NETWORK TOPOLOGY OPTIMIZATION

NETWORK TOPOLOGY OPTIMIZATION

The Art and Science of Network Design

Roshan L. Sharma

VNR VAN NOSTRAND REINHOLD
New York

Library of Congress Catalog Card Number 89-27365
ISBN 0-442-23819-3

Printed in the United States of America

Van Nostrand Reinhold
115 Fifth Avenue
New York, New York 10003

Van Nostrand Reinhold International Company Limited
11 New Fetter Lane
London EC4P 4EE, England

Van Nostrand Reinhold
480 La Trobe Street
Melbourne, Victoria 3000, Australia

Nelson Canada
1120 Birchmount Road
Scarborough, Ontario M1K 5G4, Canada

16 15 14 13 12 11 10 9 8 7 6 5 4 3 2 1

Library of Congress Cataloging-in-Publication Data

Sharma, Roshan Lal, 1927–
Network topology optimization: the art and science of network design/Roshan L. Sharma.
p. cm.
Includes bibliographical references (p.).
ISBN 0-442-23819-3
1. Computer networks. 2. Data transmission systems. I. Title.
TK5105.5.S427 1990
004.6′5—dc20 89-27365
CIP

To
Raj, Veena, Pamela, Deepak, and Ajita

Contents

Preface

Technology has changed much since the publication of the author's NETWORK SYSTEMS in 1982. The fields of telecommunications and networking have, in particular, witnessed more significant developments than any other field of human knowledge. This is primarily due to the fact that telecommunications is the engine that is driving the *information age* into our daily lives, as compared to the forces that brought us the agricultural and the industrial ages earlier.

One way to measure the progress in the field of telecommunications is to count the number of technical papers and books published during the last decade. That, however, may be an arduous task. A quick sampling of titles of these publications shows that the subject of designing private networks is still being neglected. Many of the books dealing with network design handle specialized areas such as voice and data, with hardly any information on network topology optimization. Most of the technical papers that dealt with the topological design of data networks were published during the late 1960s and early 1970s. No technical book has ever attempted to assess those papers in the light of the latest developments in the field of integrated corporate networks. A definite need for a book dealing with practical tools for the analysis and design of actual network systems, in addition to providing examples of actual network planning and design, has long existed. The author hopes that this book will meet that need.

The topology of a network system provides a great deal of insight into its behavior. To illustrate, most of the central offices of a public telephone network represents a star topology. Recently, well-publicized outages in Illinois and New Jersey disconnected many large corporations from the rest of the world. They lost millions of dollars in revenue. A redesign of their network topologies would have eliminated such catastrophies.

The recent popularity of T1 digital facilities has made strong impacts on the topologies of integrated corporate networks. Nowadays, one must be able to design mixed network topologies that allow the integration of voice, data, and video applications in a cost-effective manner. New technologies have also created the need for innovative network topologies. The introduction of Doelz's virtual packet and AT&T's ISN/Datakit network architectures requires the design of network links based on the ''directed link'' and ''interconnected star'' topologies, respectively. Many recent developments in IBM's System Network Architecture now allow an increasing amount of distributed intelligence within mixed topology networks. Several international standards related to the evolving Integrated Services Digital Network impose their own topological constraints.

It is the purpose of this book to describe well-proven algorithms for the analysis and synthesis of network topologies for voice, data, and integrated voice/data applications under one cover. To accomplish this task, the book is divided into ten chapters. The first chapter deals with basic networking concepts essential for designing cost-effective networks. Those who are well versed with these concepts may skip this chapter. The next three chapters deal with generalized tools for network analysis and design, and specialized tools for network analysis and the design of voice and data networks. These chapters provides a good deal of information for future use. The fifth chapter defines algorithms for synthesizing useful network topologies. The sixth chapter describes a state-of-the-art network design and analysis software package. The next three chapters describe the actual process of designing voice, data, and integrated voice/data network topologies, and presenting the results in a form that provides useful conclusions. The last chapter discusses how the tasks described in this book fit into the overall network planning process. Appendix B describing the evolving networking standards, is also included. The contents of Chapter 10 make copious use of the material presented in Appendix B. An undergraduate degree in telecommunications or equivalent experience in voice and data communications is recommended for full comprehension of the scope of this book.

The author is particularly indebted to a great pioneer, Mr. M. L. Doelz, for introducing the author to most of the key concepts of data communications during the development of the first modem, message switching system, economical networks, and protocol-independent, virtual packet switching. The author would also like to thank Mrs. Pamela Schemagin and Ms. Ajita Sharma for their painstaking help in performing many tedious numerical analyses, proofreadings, and improving the readability of this book.

The author can be reached at P.O. Box 822938, Dallas, TX 75382, or (214) 691-6790. Any questions or inquiries should be sent to above address.

NETWORK TOPOLOGY OPTIMIZATION

1

Basic Networking Concepts

1.1 INTRODUCTORY REMARKS

This book is primarily concerned with the design of telecommunications networks. It is the purpose of this book to use the principle of successive decomposition, or iterative modeling, as expounded in an earlier book. *Network Systems* (Sharma et al. 1982), to provide a balanced treatment of telecommunications network design.

Telecommunications deals with the transmission, emission, or reception of signs, signals, written images, sounds, or intelligence of any nature by wire, radio, optical, or other electromagnetic means. A telecommunications network consist of two or more nodes that are interconnected through transmission links based on one or several of the many available technologies and network topologies. The network may be characterized by a hierarchy of nodes and/or links.

There are basically two types of telecommunications networks—switched and nonswitched. The most common variety of nonswitched system involves communications between two or more humans. Another common variety of the nonswitched system is a broadcasting network, which primarily involves the transmission of only one type of information at any one time from a single location (or several locations repeating the transmission to different areas of reception) and the reception of the same information by a large number of listeners. Other types of nonswitched networks deal with data processing and database management, and generally involve a single service node. This book does not cover nonswitched telecommunication networks. There are many examples of switched communication systems. The best example of a switched system is that of the public telephone network that allows subscribers to talk with one another. Public data networks also provide an excellent example of the switched system used for the interchange of messages or packets of data. Such networks allow the subscribers to fully share the network facilities. This book deals with the topological design of *switched communication networks* that provide a variety of services such as voice, data, and video to subscribers. In order to accomplish that task, one must first introduce basic networking concepts such as classes of communication networks, network components, attributes of network traffic, voice and data communication, and network architecture.

1.2 CLASSIFICATION OF SWITCHED COMMUNICATION NETWORKS

Switched communication systems can be understood only in terms of their roots. One can therefore classify switched communication networks as follows:

1. geographical coverage of the switched network
2. method of accessing the switched network
3. switching technique employed in the network

1.2.1 GEOGRAPHICAL COVERAGE

There are three types of networks characterized by their geographical coverage:

a. Local Area Networks (LANs): About 60 percent of all networks fall into the category of LANs, which span one or more buildings over a campus.
b. Metropolitan Area Networks (MANs): About 22 percent of all networks cover distances of about 50 miles and fall into the catagory of MANs.
c. Wide Area Networks (WANs): About 18 percent of all networks fall into the category of WANs, which cover distances of about 500 miles (10 percent) and larger (8 percent).

Some authors have defined a MAN as an interconnection of LANs, and a WAN as simply an interconnection of LANs and MANs.

1.2.2 METHOD OF ACCESSING

Each subscriber node must be designed to access the switched network for service. Basically, there are three distinct access methods:

1. Demand Access: This is generally employed in most voice and data communication networks. According to this interrupt-based technique, the subscriber goes off-hook to demand service. The service node provides a dial tone immediately after discovering the off-hook condition. The dial tone is an invitation for a further dialogue that ultimately results in a switched path between the calling and the called subscribers.

2. Polled Access: This is another popular method for allowing the computer-controlled switching node to interrogate each subscriber regarding the need for data interchange with another subscriber.

3. Multiaccess: A few WANs employ the ALOHA multiaccess technique, whereby a source node begins transmitting whenever it has a message or packet in its memory. If a collision occurs between two messages/packets transmitted by two sources, each node retransmits after waiting for a random period. In some ALOHA systems, a source will retransmit a message/packet if it does not receive a positive acknowledgment from the destination node within a certain period. The multiaccess technique is primarily employed by LANs. Many LANs employ the *carrier-sense-multi-access* (CSMA) technique, whereby each node senses the carrier on the shared bus before sending the message or packet. Collisions are handled in the same way as for ALOHA. The remaining LANs employ the *token access*, whereby a node can only transmit when it grabs a free token.

1.2.3 SWITCHING TECHNIQUES

Switching technology employed by the network determines the manner in which the network components or facilities are shared among the subscribers. Three types of switching techniques can be enumerated, as follows:

1. Circuit Switching (CS): Provides a private, hardwired connection through the network, between two or more subscribers, for the entire duration of the call. The flow of user information is completely transparent to the network nodes in the path of the call. The connection is brought down immediately after the calling party hangs up. The circuit-switched networks have mainly been employed for voice and video teleconferencing services. The use of the public telephone network for data communications has also been increasing ever since the advent of personal computers (PCs) and low-cost modems.

2. Packet Switching (PS): Packet-switched networks permit the transfer of information between two subscribers through the routing of addressed packets of user data through the network. Unlike the circuit-switched network, each PS node employs the store-and-forward technique for switching each packet, for the purpose of routing it towards the destination on a virtual/permanent circuit established before the start of the transmission. In that sense, the PS node is not transparent to user data. Both the network links (virtual or permanent variety) and nodes are fully shared among all the subscribers. Most PS nodes use only the random-access memory, to minimize the switching delays. Some recently introduced PS systems employ very small packets and the technique of nodal bypassing (when the destination is somewhere else on the virtual circuit) to minimize the delays even further. In general, all packets are switched according to the first-in-first-out (FIFO) scheme, except when priorities are assigned to the packets.

3. Message Switching (MS): Message-switching networks employ a long-term store-and-forward (S/F) scheme for each received message. Each message is generally stored in full before forwarding it towards the destination. There are many variations of a message-switched network, depending upon the length of the message and/or the legal commitments related to the delivery of the message. Some message-switched systems provide a disk-based storage for a short-term retrieval and a magnetic tape storage for a long-term retrieval. Most message-switched networks segment the messages for high utilization of the storage media and fast retrievals from the storage media.

In general, a LAN, MAN, or WAN may employ any combination of the above-mentioned access methods or switching technologies.

1.3 COMPONENTS OF A TYPICAL NETWORK SYSTEM

The smallest switched network system must employ at least one switching (or service) node and several communication facilities known as subscriber lines (SLs) for connecting the on-net subscribers to the switching node(s). If there is more than one switching node, communication facilities known as access lines (ALs) and trunks are employed for interconnecting the various types of switching nodes. The links connecting a lower-level, privately owned, customer premise equipment (CPE) in the form of a Private Branch Exchange (PBX) or a data concentrator to tandem switching (TS) nodes are generally called access lines (ALs). The links that connect tandem switching nodes are called trunks (TKs). In case a private switched network provides communication services to the off-net subscribers (or public), one will need additional ALs for connecting a subset of its switching nodes to the Central Office (CO) switches of the public network. See Figure 1.1 for an illustration of a switched communication network and its major components, as discussed above.

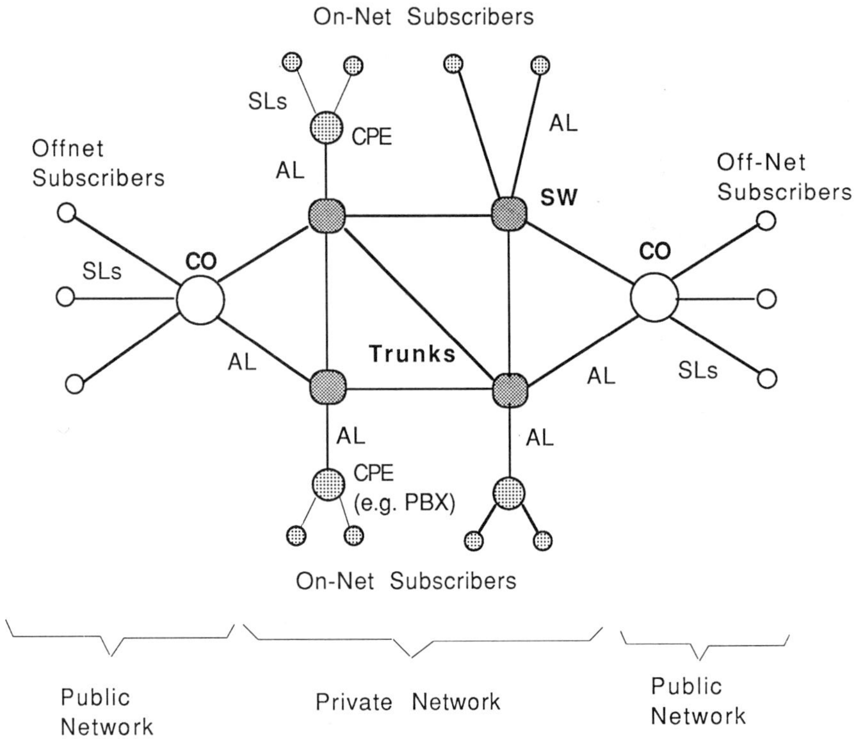

Figure 1.1. Components of a switched communication network.

1.4 BASIC TRAFFIC ATTRIBUTES

A network system exists primarily to provide useful services to a set of subscribers. These services involve many forms of information exchange between network service nodes and subscriber nodes. Several traffic attributes characterize such services. These attributes and their underlying concepts are enumerated in the following paragraphs.

A *telephone/data call* is an unstructured sequence of intelligence flowing in both directions beween the calling and the called subscribers. For circuit-switched voice networks, a call requires the establishment of a reserved, unshared path between the subscribers. A polite conversation between two persons in a voice network implies that intelligence flows in only one of the two directions. This results in a 50 percent utilization of network resources (except for the 2-wire subscriber links). A full duplex data call may not suffer from this constraint. For a packet-switched network, a shared virtual circuit is established before any data packets can be sent.

A *data message* represents the user's data directed from an originating subscriber towards a destination subscriber. Depending upon the switching technique, the message may or may not be transparent to the network node. Since a circuit-switching node does not alter or analyze the contents of the information flowing through the network, it is transparent to the users. Since a message-switching or a packet-switching node, on the other hand, must analyze a packet or a message before routing it to the destination, it therefore does not provide a transparent service to the users. An *input message* is generally called an *inquiry* and an *output message* is generally called a *response*. Many types of *control messages* flow between nodes to perform essential supervisory and signaling functions required by the various services. A message generally consists of many user-defined fields, each clearly demarcated to denote such items as originating and destination addresses, user data, and cyclic redundancy check (CRC) code for error detection. A packet may also consist of similar fields, but their size and contents (except for the user data) is generally defined by a standard or a proprietary protocol. Several packets may be required to transmit a message.

A *transaction* generally involves a series of messages in both directions relative to the user and the network node. A transaction is simply a characteristic of the particular application or service. For example, a Binary Synchronous Communication (BSC) protocol involves a sequence of messages between the data terminal and a host before receiving a response message from the host.

A *session* is almost identical to a transaction except that it may involve one or more transactions. A transaction or a session can not commence until a permanent or a temporary virtual circuit has been established.

The *expected call duration* (ECD) is defined by the duration of a single voice session. Some analysts substitute call holding time for ECD. A *message length* is generally determined by the number of bits or characters in the message. The *expected message duration* (EMD) can also be defined as the message length divided by the link capacity denoted in terms of bits or characters per second. Call or message duration may be characterized by randomness. For that case, a call/message duration distribution with at least a mean (μ) and a standard deviation (σ) must be defined before computing system resource utilizations and response times. A random distribution of call/message duration also results when many types of fixed length messages travel randomly over a shared virtual circuit, each consisting of one or more physical links.

Call or message arrival rate (λ) defines the rate at which the calls or messages are presented to a network resource/server. Traffic intensity (A) is generally defined as the product of arrival rate and the expected call duration of ECD. Therefore, $A = \lambda * \text{ECD}$ erlangs. Call intensity is measured in terms of erlangs, named after the famous Danish mathematician. Call intensity determines the average number of servers (e.g., links) concurrently occupied. In some quarters, hundred call seconds (CCSs) determines the call intensity. However, it does not have any physical meaning. CCSs can easily be divided by 36 to compute erlangs. Call intensity is an important concept that is employed for computing the number of servers (or links/trunks) required to provide a desired performance level.

Traffic flow is a measure of the traffic intensity on any directed path consisting of an order sequence of network nodes. A sequence of two nodes implies a single circuit bundle connecting the two nodes. A sequence of N nodes implies a directed path of ($N - 1$) intervening links, each consisting of several physical circuits. Most well-designed network systems provide some traffic flow control to prevent congestion that may result in system failures. There are many ways to achieve flow control. Access limiting and dynamic routing of traffic are some of the many popular methods.

The concept of switched systems and the associated traffic attributes, as introduced above, can be further expanded by discussing the well-known voice and data communication systems.

1.5 CONCEPTS OF VOICE COMMUNICATIONS

Voice communications are synchronous with telephony, which began with Alexander Graham Bell's invention of the telephone in 1876. The first mechanical circuit switch was installed in 1878 in New Haven, Connecticut. The first automatic, step-by-step Strowger switch was installed in 1892 in La Porte, Indiana. The first crossbar-type electromechanical switch was installed in 1938 in the U.S.A. Direct distance dialing (DDD) as based on the 7 digit code was introduced in 1947 in the U.S.A. The first stored program controlled (SPC) No. 1 ESS circuit switch (CS) was installed in 1965 in the U.S.A. A deregulation of the customer-premise-equipment began in 1968, as a result of the Carterfone decision. This gave impetus to the development of private automatic branch exchange (PABX) and privately owned tandem switches required for a private network. The first digital circuit switch employing pulse code modulation (PCM) and 255 μ-Law was commercially announced by Collins Radio Company in 1971. The divestiture of AT&T took place in 1984, which created about 250 Local Access and Transport Areas (LATAs) and resulted in seven Regional Bell Holding Companies (RBHCs) to handle local access and intra-LATA services, and many other common carriers (OCCs), such as the new AT&T, MCI, and U.S. Sprint, to handle inter-LATA and long-distance services and the leasing of long-haul analog/digital circuits. At present much activity is taking place in introducing No. 5 Electronic Switching System (ESS) Central Office (CO) switches capable of handling Integrated Services Digital Network (ISDN) based services. See Appendix B for a description of the ISDN-related standards and network topologies.

Voice communication networks are generally classified by the type of switching nodes employed. The first generation of voice switches used dumb Strowger switches. In the second generation of voice switches, SPC-type circuit switches were employed. The third generation of voice switches generally used digital CSs. In the fourth generation of voice switches, each switch can handle both voice and data. That, in turn, implies the capability to provide concurrent access to both voice and data terminals at each subscriber location and to share a modem pool for WAN applications. In addition to providing a traditional integrated voice and data (IVD) capability, an ISDN switch provides higher data rates and greater intelligence than the fourth generation circuit switch. For that reason, an ISDN switch may be called the fifth generation switch. In short, the various generations of voice switches can be characterized by an increasing amount of nodal and/or network intelligence required for cost effective network management and technical control (NM/TC).

Voice communication networks were originally characterized by analog transmission facilities with only space division multiplexing (SDM). With time, frequency division multiplexing (FDM) became the most cost-effective method of carrying a large number of voice conversations over a coaxial cable. For example, the L1, L3, L4, and L5 type carrier systems could carry 600, 1860, 3600, and 10800 voiceband channels, and were introduced in 1946, 1953, 1967, and 1974, respectively. Later analog and digital microwave systems (also called radio systems) became cost effective for long- and medium-haul routes for a medium number of voice channels. The T-carrier systems employing digital PCM-based multiplexing techniques became cost-effective in the early 1960s for short-haul routes. With the advent of fiber optics, the repeaterless distance and the number of voice channels carried by digital transmission systems, known as T-carriers, has increased rapidly. One can now state without equivocation that digital transmission is the only cost-effective means of carrying a large number of voice conversations and/or data traffic over all ranges of distance.

At present, the availability of cost-effective digital switching and transmission techniques is quickening the pace towards a fully integrated digital network (FIDN). Since ISDN still employs CS for voice and CS and/or PS for data, we are still far from realizing an FIDN. But with the recent development of fast-packet switching technology (using small packets, fast routing analysis, and nodal-bypass mechanism), packetized voice is not far away. When packetized voice becomes economically feasible, the FIDN will finally become a reality, resulting in large economies-of-scale and a fully integrated network.

1.6 CONCEPTS OF DATA COMMUNICATIONS

Data communications is much older than voice communications, despite notions to the contrary. The earliest versions involved coded messages using drum sounds, smoke signals, and semaphore devices. The modern variety of data communications began with the invention of telegraphy in 1835. However, the biggest impetus to data communications came from the development of the first commercial modem by the Collins Radio Company in 1955. Such a modem soon became the foundation for several coded-feedback communications systems used for tape-to-tape and card-to-card data transport. A portion of the data was first divided into blocks, before their transmission. Each block had an overhead code for detecting any errors caused through transmission. A number of data blocks could be outstanding until feedback information regarding their correctness was received and before continuing the transmission of additional blocks. The next epoch occurred with the development of the first microprogrammable front-end (FE) communications processor in 1960. Later, Collins Radio Company made this processor the basis of the first SPC message switching system, ultimately replacing all of the existing torn-tape teletype systems by 1964. In 1968, Collins Radio Company also developed the C-System, which was a large LAN consisting of mainframes, minicomputers, and input/output devices, interconnected via time-division multiplexed (TDM) loops, capable of providing both message and packet-switching services to a large number of subscribers. The first Advanced Research Project Agency (ARPA) sponsored packet-switched network became operational in 1969 with only four nodes. The first public PS data network, as operated by Telenet Corporation, became operational in 1975.

The data switching nodes can also be characterized by several generations of development. The first generation switches were those that provided interchange of data between the host and terminals using the mechanism of polled access. The second generation of data switches were the SPC message switches nodes. The third generation of data switches were based on packet switching. The fourth generation of data switches attempt to integrate voice and data on transmission links only.

The technologies related to voice and data are becoming integrated, first in terms of transmission, and eventually switching technologies. A complete integration will imply a great deal of savings to the users. A fully integrated digital network will eventually become as basis of an emerging utility that will provide unfettered sharing of information to all subscribers with diverse bandwidth requirements. This would be very similar to the existing electric and water utilities, which provide any amount of utility through a simple wall outlet or a faucet. Unfortunately, it will take some time. Experience shows that it took six years to make packet switching commercially available after one could prove its cost-effectiveness. Several products based on packetized-voice finally appeared in late 1986. This came almost ten years after packetized voice was shown to be commercially feasible. The more complex the system, the longer is the delay between those two events. Based on that assumption, FIDN-based information utilities should be available by 2000 A.D. FIDNs will be fully integrated in terms not only of services, but also network management and control (NMC) functions. They will employ expert systems (ESs) to achieve self-healing.

1.7 CONCEPTS OF NETWORK ARCHITECTURE

A network architecture is generally characterized by three networking attributes:

1. Network Topology. This defines the manner in which the network nodes are interconnected. Generally, there is a hierarchy associated with the network nodes and/or links.
2. Transmission and Multiplexing Technique. This defines the modulation, encoding, transmission/reception, and multiplexing techniques employed to synthesize links for interconnecting the various network nodes. Issues such as analog or digital transmission, coaxial cable or optical fibers, and single or multiplexed streams of data are resolved.
3. Network Management and Control (NMC) Technique. This technique is involved with the switching technology, traffic flow control, and methods of distributing and controlling intelligence in the network.

Each of these concepts is elaborated upon in the following sections, using the principle of successive decomposition.

1.7.1 NETWORK TOPOLOGIES

The basic topologies that have proven useful for telecommunications networking can be classified as follows:

1. mesh topology—fully connected or partially connected varieties
2. minimal spanning tree or multidrop topology
3. directed link topology
4. star topology

5. bus topology
6. ring topology

Additional topologies can be obtained by mixing the above basic topologies, through the process of superposition and interconnection. Since most of the cost of a communication network is dictated by the cost of communication links, network topology must be optimized for each application. The goal of this book is to address that need.

1.7.1.1 Mesh Topology

A fully connected mesh topology of Figure 1.2(a) provides a direct link or path between every pair of network nodes of a given level of hierarchy. Such a topology is economically feasible only when the number of nodes at a given network hierarchy is very small. The communication paths in a fully connected mesh are reduced to a single-hop variety. Since the cost of transmission in most existing networks is predominantly influenced by link costs and not switching, a fully connected mesh topology is only employed at a very high level network hierarchy. Even for those cases, the partially connected mesh topology of Figure 1.2(b) is preferred over the fully connected topology when system costs are considered. A partially connected topology is recommended when the traffic flow between certain nodes is negligible and it can be switches on longer paths while resulting in a better economy-of-scale.

1.7.1.2 Minimal Spanning Tree (MST) or Multidrop (MD) Topology

An MST topology is a subset of the generalized tree or MD topology, designed with the constraint that the sum of all link lengths in the network is minimal. Figure 1.3(a) represents an MST topology. Assume that node 6 of Figure 1.3(a) was a switching node. Using that information, a 4-level hierarchy for Figure 1.3(b) can be created according to the number of links between any pair of nodes. Chapter 5 will discuss algorithms for designing an economical MD topology.

1.7.1.3 Directed Link (DL) Topology

Several new network architectures allow the synthesis of networks with only directed network links (also called the netlinks) that connect a subset of nodes, in the manner similar to the way a salesman travels. See Figure 1.4 for an illustration of several netlinks, each based on the directed link topology.

1.7.1.4 Star Topology

A network consisting of a single switching node and directly connected subscribers can be represented by a star topology. See Figure 1.5. For that case, each subscriber is connected to the switching node directly through a subscriber or user line. Such a topology is commonly used for connecting all of the subscribers in a building or on a campus to a voice switch known as the PABX or a data switch. The star topology is also used for connecting the local subscribers to a CO of a BOC.

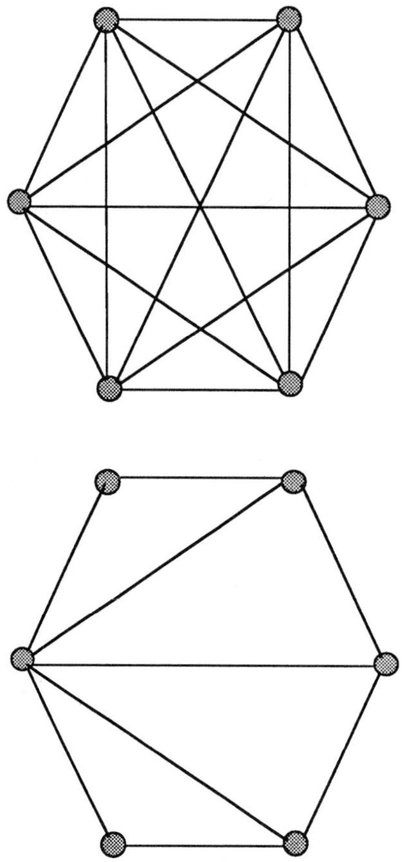

(a). Fully Connected Mesh

(b) Partially Connected Mesh

Figure 1.2 (a, b). Mesh topology.

1.7.1.5 Bus Topology

The bus topology employs a fully shared broadcast transmission facility, called a bus, for exchanging information between nodes. Such a topology, as shown in Figure 1.6, is generally the basis for a high-speed, distributed system used for interchange of information between computers and computer-controlled devices located within a building or a campus. The bus topology is thus very popular in synthesizing high-speed LANs that employ either the token or the CSMA access methods.

1.7.1.6 Ring Topology

When the two ends of a shared transmission media of bus topology are closed to form a loop, the ring topologies of Figure 1.7 results. Whereas Figure 1.7(a) represents the S/F-type ring topology, Figure 1.7(b) represents a broadcast-type ring topology. Many popular LAN architectures employ the ring topology, using the token access.

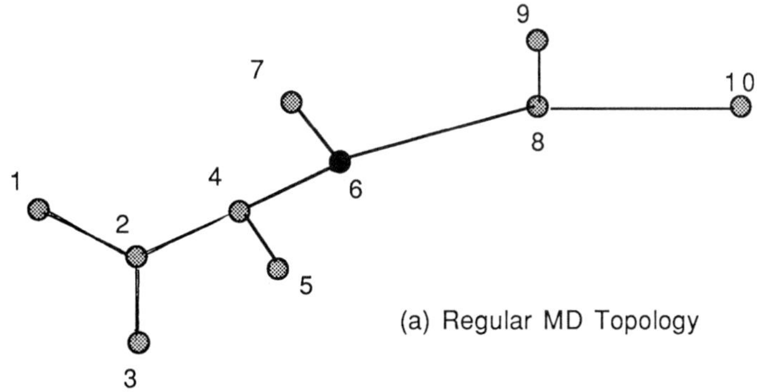

(a) Regular MD Topology

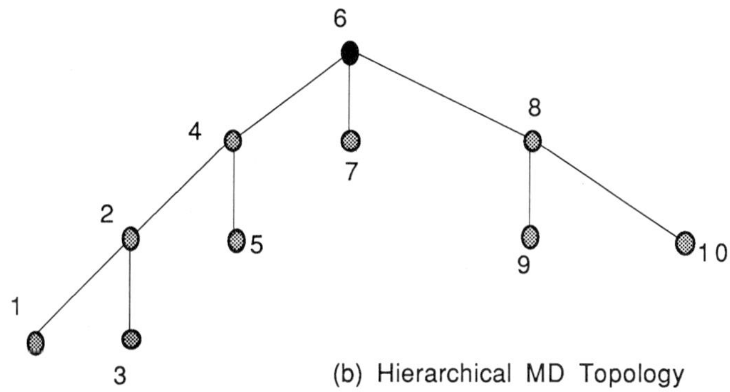

(b) Hierarchical MD Topology

Figure 1.3 (a, b). Multidrop topology.

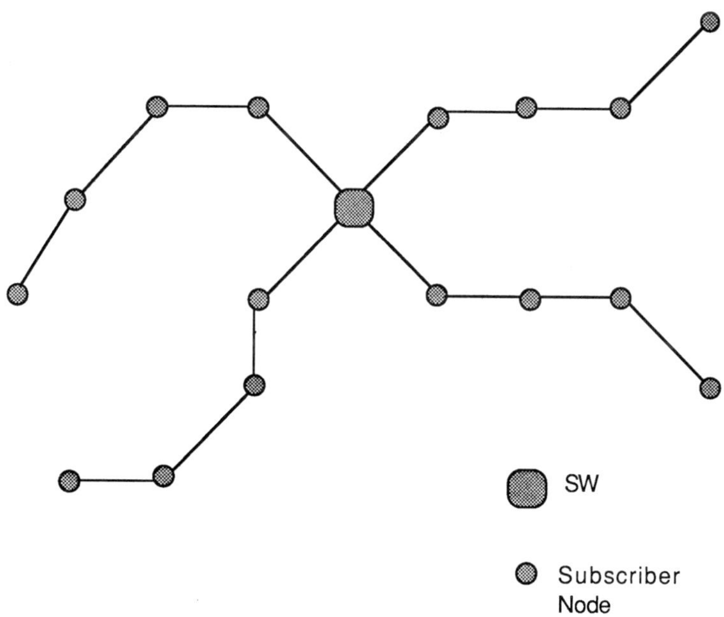

Figure 1.4. Directed link topology with four netlinks.

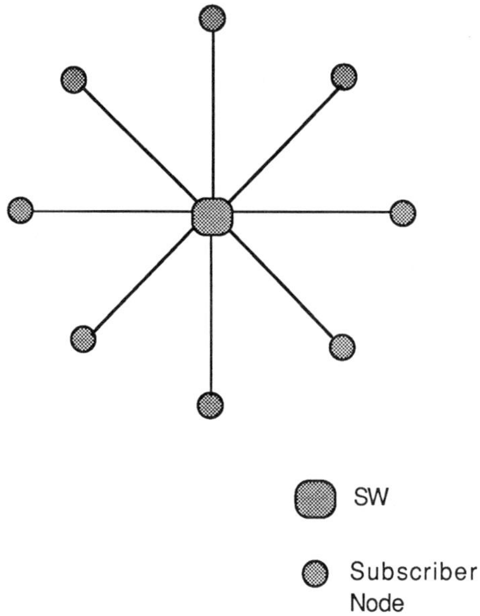

Figure 1.5. Star topology.

1.7.1.7 Mixed Topologies

Many mixed topologies can be derived from the above-mentioned basic topologies through the process of superposition or interconnection. For example, the multicenter, multistar topology of Figure 1.8 consists of the mesh and the star topologies. This topology is commonly used for corporate voice and PS networks. Similarly, the multicenter, multidrop topology of Figure 1.9 consists of the mesh and the MST/MD topologies. This mixed topology is the basis for many Systems Network Architecture (SNA) based data communication networks. The network of Figure 1.10, as based on the Doelz™ Extended Slotted Ring (ESR) Architecture, uses the mesh and the DL topologies. The IBM Series/1 Chat Ring as based on the token ring architecture illustrates a mixture of the star and the ring topologies, as shown in Figure 1.11. Whereas each local communication controller (LCC) can serve a mixture of personal computers (PCs), printers,

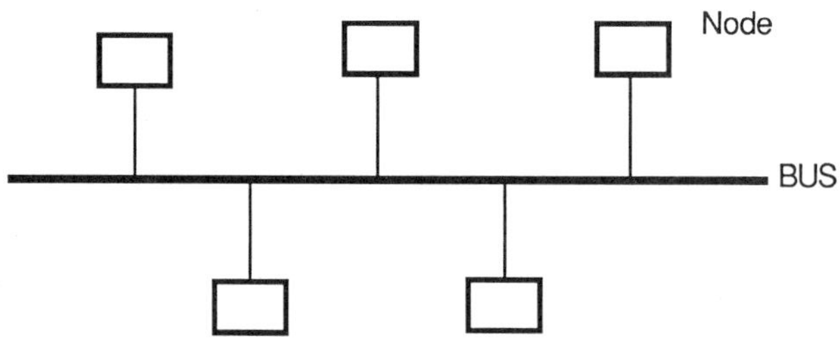

Figure 1.6. Bus topology.

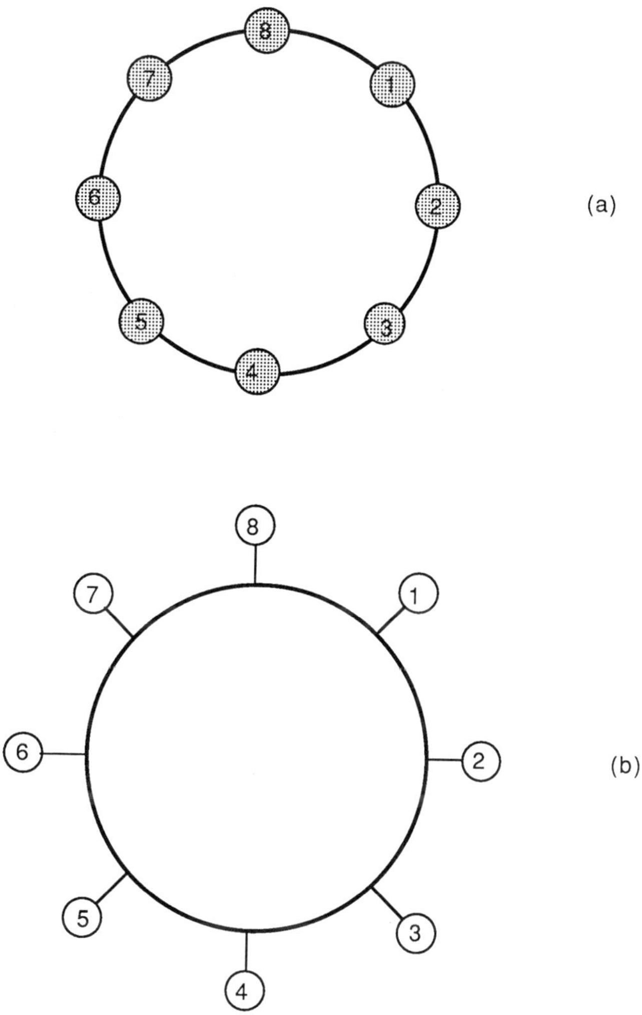

Figure 1.7 (a, b). Ring topology.

and file servers, the LCCs on the ring provide information exchange among a large number of computer-based terminals. A typical implementation of Datapoint's ARCnet LAN, as shown in Figure 1.12, employs a mixture of bus and star topologies. Such a network topology is sometimes also called the starburst topology. A network that handles an IVD application generally employs the star, MD, and mesh topologies.

1.7.2 TRANSMISSION AND MULTIPLEXING TECHNIQUES

A balanced design of a network system requires an integrated plan for transmitting both the user and control data from one node to another with the network. A knowledge of transmission and multiplexing techniques is essential. Whereas the transmission tech-

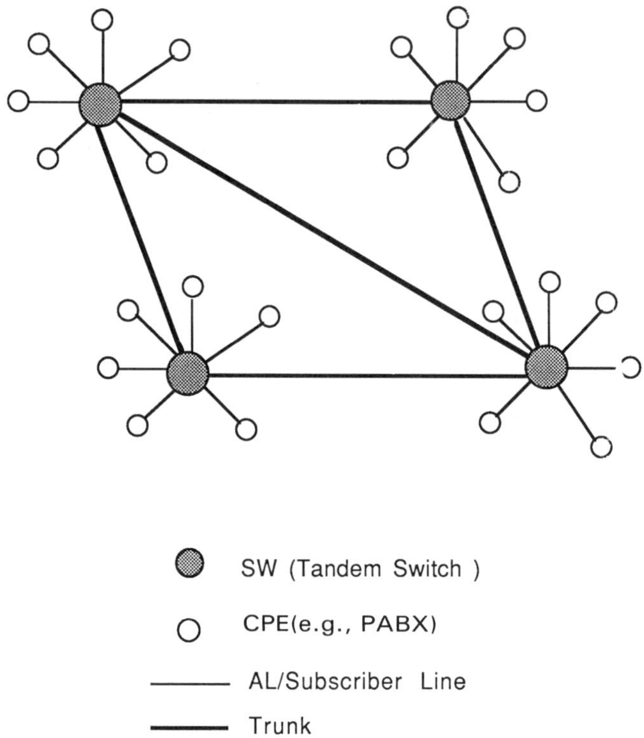

Figure 1.8. Multicenter, multistar topology.

niques usually involve the method of encoding each basic symbol prior to transmission over the channel, the multiplexing technique defines the way multiple streams of user and control data can be combined together to share a single physical link.

1.7.2.1 Transmission Techniques

Discrete information can be transmitted in the form of either analog or digital signals.

Analog Transmission Techniques. An analog signal is realized by multiplying a continuous modulating function (usually the information) with a high-frequency carrier waveform centered at F_c. Figure 1.13 illustratres the use of amplitude modulation (AM) and frequency modulation (FM) for producing continuous analog signals. Figure 1.14 illustrates the use of AM, FM, phase modulation (PM), pulse amplitude modulation (PAM), pulse width or pulse duration modulation (PWM or PDM), and pulse position modulation (PPM) for producing discrete analog signals employed in data communications. These techniques are the basis of modulation and demodulation (MODEM) devices.

Digital Transmission Techniques. A digital signal is simply a baseband signal in the form of a digital pulse train that represents a coded version of the original information.

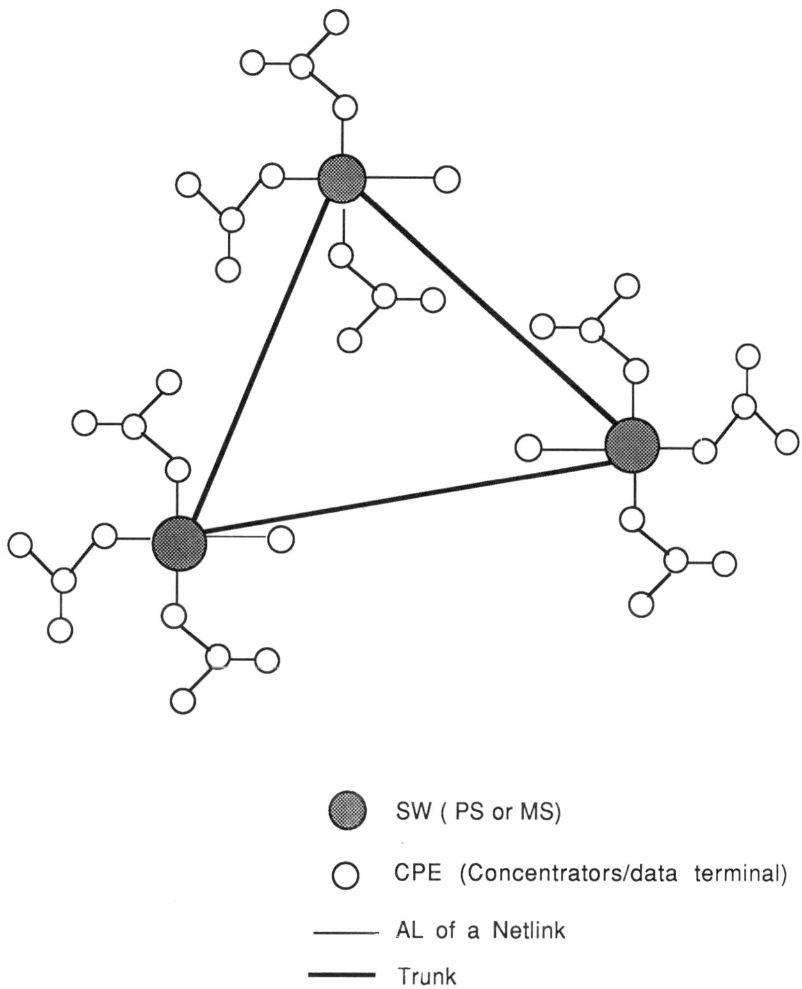

Figure 1.9. Multicenter, multidrop topology.

Pulse code modulation (PCM) and delta modulation (DM) are two popular techniques for producing digitized voice signals.

Pulse code modulation employs the principle of sampling a continuous signal, such as voice, quantizing the sampled value into a discrete level, and encoding this level into a fixed sequence of binary pulses. The larger the number of quantization levels, the higher the fidelity of the received signal. Figure 1.15 illustrates the manner in which a piece of voice signal is converted into a sequence of binary pulses as a part of the entire PCM pulse train. According to Figure 1.15, 8 quantization levels (equal to 2^3, called 3-bit code) are employed, thus resulting in a total bandwidth (BW) requirement of 2 * signal BW * number of pulses per sample (equal to 3). Assuming a typical voice signal of 4000 Hz, the PCM bit rate required is therefore 2 * 4000 * 3 = 24000 bps (bits per second). The telephone industry standard throughout the world is a PCM system with

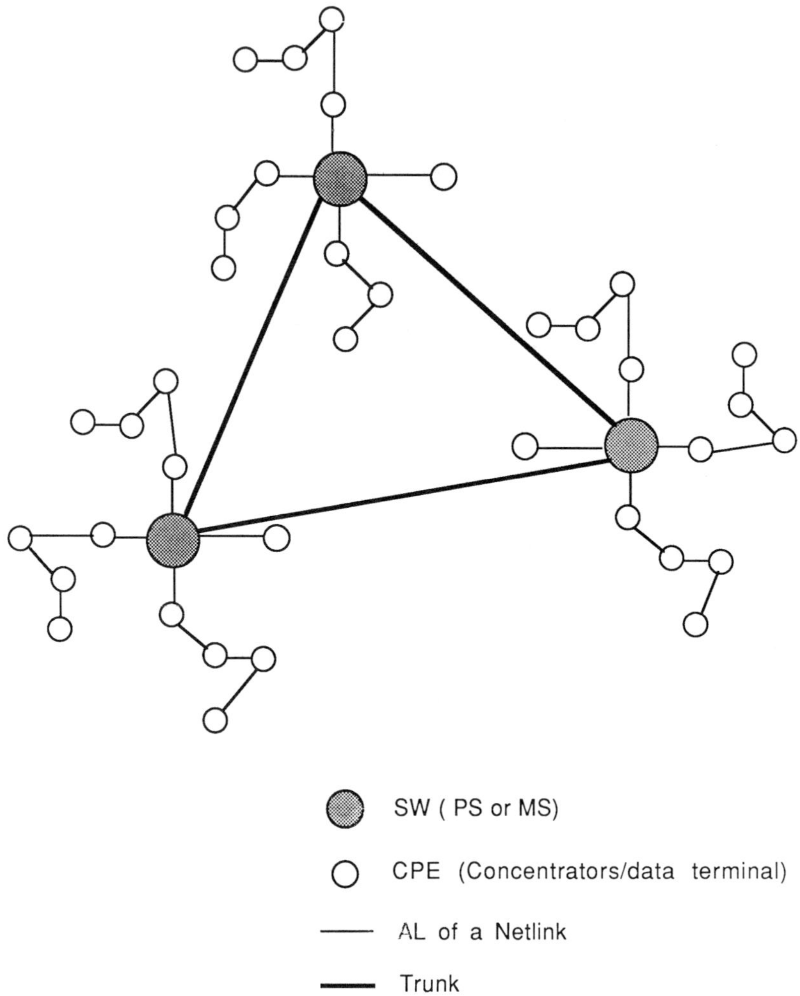

Figure 1.10. Multicenter, multi-DL topology.

256 quantization levels (equal to 2^8, called 8-bit code), which requires a PCM bit rate of $2 * 4000 * 8 = 64000$ bps. A new technique called the adaptive differential PCM (ADPCM) is being introduced that requires a PCM bit rate of only 32000 bps. It is based on the fact that a highly correlated signal (e.g., voice) changes slowly from one Nyquist sample to another, and therefore it sends only the difference. The sent information is further reduced by adapting the step size (or delta) to provide increased dynamic signal ranges. Although the BW requirements of ADPCM look very attractive, one must note that 9600 bps data calls cannot be handled by ADPCM. Furthermore, the cost of transcoders for interfacing ADPCM to existing network facilities must also be considered. In other respects, ADPCM is quite comparable to the traditional PCM.

Delta modulation (DM) is another form of the sample data technique that produces a

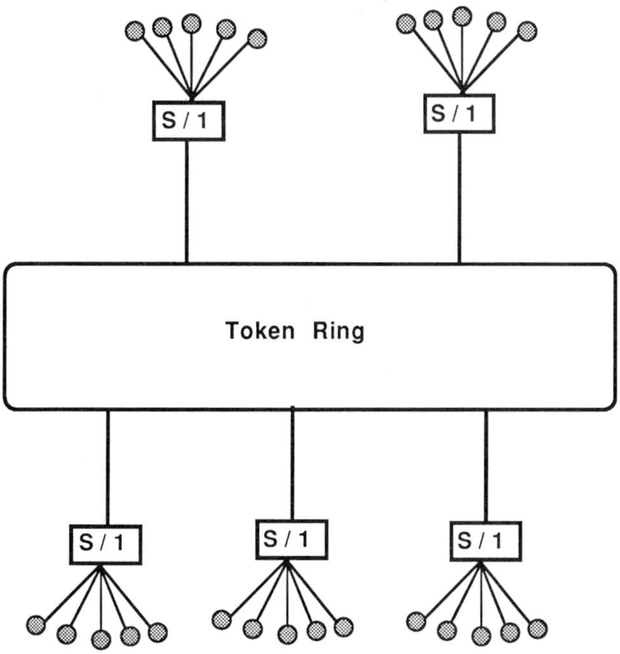

Figure 1.11. Mixed topology for IBM's Chat Ring.

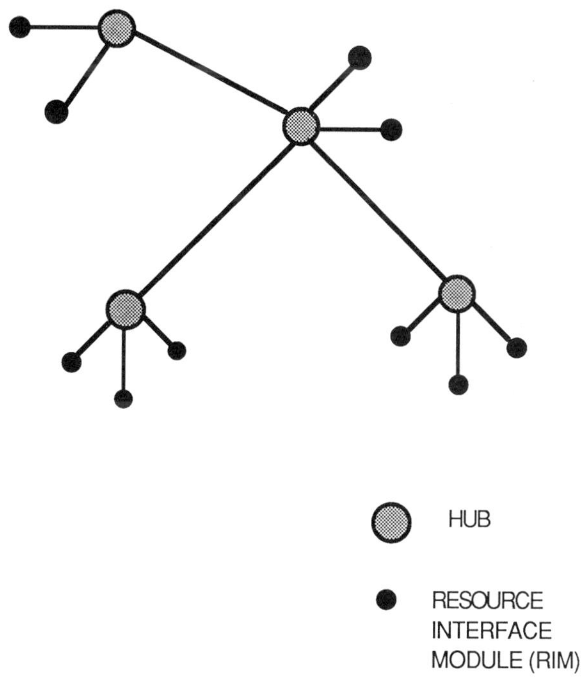

Figure 1.12. Mixed topology for Datapoint's ArcNet LAN.

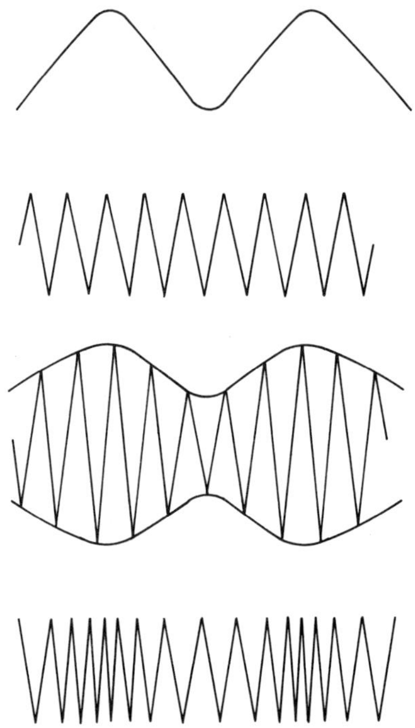

Figure 1.13. Modulated analog signals for a continuous message.

1 or 0 when the sampled value is higher or lower in amplitude than that for the previous sample. An alternating 1 and 0 represents a constant signal value. DM requires very little hardware complexity when compared to the PCM technology. For that reason, it is mainly used for mobile/military systems. Although it requires less BW than a PCM system, it requires a higher sampling rate to achieve the same quality of received signal. Some new DM-based systems are based on the continuously variable slope DM (CVSDM) concept, which is generally superior to the older DM and compares favorably with the PCM technique.

The pulse trains resulting from PCM or DM are transmitted in either the unipolar, polar, or bipolar form, as shown in Figure 1.16. Figure 1.17 shows how the clear channel capability (CCC) is achieved for an extended super frame (ESF, as discussed later) using the bipolar, 8-zeroes suppression (B8ZS) encoding scheme.

Why Digital? Analog transmission was always beset with the effects of nonlinear distortion, attenuation, and additive noise. Furthermore, the high cost of multiplexing and repeatering eventually caused its replacement by digital transmission techniques.

Digital transmission allows (1) simplicity in multiplexing and signaling, (2) lowering of costs due to the use of Very Large Scale Integrated (VLSI) circuits, integration with digital switches and fewer repeaters, (3) higher fidelity of information due to low distortion and additive noise, (4) integration of many services with and without encryption, and (5) an integrated management and control of the entire network.

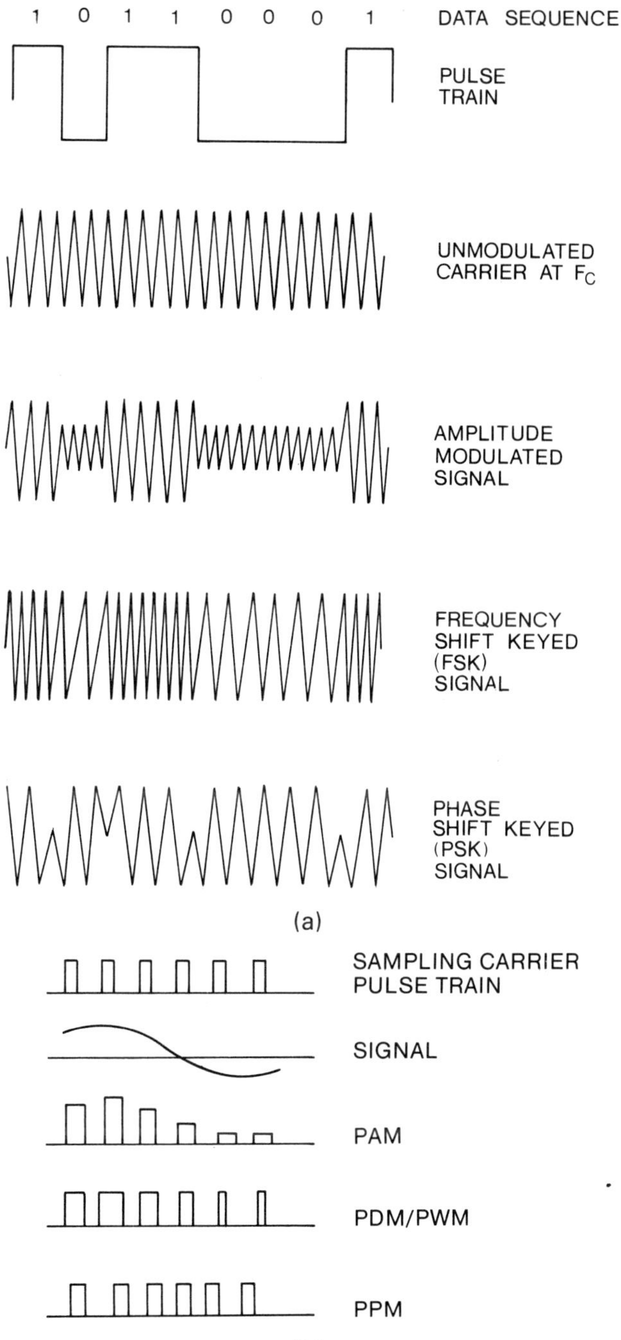

Figure 1.14 (a, b). Modulated analog signals for analog (a) and discrete (b) messages.

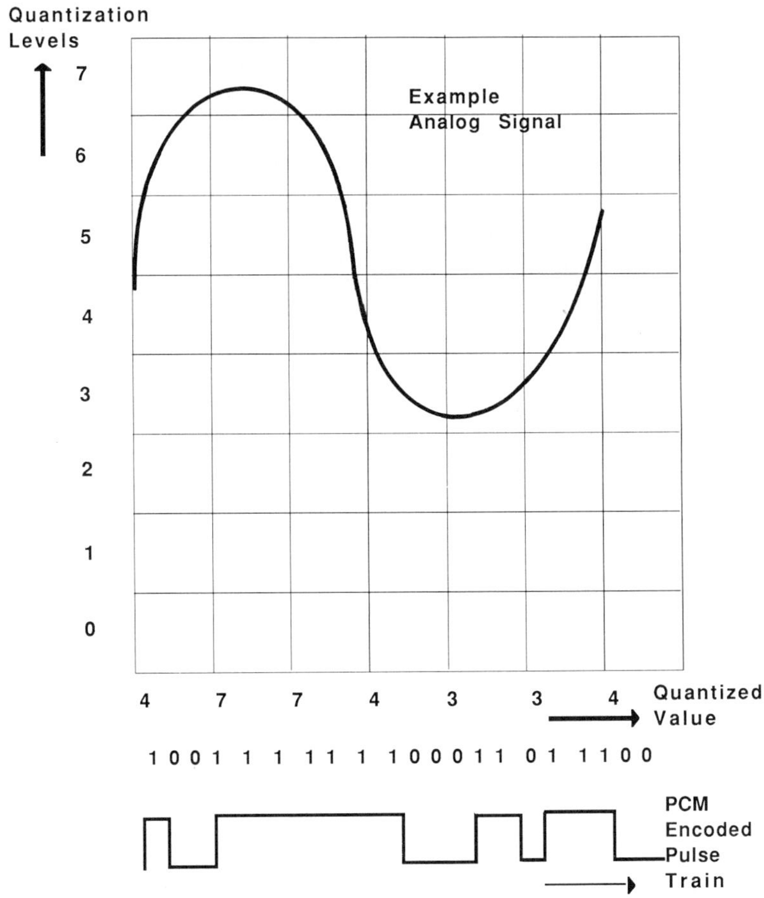

Figure 1.15. Pulse code modulated digital signal for an analog message.

At a public forum, it is hard to find anyone who will even dare to champion analog transmission. Before 1980, it was almost dangerous to speak in favor of digital transmission.

1.7.2.2 Multiplexing Techniques

Most transmission links employ some form of multiplexing of individual user and control channels to utilize the inherent structure or bandwidth of the physical link, hence reducing the cost of per-channel transmission. Special multiplexing techniques are used for both the analog and digital transmission systems.

Multiplexing Techniques for Analog Transmission. The first multiplexing technique employed for analog transmission was called space division multiplexing (SDM), which implies the bundling of individual voice-grade channels in a physical/spatial manner. This method allows the sharing of the common conduit, and hence the cost of cable

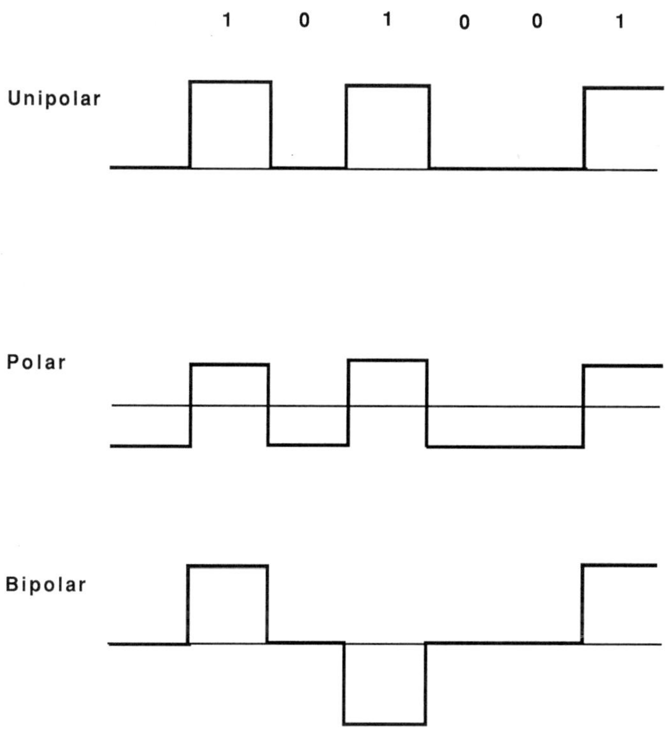

Figure 1.16. Digital signal transmission wave forms.

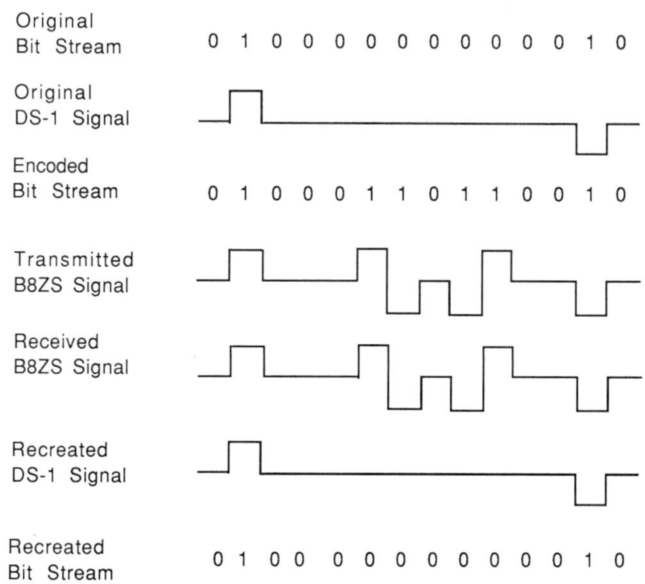

Figure 1.17. Digital signal formation for clear channel capability.

laying. However, this scheme is beset with the effects of nonlinear distortion and attenuation as the number of bundled channels or the distance increases. For that reason, the SDM technique is mainly used for local loops and the LAN environments.

Frequency Division Multiplexing (FDM) combines the spectrums of individual channels to form a large spectrum, if allowed by the physical link. It requires the use of guard bands at the transmitter to separate the individual channels, and the use of filters to separate the individual channels at the receiver. FDM-based systems tends to be quite expensive and bulky, due to the excessive amount of analog signal processing required. For a long time, FDM was a better method than the only other available technique of SDM. An interesting application of FDM was to send data under voice (DUV), a technique used for combining DDS channels with FDM voice channels.

Multiplexing Techniques for Digital Transmission. Time-division-multiplexing (TDM) is the most popular technique for combining several digital channels into a single channel. PCM–TDM is the most popular technique for combining 24 PCM channels to create a T1-carrier rated at 1.544 Mbps. See Figure 1.18 for the T1 carrier structure that includes a framing bit for each 8 ∗ 24 (equal to 192) bits of data. The 193-bit frame consists of user data, signaling data, and one framing bit used for synchronization. A

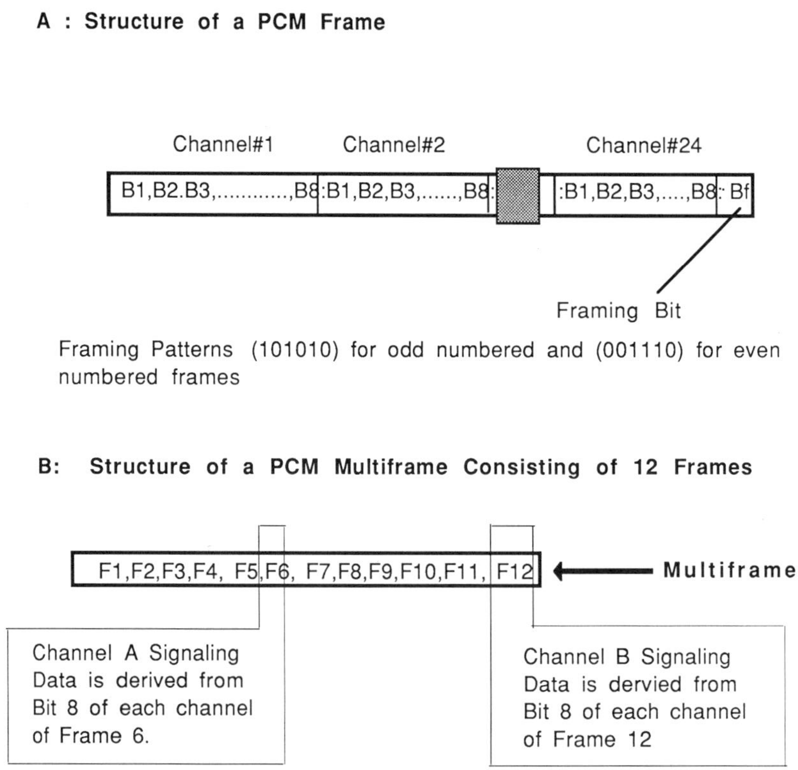

Figure 1.18. The PCM time division multiplexing technique.

better understanding of the PCM–TDM technique can be achieved by considering a set of 12 frames. The odd-numbered frames (1, 3, 5, 7, 9, 11) should yield the multiframe synchronization pattern of 101010, and the even-numbered (2, 4, 6, 8, 10, 12) frames should provide a multiframe synchronization pattern of 001110. The signaling data (on-hook, off-hook, dial-digits, etc.) is derived from the eighth bit of each channel of the sixth and twelfth frames (Channels A and B). This is also known as the bit-robbing technique.

T1-carrier is the first level of digital multiplexing hierarchy in North America and Japan. The international standards organization, Consultative Committee International Telegraph and Telephone (CCITT) has another standard for Europe and other countries. That involves 30 channels of user data and 2 channels (channels A and B) of signaling channels to create a 2.048 Mbps (32 * 64 Kbps) carrier. This eliminates the need for bit-robbing, resulting in simpler hardware. See Figure 1.19 for the digital PCM hierarchies for the above two standards. Figures 1.20 and 1.21 illustrate additional nomenclature associated with the North American PCM–TDM hierarchies. Table 1.1 compares several generations of the well-known D-Type multiplexer. At present, there are no standards for fiber optical facilities. The Synchronous Optical Network (SONET) is a proposed standard for a new family of transmission rates ranging from 51.84 Mbps to 13.22 Gbps. The SONET format will represent a new transport and multiplexing hierarchy above the DS-3 level.

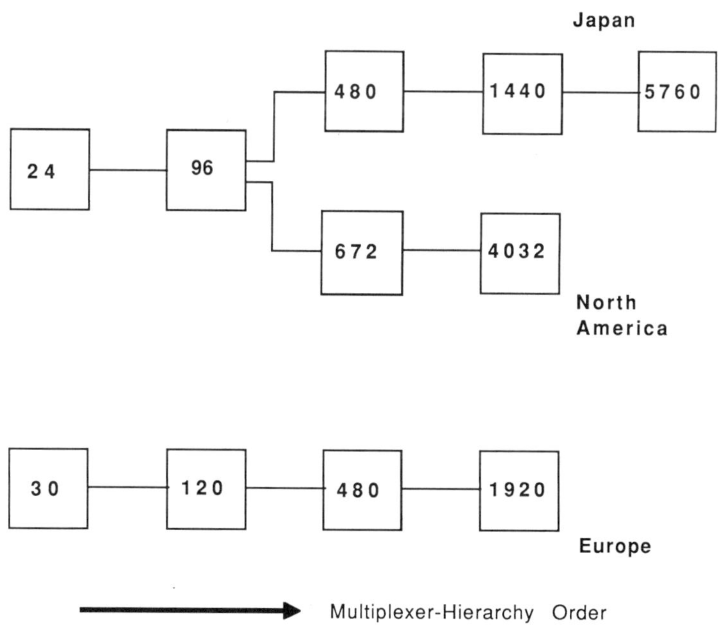

Note: The integer within the box denotes number of 64Kbps channels

Figure 1.19. PCM time division multiplexing hierarchies.

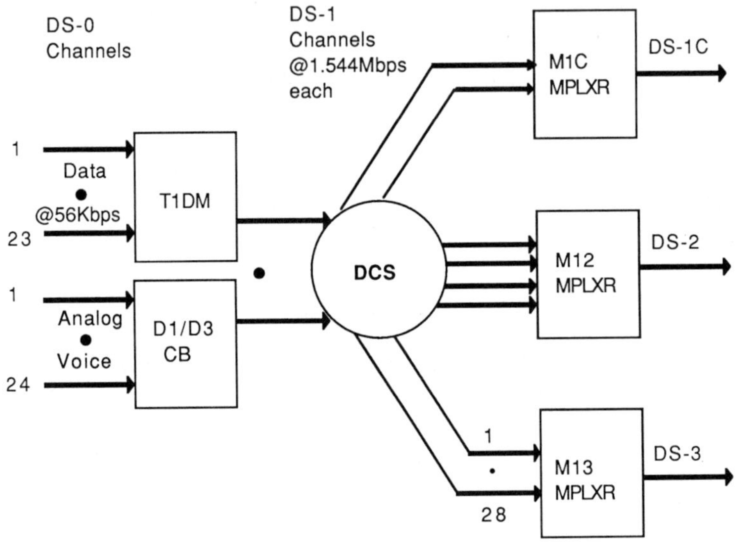

Notes: CB = Channel Bank
DCS = Digital Crossconnect System
MPLXR = Multiplexer
DS-i = ith Hierarchy Digital Signal

Figure 1.20. North American time division multiplexing hierarchy.

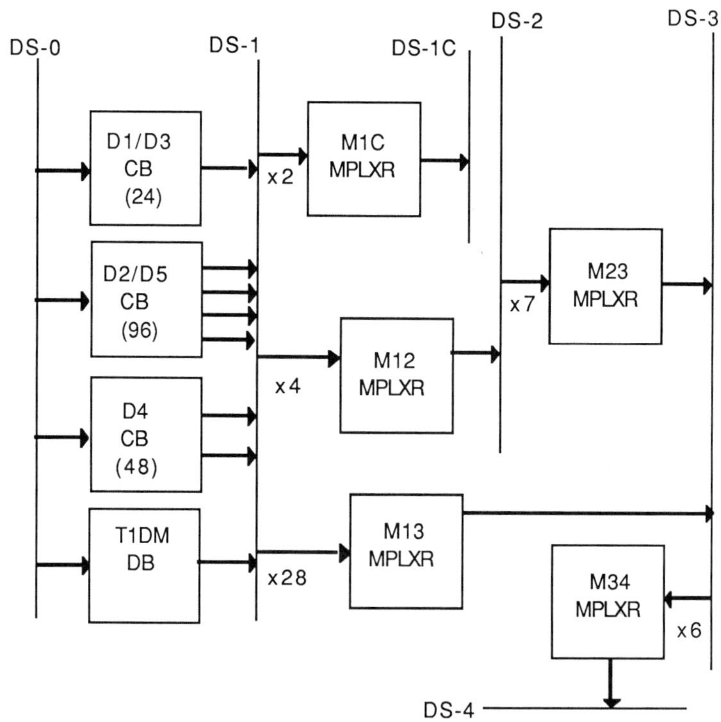

Figure 1.21. North American time division multiplexing hierarchy—second view.

Table 1.1 Comparison of PCM Channel Banks

Design Parameter	D-Channel Type D1	D2	D3	D4	D4E	D5
No. VG Ckts.	24 72,96	24,48,	24	24,48	30 72,96	24,48,
Sampling Sequence	1,13,2,14, 3,15...... 12,24.	12,13,1,14, 5,21,9,15.... 8,24.	1,2,3,...	Options for D1,D2, D3	1,2,3,... 29,30.	Options for D1,D2, D3,D4
No. of Bits per Sample	7	8	8	8	8	8
Signaling Bit in Channel	Every 8th Bit	8th Bit of 6th Frame	Same as D2	Same as D2	16th Time Slot	Same as D2
Companding Parameter*	μ=100	μ=255	μ=255	μ=255	A=87.6	μ=255

Notes (*)

1. Uniform PCM is inherently inefficient since strong signals do not occur as infrequently as the weak ones. There are two companding techniques employed to enhance the weak signals and deemphasize the strong ones: μ-Law is used in Japan and North America, A-Law is employed in Europe.
2. The μ-Law companding transforms the input signal (x) into an output signal as follows:

$$\text{Output signal} = \text{sign}(x) \left\{ (\ln(1 + \mu*[x]/V)) \,/\, (\ln(1+\mu)) \right\} \ldots.\ 0 <= [x] <= V$$

where

sign(x) is the polarity of x

[x] is the absolute value of x

ln(k) = natural logarithm of any real number k

μ = μ-law parameter value (usually = 255)

Table 1.1 *(Continued)*

3. The A-Law companding transforms the input signal (x) into an output signal as follows:

$$\text{Output Signal} = \text{sign}(x) \left\{ (A^* [x]) / (1+\ln(A)) \right\} \ldots\ldots\ 0 <= [x] <= 1/A$$

$$= \text{sign}(x) \left\{ (1+\ln(A^* [x])) / (1+\ln(A)) \right\} \ldots\ 1/A <= [x] <= 1$$

where

A =A-law parameter value (usually=87.6)

The bit-robbing technique, as discussed above, adds a constraint to the 8 bits of user data for certain frames. Bit-robbing always resulted in some voice quality degradation. To remove this constraint, a multiframe consisting of 24 PCM-frames (instead of the 12 frames used traditionally) will be employed in conjunction with some new ways of encoding signalling data. Such a frame is called the extended super frame (ESF). The multiframe (ESF) synchronization will now be derived from a single pattern, ''001011,'' derived from the framing bits from frames 4, 8, 12, 16, 20, and 24. A cyclic redundancy check (CRC) code will be imbedded in the framing bits of frames 2, 6, 10, 14, 18, and 21. This is a new scheme for providing an end-to-end checking of a link for diagnostics. A great deal of network management data dealing with items such as alarms, supervision, status, network configuration, performance indication and maintenance will be imbedded in the framing bits of the remaining frames, i.e., 1, 3, 5, 7, 9, 11, 13, 15, 17, 19, 21, and 23. The traditional signaling data dealing with on-hook, off-hook, dialing digits etc. will be sent on a separate common channel signaling (CCS) network. With this approach, the users will be able to get the full 64 Kbps capacity out of each PCM channel, instead of the existing 56 Kbps capacity. In order to implement the ESF concept and use the preferred bipolar digital transmission technique, the bipolar, 8-zeros suppression (B8ZS) technique must be employed to prevent the formation of long null sequences. Refer back to Fig. 1.17 for an illustration of the way that the B8ZS technique works to achieve an ESF that provides the clear channel capability (CCC) in digital systems of the future for North America and Japan. The multiplexing techniques used in Europe and other countries never had the problems created by bit-robbing.

1.7.3 NETWORK MANAGEMENT AND CONTROL (NMC) TECHNIQUES

One way to classify a network is by the amount of intelligence distributed throughout the network. Network links are assumed to possess only a temporary storage dealing with the information being transmitted. If a repeater does any amount of information processing to keep the link operating, it should be considered as an intelligent network node. Nodal intelligence is determined by the amount of random access memory (RAM) provided, the number of million instructions per second (MIPS) that can be executed by the nodal processor(s), the total number of instructions in all of the software/firmware modules available, and the type/usefulness of the various software/firmware modules.

The second way to classify networks is the extent to which the nodal intelligence is used to manage and control the performance and operation of the entire network. In some systems, each node may be very intelligent, but it cannot communicate with the other nodes. Such a state can occur when the network nodes are manufactured by different vendors, and when each vendor employs a proprietary scheme for internodal communication. This situation will prevent the realization of an integrated network management and control philosophy. NMC integration implies the extent to which network nodes communicate with one another to manage and control the entire network. The network is as intelligent as the amount of NMC integration employed by it.

NMC functions can be classified into four distinct categories:

1. Routing of network traffic;
2. Data base related services;
3. Maintenance and repair management services; and
4. Network operation services.

We will now describe each of these NMC functions in a little more detail.

1.7.3.1 Routing of Network Traffic

Allowing the traffic to flow on alternate routes results in a better use of network facilities and better performance to subscribers. Three methods of traffic routing are generally employed by a system with alternate routing capabilities.

Deterministic Method. This is the simplest routing control, involving the use of fixed primary (also called the high usage, or HU) routes and a fixed number of alternate (also called the final) routes. Such a method was employed in the early versions of electromechanical switches, which stored the routing information in hardware. Since each switching node made its own decision as to the link it chose for the path, the originating or any other node did not know the exact path taken by the information. Such a method was called the successive node route control (SNRC). The public voice networks have employed this method until recently, when the common channel signaling (CCS) was introduced.

Synoptic Method. The network management center (NMC) samples the traffic flows in the entire network frequently and, based on the information collected, it decides which path the next call or message will follow. Sometimes routing tables are dependent upon the time-of-day. In some systems, the cost of a call determines the route taken. This approach is similar to that in an ISDN employing the CCS system. Using this scheme, the network becomes more intelligent in terms of the paths employed for routing traffic. Since most of the networks of the future will employ a NMC, the synoptic method will become commonplace.

Adaptive Method. Each network node gathers information from the neighboring nodes on a periodic basis and decides how calls or messages are to be dynamically routed towards their destination. Some control information must be collected (e.g., in the form of data stamps, as it passes through different nodes) to gather information regarding the paths used. Additional care must be taken to avoid looping. This approach is generally employed in fully distributed packet-switching systems (PSSs).

During some holidays and periods of network outages, it may also be necessary to deny service to subscribers. A voice network achieves this goal by either not providing a dial tone or by giving a busy tone to the calling subscriber. The data networks achieve this goal through flow control. This may involve a simple form of feedback to the subscriber or a complicated set of rules dealing with permission to send data.

1.7.3.2 Data Base Related Services

A properly designed NMC system must maintain a relational data base that provides the following essential capabilities:

a. Network directory service;
b. Network component inventory;
c. Traffic/performance statistics gathering; and
d. Billing.

1.7.3.3 Maintenance and Repair Management Services

A properly designed NMC system should provide the following capabilities required to maintain an almost fault-free system:

a. Collection of system alarms;
b. Converting the system alarms into trouble tickets and sending these to qualified persons;
c. Scheduling diagnostics and repairs; and
d. Predicting impending failures and taking evasive actions (self-healing).

1.7.3.4 Network Operations Services

A properly designed NMC system should provide the following functions to help the corporation run a smooth operation:

a. Scheduling training classes for the NMC-related personnel;
b. Staffing of the NMC personnel; and
c. Ordering new network components or retiring other components for the purpose of maintaining an optimum network design on an ongoing basis. This effort will require the use of analysis and design tools described in this book.

2

Generalized Tools for System Analysis and Design

2.1 INTRODUCTORY REMARKS

Any network system can be modeled at any level of detail. Even the largest network system can be modeled as a single node that handles a certain mix of user transactions per second. As the model gets more complicated, one has to define the performance of each network system in terms of a larger number of transactions and different mixes. User transactions become mixed with other transactions required to maintain a given level of network system integrity, while providing an acceptable level of services to the subscribers.

The task of modeling is simply that of representing the entire system in terms of a network of nodes belonging to certain levels of a hierarchy. Sharma et al. (1982) provide additional understanding of this. It will suffice to say here that a successive decomposition (each resulting in a different model) is continued until no more useful understanding results.

In this chapter, we will introduce the concepts of systems analysis, design, and performance in a way that will be useful to readers for understanding the material of subsequent chapters.

2.2 CONCEPTS OF SYSTEM ANALYSIS AND DESIGN

Figure 2.1 illustrates the relationship between modeling, analysis, and design. It shows that it is very hard to separate the efforts generally associated with either modeling, analysis, or design. Such efforts are completely intertwined. In general, an analysis effort should provide a prediction of system performance, assuming correct distribution of transaction arrivals and their holding times. A system design effort, on the other hand, should yield a network system architecture and topology that will provide the desired performance. But in all cases, many passes will be needed to arrive at the correct system architecture and topology. The iterative process, as illustrated in Figure 2.2, is probably the least understood process by the telecommunications professionals. They still assume that network design is a one-time effort executed on a large mainframe at the beginning of a network planning cycle. Many of them expect the use of detailed tariff-related data bases for each resulting topology during the above-mentioned iterative process. Such an approach will always slow down the design process, even on a mainframe. Since the

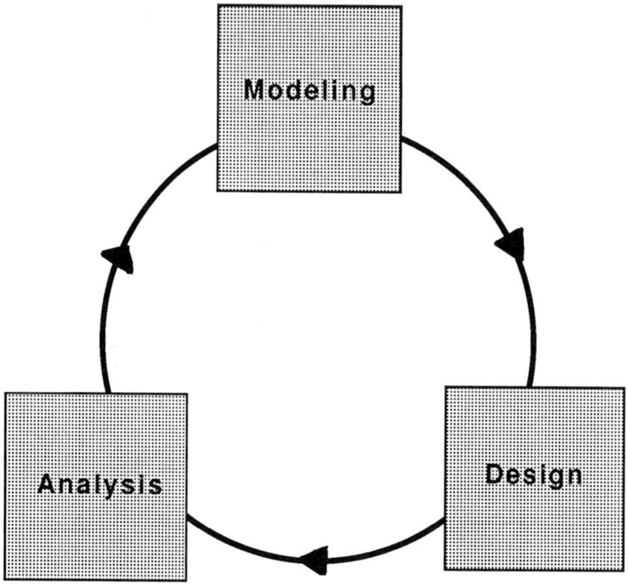

Figure 2.1. Relationships between modeling, analysis, and design techniques.

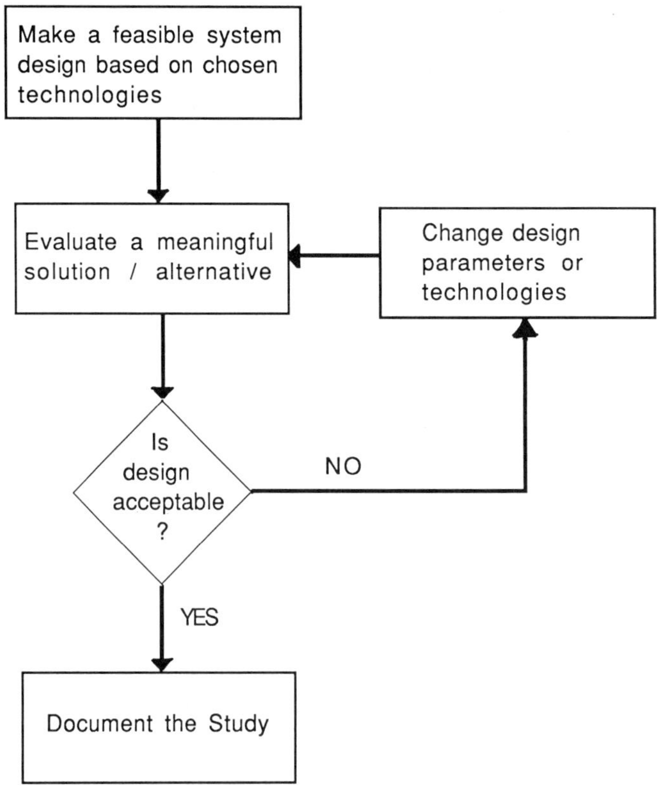

Figure 2.2. Iterative network design methodology.

topology of a network is affected by architectural/technological choices, and never by tariffs (consult Chapters 7, 8, and 9 for justification), one should only employ simple models of tariffs, in order to make the design process of Figure 2.2 a truly *interactive* one. See Chapter 6 for a discussion of a software package that achieves this goal while providing a user-friendly interface. Since this package runs on a desktop workstation, it also encourages ongoing use for studying the effects of new tariffs and technologies during the entire life cycle of each network system.

Figure 2.3 illustrates the life cycle of a properly designed network system. Ten tasks are executed in series. Modeling of existing environs, user requirement modeling (existing traffic intensities and performance), user requirements growth modeling, analysis and design of all of the alternatives system cutover models, return-on-investment (ROI) modeling, choosing of the most cost-effective system, system implementation, system performance measurements after the cutover, and documenting the results. But the most important event is the ongoing collection of analysis and design tools required to systems

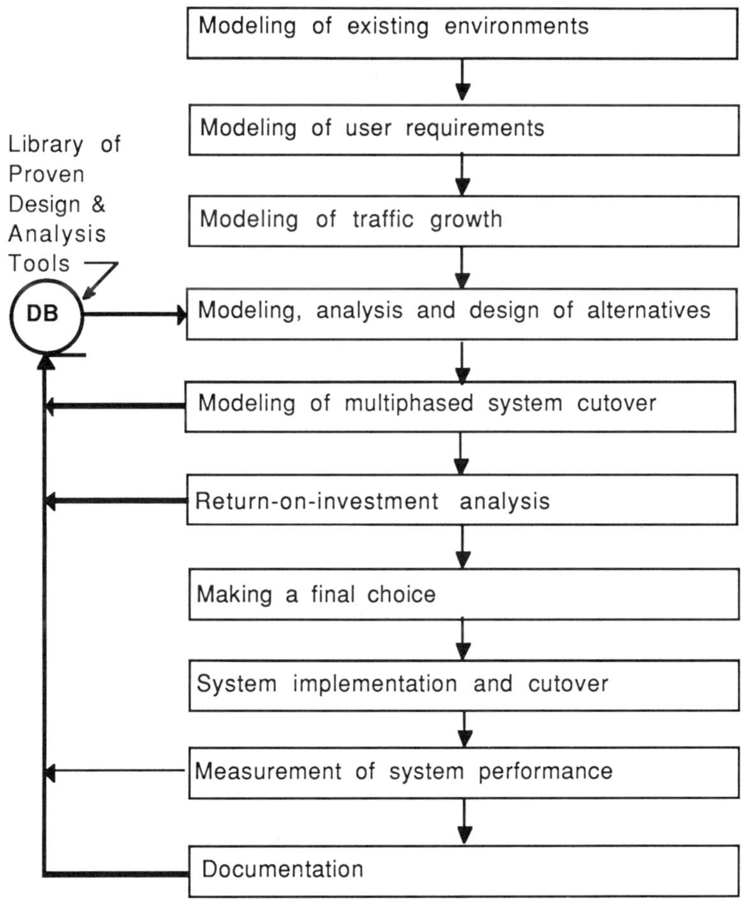

Figure 2.3. Life cycle of a network system.

engineer this and all other previous network systems. The data base of all of the tools represents the art and science of networking, if these tasks are performed with diligence and patience for several complex network systems.

There are three basic methods of analyzing system performance:

1. Physical simulation, which requires the availability of the final system under investigation. Tests are performed under typical traffic loads to obtain all aspects of the performance. This is by far the most expensive and time-consuming approach.
2. Computer simulation which requires a software modeling of all operations and running the program until one achieves a statistical equilibrium. The development of the simulation software may take a sizeable amount of time. Furthermore, the use of a mainframe for predicting the performance of every interesting configuration may cost a great deal of money.
3. Analytic simulation, which employs well-proven analytical techniques for modeling and predicting the performance of a network system or its elements. This method is by far the most elegant, the quickest, and the cheapest.

The author's experience has shown that even the most complex system can be broken into smaller and smaller components until analytical techniques can be employed, and then the results are combined to predict the performance of the entire system. It must be emphasized that all methods yield approximate results, since no one can simulate all possible environs. Large discrete systems are afflicted with too many design and traffic variables. The number of possibilities generally exceeds the number of electrons in the universe. In most cases, approximate analytical tools will generally yield more meaningful results than other methods, considering cost and time constraints.

2.3 CONCEPTS OF SYSTEM PERFORMANCE

It is a sad fact of life that very little concern for performance is being given to most of the network systems that are being implemented, or for which standards are being developed. The simple answer lies in the complexity and high costs related to the time-consuming tasks of predicting the performance of any large network system. These observations apply both to public and private network systems. Performance of a network system generally deals with costs, throughputs, quality-of-service (QOS), and grade-of-service (GOS). These are described in the paragraphs that follow.

2.3.1 SYSTEM COSTS

This criterion is probably the most critical performance parameter, since it determines the cost of service provided to the users. The cost of a call made on the public telephone network is a prime example of a system cost. Most system designers talk of system throughput in terms of the telephone call handling capacity, as an example. Very little attention is given to the system costs and the efforts to lower those costs through network optimization. As a result, the users continue to pay higher rates for the services provided by the public networks.

Costs of a system should be divided into major components such as transmission,

NMC, and hardware. Experience shows that the transmission costs are generally the predominant part of the total cost. For most private networks, the costs related to transmission may constitute as much as 70 percent. The costs incurred for network management, operations, and control may be 20 percent of the total costs. A common practice is to show the costs on a monthly basis, since all carriers bill the users each month. The costs of NMC are also natural for a monthly exhibit. It is also easy to spread the costs of hardware for the entire life cycle of the system, while considering the costs of money (e.g., interest, depreciation, etc.), and obtain an equivalent monthly cost. See Subsection 2.4.4 of this chapter for a methodology to convert fixed hardware costs into equivalent monthly expenses.

2.3.2 SYSTEM THROUGHPUTS

System throughput measures the overall capability of the system in terms of the maximum number of transactions that can be handled per unit of time.

The throughput of a circuit-switched system is represented by the maximum number of call attempts and calls handled per average second of a busy hour by each network node and by the entire network system. The throughput of the entire system is also determined by the maximum number of calls that can be handled concurrently during the busy hour. The latter value is affected by the call duration and the number of nodes in the call path.

The throughput of a packet-switched system is represented by the number of input packets or messages that can be handled per average second of a busy hour. Some vendors represent the throughput of their packet switch node by the maximum bit rate (input rate plus the output rate) handled during a second. Another useful measure of system throughput should be the maximum number of packets or messages handled by the entire network during a second or busy hour.

2.3.3 QUALITY-OF-SERVICE (QOS)

Quality-of-service deals with performance issues such as the transmission quality, voice quality at the receiver, length of error-free periods, average bit-error-rate (BER), system reliability, as determined by such measures as mean-time-between-failures (MTBFs) for each node and link and mean-time-between-system-failures (MTBSFs), and system connectivity (a measure of the capability to provide service to all subscribers during component failures). In order to study system connectivity, one must first study the capability of each node to connect to a large number of subscribers and/or access lines or trunks.

2.3.4 GRADE-OF-SERVICE (GOS)

Grade-of-service criteria deals with the degradation in service caused by contention for critical resources when all of those resources are functioning.

The GOS for a circuit-switched system deals with the statistical distributions of the following design parameters:

1. System response time (elapsed time measured from the moment a user goes off-hook to the moment the user hears a dial tone);

2. Call connection time (elapsed time measured between the moment the last digit is dialed and the moment connection is made between the calling and the called subscriber); and
3. Call set-up time (sum of the above two quantities plus the time to dial all of the digits).

The GOS of a CS system is also determined by the probability of a call loss or blockage caused by the unavailability of a path through the network during a busy hour. The GOS of a CS system can only be computed by taking into account the actual network topology and the blocking probability experienced by each network node-link bundle pair.

The GOS for a packet-switched system deals with the statistical distributions associated with the following design parameters:

1. Nodal response time (elapsed time measured from the moment that the last byte of the packet is received by the node to the moment that the first byte of the same packet is transmitted), and
2. Call/session set-up time (elapsed time between the moment that the last bit or byte of the call-setup-packet is transmitted and the moment that the first byte of the acknowledgment-packet is received by the packet mode terminal).

The GOS for a message-switched system deals with the statistical distributions associated with the following design parameters:

1. Nodal response time (elapsed time measured from the moment that the last byte of the message is received by the node to the moment that the first byte of the same message is transmitted by the node), and
2. Message retention time provided by the system (elapsed time measured from the moment that the last byte of the message is received by a network node to the moment that the system puts the message into a storage media requiring manual handling).

In most cases, the exact distributions for system performance measures are not available. But it is easy to obtain their average, and 90 percentile values, using the central limit theorem that is applicable for the case when a large number of random variables determines the system performance measure. The average and the 90 percentile values of any system performance measure are quite adequate for studying improvements in a system design or for analyzing the performance of a given system. See Chapters 3 and 4 for the tools employed to analyze the performance of a given or a new system.

The GOS for each network node must be studied before computing the GOS of a system. Vendor data and technical proposals are excellent sources for nodal GOS data in the form of a termination (via subscriber line (SL), access line (AL) or trunk) cost, maximum number of terminations allowed, the type of modularity associated with line/trunk terminations, maximum traffic handling capacity in erlangs for each type of termination, nodal switching network architecture (e.g., the number of stages and type of transmission interfaces), nodal blocking/delay characteristics, CPU type and capacity in MIPS, maximum size of RAM, types of secondary storage devices allowed, CPU word length, capacity to handle busy hour call attempts or nodal throughput, the number and type of services/features handled, NMC capability matrix and equipment (switching, air conditioning, power, etc.), footprints, and so forth.

2.4 GENERAL-PURPOSE ANALTYICAL TOOLS

Experience has shown that a large number of tools are often needed during the network design phase. The absence of such tools can make the designer's task quite difficult. Easy access to such tools can provide the designer with a keen insight into the operation of the network system or provide ''what if'' type solutions to many design alternatives. A set of generalized tools can be divided into the following categories:

1. Combinatorics and error analysis;
2. Data stream analysis;
3. System reliability analysis; and
4. Return-on-investment (ROI) analysis

The textbooks of Feller (1957) and Parzen (1965) may be consulted to get a deeper understanding of the actual tools described in this and the next two chapters.

2.4.1 COMBINATORICS AND ERROR ANALYSIS

Combinatorics is a field dealing with the possible combinations and permutations of limited items taken out of certain available items. As a result, this field deals with the set/group theory that deals with the enumeration of special patterns. It is not surprising that combinatorics is quite often used for applications dealing with computers and communications that involve 0's and 1's.

2.4.1.1 Number of Combinations or Permutations of x Out of N Items

Consider N items, empty or occupied, as represented by 0's or 1's. It may be interesting to compute the number of combinations of x occupied items taken out of N items while ignoring the order into which the x items are occupied. This can be expressed as follows:

$$C(N, x) = N!/[x!(N - x)!] \tag{2.1}$$

where

$N! = 1 * 2 * 3 \ldots * N = N$-factorial

Some authors also equate C_x^N and $\binom{N}{x}$ with $C(N, x)$.

The number of permutations of x occupied items taken out of N items requires the knowledge of the order in which the items are occupied. $P(N, x)$ can be expressed as follows:

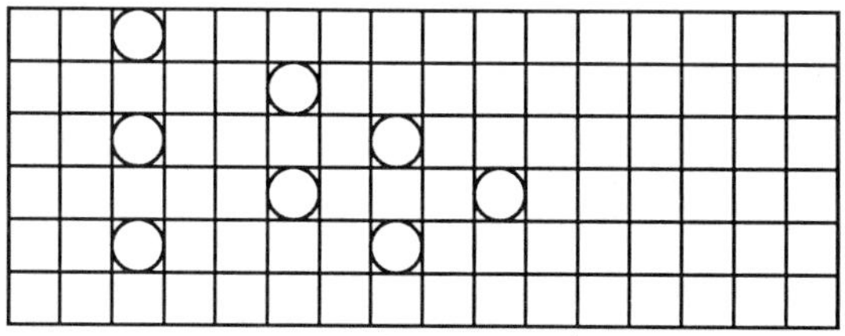

Figure 2.4. Cellular array pattern {3,2,2,1} of occupied cells.

$$P(N, x) = N!/(N - x)! \tag{2.2}$$

The probability of a pattern $\{j, k, \ldots, l\}$ of occupied cells (See Figure 2.4 for an example pattern with $L = 6$ and $N = 16$) anywhere in the cellular array can be expressed as follows:

$$\begin{aligned} &\text{Prob}\,\{j, k, \ldots, l\} \\ &\quad = [C(L, j) * C(L, k) * \ldots * C(L, l)][P(N, x)]/ \\ &\quad\quad [C(NL, m)][u!v! \ldots w!] \end{aligned} \tag{2.3}$$

where

u, v, and w are the number of vertical columns that have an identical number of occupied cells (e.g., u, v, w are 1, 2, 1 for Figure 2.4)

$j + k + \ldots + l = m$ (e.g., $m = 8$ for the pattern 3, 2, 2, 1 of Figure 2.4)

$u + v + w = x$ (e.g., $u + v + w = 4$ for Figure 2.4)

2.4.1.2 Some Applications of Combinatorics to Error and Throughput Analysis

Every system analyst is occasionally called upon to compute the output data error rates in a data communication system as a function of the raw error rate, and the error detection or error correction technique employed on the input data. The analyst may also be interested in the optimum sizes of the transmitted block for maximizing the utilization of the limited storage or limited bandwidth of the output tranmission channel. We will now describe some typical applications that require the use of several effective tools provided here.

Application 1. A given binary array of information of size N by L, consists of N bytes, each consisting of seven information bits and one parity bit. This represents a 1-dimensional parity scheme. Such a scheme will detect all odd numbers of errors (most probably only one error) occurring in each byte. It will generally allow a pattern of two errors in a byte go undetected. The designer can add the last byte to provide a horizontal parity bit to each row of bits in the array consisting of $N - 1$ bytes. This represents a 2-dimensional parity scheme. Such a scheme will generally allow a pattern of four errors to go undetected. Given the raw probability of bit error, P_e, the undetected bit-error-rate

(UBER) can be computed. UBER can be expressed as the product of two probabilities: the probability of a given pattern of errors occurring anywhere in the array, as defined earlier, and the probability of exactly "x" errors occurring anywhere in the array for a given value bit-error-rate (BER) specified by the value P_e. Assuming only random errors and $P_e \ll 1$, the value of UBER for patterns of errors defined as $\{2\}$ and $\{2, 2\}$ occurring anywhere in the array can be obtained and expressed as follows for the 1- and 2-dimensional parity schemes, respectively:

$$\text{UBER1} = \left\{[C(L, 2)] * [P(N, 1)]/C(NL, 2)\right\} * \left\{C(NL, 2)\, P_e^2 (1 - P_e)^{NL-2}\right\} \tag{2.4a}$$

$$\text{UBER2} = \left\{[C(L, 2) * C(L, 2)] * [P(N, 2)]/2C(NL, 4)\right\} * \left\{C(NL, 4)\, P_e^4 (1 - P_e)^{NL-4}\right\} \tag{2.4b}$$

where $C(N, x)$ and $P(N, x)$ are the same as defined before.

Application 2. A block of data with N bits can employ a cyclic redundancy check (CRC) code of R bits long that is imbedded in the N bits. Table 2.1 shows the relation-

Table 2.1 Number of information Bits, k, Per Block of Length n Versus Hamming distance d and number of Detectable Errors, d-1 for Bose-Chaudhuri-Hocquenghem (BCH) Codes

Fixed Integer	Block Length	d (Hamming distance in bits)					
		3	5	7	9	11	13
m	n	k (Number of information bits)					
4	15	11	7	3	-	-	-
5	31	26	21	16	11	6	1
6	63	57	51	45	39	33	27
7	127	120	113	106	99	92	85
8	255	247	239	231	223	215	207
9	511	502	493	484	475	466	457
10	1023	1013	1003	993	983	973	963

Notes:

(1) $n = 2^m - 1$ = block length

(2) d=2t+1 where t is the number of errors that can be corrected

(3) d-1=number of errors that can be detected

(4) BCH Codes are a subset of the Cyclic Redundancy Codes

ship between the Hamming distance, d, or the number of errors that can be detected, $d - 1$, to the information bits, k, and the block length, n. The CRC code is determined for each information block of k bits and added to it just before it is transmitted. Experience has shown that the Bose–Chaudhuri–Hoquengham (BCH) codes that are a subset of the CRC codes are easier to implement than the 1- or 2-dimensional parity schemes. Furthermore, the BCH codes are more efficient than the 1- or 2-dimensional parity schemes, since these also detect most of the bursts of two and higher bit errors. One can compute the UBER for a known raw bit-error-rate of P_e for the most likely undetectable number of errors (i.e., $q = d$, where d is the Hamming distance), as follows:

$$\text{UBER3} = C(N, q)\, P_e^q (1 - P_e)^{N-q} \tag{2.5}$$

See Figure 2.5 for some useful curves showing the benefits (in terms of UBER) and the penalties (e.g., in terms of redundancy) of the old 1- and 2-dimensional error checking and the modern CRC codes. The values of UBER3 will be lower than those plotted, for the reasons already discussed above.

Application 3. Another interesting application addresses the computation of an optimum block size for storing long messages with unknown lengths. Wolman (1965) developed an interesting relationship that defines the optimum block length for a large

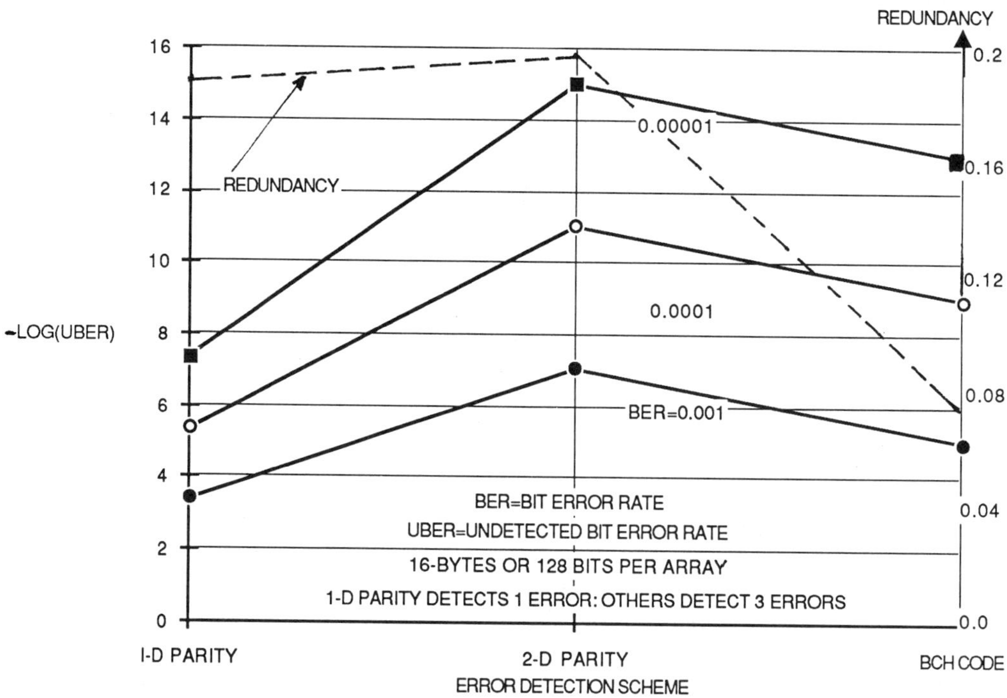

Figure 2.5. Probability of undetected error versus the error detection scheme employed.

number of message length distributions, as follows:

$$L_{opt} = \text{SQR}[2 * \text{BR} * L_m] \text{ bits} \quad (2.6)$$

where

BR = block redundancy due to overhead (in bits)
L_m = average message length in bits

Application 4. The Equation (2.6) above deals with a situation where an optimum use of the storage media is required for the case of no media errors. A closely associated problem deals with computing the optimum block size for a transmission channel characterized by random errors. Error control (implying almost no undetected errors) in transmission systems generally deals with blocked transmission and error correction through retransmission via feedback control. Each block contains a CRC code. A certain number (N) of blocks are transmitted before looking for a feedback (also called the Go-Back N scheme) message stating whether all blocks were received correctly or not. If one makes the block too long, each block may be received in error and must be retransmitted. If the block is too small, the overhead due to redundancy may be too high. Therefore, the designer is frequently called upon to solve for the optimum block length for a given raw binary-error-rate, P_e.

The optimum block length (see Schwartz, 1987) can be expressed as follows:

$$L_{opt} = \text{SQR}[\text{BR}/P_e] \ldots \text{ for BR} * P_e \ll 1 \quad (2.7)$$

where

BR = overhead redundancy per packet or block (=48 for HDLC)

There is also some interest in computing the efficiency of transmission. It is expressed in terms of P_e and a, which is the ratio T_t/T_b (T_t is the sum of block transmit time and the two-way turnaround propagation time, and T_b is block transmit time), as follows:

$$\zeta = \{\text{BI}/[\text{BI} + \text{BR}]\} * \{[1 - p]/[1 + (a - 1)p]\} \quad (2.8)$$

where

BI = information bits per block
BR = overhead redundant bits per block
$p = 1 - (1 - P_e)^{[\text{BI} + \text{BR}]}$

See Figure 2.6 for a plot of some useful curves relating ζ to block lengths for typical terrestrial and satellite channels with $P_e = 0.00001$, and for two-channel capacities.

2.4.2 DATA STREAM ANALYSIS

There are times when studying the manner in which the packets of a message become distributed in time over a shared input/output transmission line when several local ports

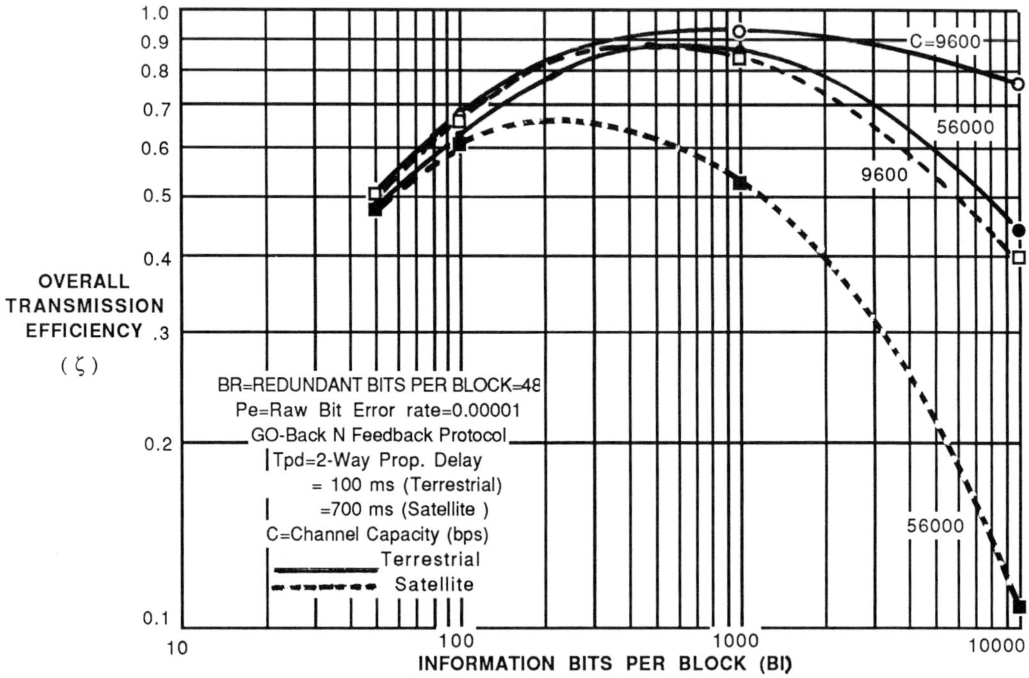

Figure 2.6. Transmission efficiency versus block size for typical terrestrial/satellite channels.

act as sources of message streams in a packet assembler and dissembler (PAD) is required. See Figure 2.7.

Assuming a first-in-first-out (FIFO) mechanism, identical speeds at both the input and output channels, and effective channel utilization less than unity, one can show that each message on the channel will experience some elongation in time. The average message expansion factor (AMEF) is related to the average number (N) of slots (or packets) per message and the number (S) of active sources (or local ports). An analytical solution was obtained through using computer simulation. The resulting expression for AMEF is as follows:

$$\text{AMEF} = 1 + [1 - 1/N - 1/S + 1/(NS)]*\rho \tag{2.9}$$

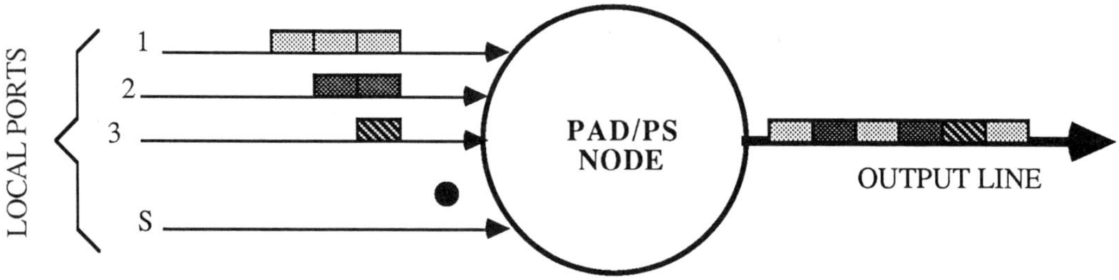

Figure 2.7. Merge operation involving several data streams.

where

N = average number of packets (or slots) per message
S = number of active sources of messages
ρ = utilization of the shared channel

Figure 2.8 plots several useful curves relating AMEF to N and S and for ρ equal to 1. It is interesting to observe that the maximum value of AMEF is 2 for an infinite number of active sources and large values of N. That is not intuitively obvious. Of course, the value of AMEF is reduced to 1 for $N = 1$ and any S. This value is intuitively obvious. The result of Equation (2.9) can be used to estimate the delays in receiving the entire length of a multipacket message at the receiver when many sources are vying for the same destination over a large number of virtual circuits. There are many applications where the receipt time of the entire message is required.

2.4.3 SYSTEM RELIABILITY ANALYSIS

There are many occasions when it is necessary to compute the reliability of a nodal subsystem or the entire network system. In order to achieve that goal, we must define several quantities, since reliability can be expressed in several ways.

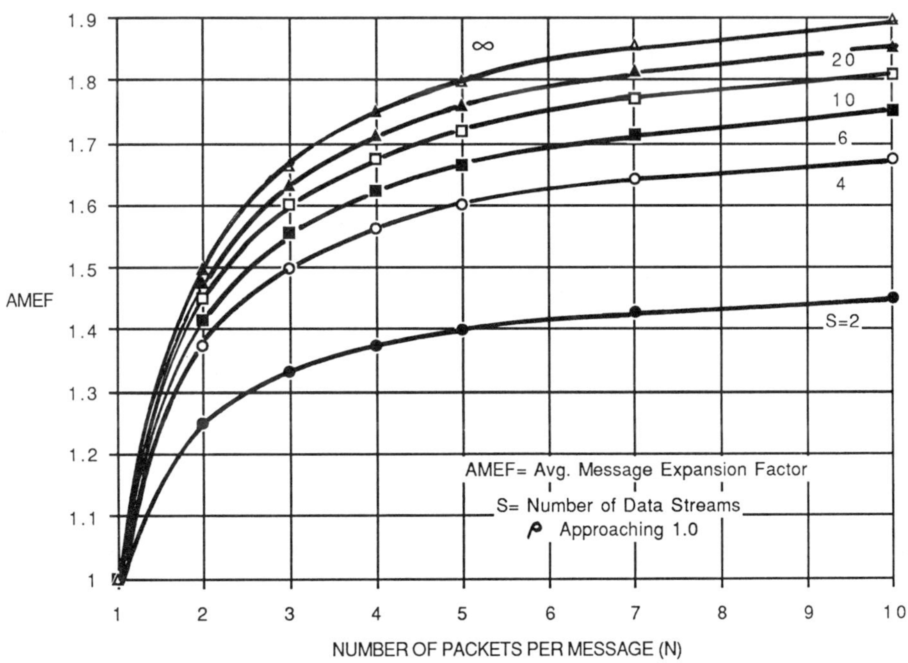

Figure 2.8. Message expansion factor versus the number of packets per message and the number of datastreams.

2.4.3.1 System Availability

System availability is defined as the percentage of time that the system is available for its assigned mission. It can be expressed as follows:

$$A = [\mathrm{MTBF}]/[\mathrm{MTBF} + \mathrm{MTTR}] \tag{2.10}$$

where

MTBF = mean time between system failures
MTTR = mean time to repair a system failure

The system can possess any arbitrary scope or size. It could be a nodal subsystem or the entire network system. The value of MTBF for nodal products is generally supplied by the vendor. The MTBF of a network system is generally estimated by the vendor for a proposed system, if required by the user. The value of the MTBF of a well-defined product is generally computed by using one of many methodologies. Each large vendor has a reliability department that dictates the methodology. In every case, the product is modeled as a collection of unique indivisible parts, each part associated with a unique failure rate. Military standards exist to aid in the selection of such failure rates. Using such models, the MTBF of a product is determined. Some of the models are discussed in the following paragraphs. The values of MTTR are estimated, using experimental data and some extrapolations based on product maturity.

The probability of successful mission (or mission reliability) during an observed mission time, T, can be stated as follows:

$$R(T) = \mathrm{Exp}\,[-T/\mathrm{MTBF}] \tag{2.11}$$

where

$\mathrm{Exp}\,[x]$ = exponential function of $x = e^x$

Probability of K failures during a mission time T can be expressed as follows:

$$P[K, T] = \mathrm{Exp}\,[-N_f] * [(N_f)^K]/[K!] \tag{2.12}$$

where

N_f = average number of failures during a mission time T
= T/MTBF
$K!$ = K factorial = $1 * 2 * 3 \ldots * K$

Each large system can be represented as a set of connected subsystems. The reliability of the entire system is a function of the way that the subsystems are connected together and the manner in which the system functions together. There are three ways that the subsystems can be connected together to allow the system to operate as a whole:

1. Series connection, whereby the system requires all subsystems to function concurrently;

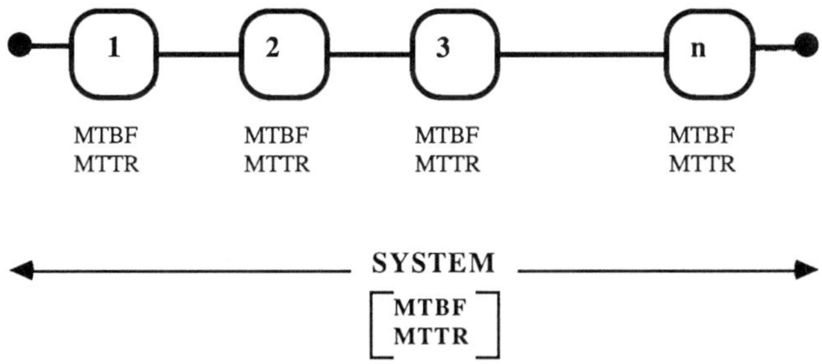

Figure 2.9. System with series-type redundancy.

2. Parallel connection, whereby at least one subsystem needs to function properly; and
3. *K*-out-of-*N* type connection whereby at least *K* subsystems must operate for the entire system to provide a successful mission.

2.4.3.2 Reliability of a Series-Type Redundant System

A large system can be modeled as a set of subsystems connected together in a series as shown in Figure 2.9.

The reliability of the entire system can be expressed as follows:

$$A_s = \prod_{i=1}^{n} A_i = A_1 * A_2 * \ldots * A_n \tag{2.13}$$

$$(\text{MTBF})_s = \left[\sum_{i=1}^{n} 1/\text{MTBF}_i \right]^{-1} \tag{2.14}$$

$$(\text{MTTR})_s = \{[1 - A_s]/A_s\} * (\text{MTBF})_s \tag{2.15}$$

2.4.3.3 Reliability of a Parallel-Type Redundant System

A system characterized by a parallel-type redundancy is shown in Figure 2.10. The reliability of a system employing parallel-type redundancy can be expressed as follows:

$$A_s = 1 - \prod_{i=1}^{n} (1 - A_i) \tag{2.16}$$

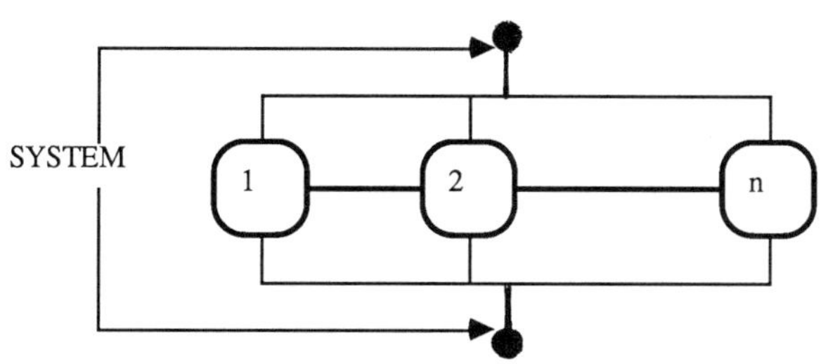

Figure 2.10. System with parallel-type redundancy.

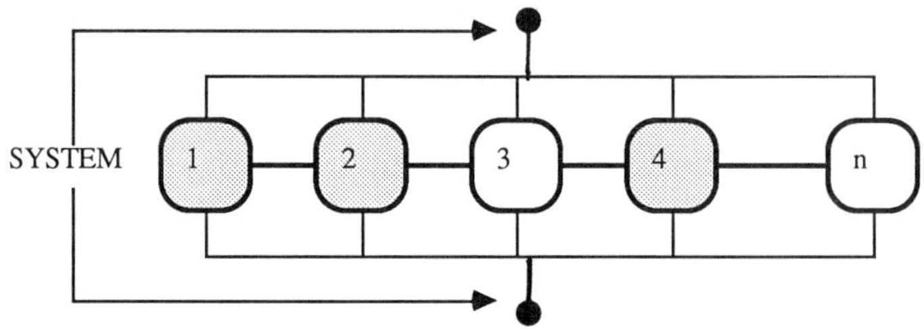

Figure 2.11. System with k-out-of-n type redundancy.

$$(\text{MTTR})_s = \left[\sum_{i=1}^{n} 1/\text{MTTR}_i\right]^{-1} \qquad (2.17)$$

$$(\text{MTBF})_s = [A_s/(1 - A_s)] * (\text{MTTR})_s \qquad (2.18)$$

2.4.3.4 Reliability of a *K*-Out-of-*N*-Type Redundant System

See Figure 2.11 for an illustration of a system employing *K*-out-of-*N* type redundancy. The reliability of such a system can be expressed as follows:

$$(\text{MTTR})_s = (\text{MTTR})_i/[N - K + 1] \qquad (2.19)$$

$$(\text{MTBF})_s = (\text{MTBF})_i[(\text{MTBF})_i/(\text{MTTR})_i]^{N-K} * [(N - K)!(K - 1)!]/N! \qquad (2.20)$$

$$A_s = (\text{MTBF})_s/[(\text{MTBF})_s + (\text{MTTR})_s] \qquad (2.21)$$

These above models can be used to compute the reliability of a composite nodal system or the entire network system. An experienced network designer generally reduces the above computational models into easily accessible computer programs. Using this methodology, it is an easy task to compute the reliability of a very large system. The major challenge lies in obtaining the correct values of MTBF and MTTR for individual subsystems (or products).

2.4.4 RETURN-ON-INVESTMENT ANALYSIS

A very cost-effective synthesis for a network system can be produced using the algorithmic methods described in this book. But, if the designer cannot present ideas to the upper management of the user community in a manner to which they are accustomed, the entire network design may come to naught. A common approach to making an effective presentation entails the following steps:

1. Make a model of the 5–10 year life cycle of the network system;
2. Determine the cost of deploying the current system during the life cycle;
3. Determine the cost of deploying the proposed system during the life cycle;
4. Determine the savings achieved during each year and during the life cycle; and
5. Compute the payoff period for the new hardware.

In order to succeed in the above effort, one must create economic models that consider the cost of money. A network system entails two types of costs: (1) transmission costs, which are usually incurred each month, and (2) hardware costs, which are usually incurred at the beginning of the life cycle. Of course, there are exceptions. Some users own their own microwave transmission facilities. Some users lease their hardware in order to avoid the one-time cost upfront. For that case, it is usually sufficient to complete Step 4 by showing the total savings achieved in each year of the life cycle and during the entire life cycle. But in every case, one must present several alternatives to the management, in order to avoid any surprises.

In order to compute the payoff period, one must first compute the life cycle costs of hardware by considering the cost of money. A simple technique is to compute the monthly payment (MP) for a principal value (PV) and a yearly interest rate (IR), and the total number of monthly payments (n). This value can be computed as a ratio of PV and a dividing factor (DF), as shown below:

$$\text{MP} = \text{PV}/\text{DF} \tag{2.22}$$

where

$$\text{DF} = \left\{[1 - 1/(1 + i)^n]/i\right\}$$
$$i = \text{IR}/12 = \text{monthly interest rate}$$
$$n = 12 * (\text{life cycle in years}) = \text{total number of payments}$$

The values of DF for IR = 10% are listed in Table 2.2 for some typical life cycle values.

Using this approach, the total cost of the hardware can be computed for the life cycle, as follows:

$$\text{Total life cycle cost} = \text{MP} * 12 * \text{Life cycle} = \text{MP} * n \tag{2.23}$$

Several system alternatives can thus be compared easily by changing the products and/or varying the values of interest rates or life cycle. Using this approach, all of the five steps mentioned earlier can thus be accomplished. These results should then be reduced into effective graphs for final presentations to the upper management.

Table 2.2 Values of DF Vs. Lifecycle Durations

Lifecycle (Years)	n (months)	DF
5	60	47.065
7	84	60.237
10	120	75.662
15	180	93.058

2.4.5 *COMPUTING DISTANCES AND TRANSMISSION LINE COSTS*

The costs of any transmission link are generally proportional to its length. Furthermore, the task of optimizing a network topology requires a large number of computations related to finding the closest and the farthest nodes from a given node. The execution time of a network design program will therefore be largely dependent on the distance and link-cost evaluation technique. Alternatively, the designer will always need an efficient technique for computing the distance between any two given points represented with vertical (V) and horizontal (H) coordinates. For North America, one can designate the location of any one given point by the readily available V and H coordinates. Data bases that relate a telephone number to the associate V and H coordinates can be purchased. The three-digit NPA (commonly known as the area code) and the NXX (commonly known as the CO number) codes for a customer premise (CP) location determine the V and H coordinates of the associated wire center of a BOC. To illustrate, the NPA-NXX equalling 213-390 for a CPE located in the Southern California region corresponds to V and H coordinates of 9227 and 7920 for a wire center (WC) named as LAS.

The V and H coordinate system is based on the Donald's projection of the North American continent, which includes Canada, the U.S.A., and Mexico. Such a projection system was derived to yield a simple method of computing straight-line distances between any two wire-centers of the Bell System and also thc associated costs of leasing a circuit. The distance between any two locations can be expressed as follows:

$$D_{12} = \text{SQR}\left\{0.1 * \left[(\text{V1} - \text{V2})^2 + (\text{H1} - \text{H2})^2\right]\right\} \tag{2.24}$$

where

V1, V2 are the V-coordinates of the two locations, and
H1, H2 are the H-coordinates of the two locations.
D_{12} equals the straight-line distance between the two locations in miles.

For countries without such a coordinate system, the value of D_{12} can be expressed in terms of the readily available longitudes and latitudes as follows:

$$D_{12} = \arccos\left[\sin(\text{LAT1}) * \sin(\text{LAT2}) + \cos(\text{LAT1}) * \cos(\text{LAT2}) * \cos(\text{LNG2} - \text{LNG1})\right] * 6371.256 \text{ Kmeters} \tag{2.25}$$

where

LAT1 and LAT2 are latitudes of the two locations, and
LNG1 and LNG2 are the longitudes of the two locations.

The author employed Equation 2.25 during the early 1960s, until someone mentioned the availability of the V and H coordinates for North America. A comparison of Equations 2.24 and 2.25 should easily convince the reader that the use of longitudes and latitudes will consume an inordinate amount of computer time when designing a large network. See Chapter 6 for an approximate method of obtaining equivalent V and H coordinates for countries situated in other continents.

One-time costs (or derived monthly costs for a given life cycle and cost of money) for a fully owned transmission line can be computed by constructing analytical models dealing with wideband (WB) modems, antennas/towers, media type/length, repeaters, trenching, and media protection devices. The monthly cost of leasing a circuit will require the use of many data bases.

For the predivestiture period, one required the knowledge of high-density (HD) and the low-density (LD) rate centers as defined by the Tariff 260 for computing the monthly lease of a multipoint voice grade circuit. The customer had no choice as to the route of the leased line. The monthly cost of a leased line depended on three schedules: Schedule A, corresponding to the assocated two HD rate centers, Schedule B, corresponding to one HD and one LD rate centers, and Schedule C, corresponding to two LD rate centers at the two ends of the circuit. The monthly cost per mile for each of the schedules and for each of the applicable mileage band are listed in Table 2.3. Similar tariffs were available for other services, such as DDS, and for other common carriers. The predivestiture tariffs required much less description than their counterparts for the post-divestiture period. This can be attributed to the numerous service options that are now available to the users.

Since the 1984 divestiture of AT&T, different types of data bases are required. One data base lists, for each rate center, the associated LATA (Local Access and Transport Area) number, name of the rate center, the state abbreviation, the CLLI (Common Lan-

Table 2.3 Pre-Divestiture Multipoint Private Line (MPL) Tariffs

Mileage Range	Monthly charges per mile		
	Schedule A	Schedule B	Schedule C
1- 15	$1.89	$3.47	$4.63
16 - 25	1.58	3.26	3.99
26 - 40	1.18	2.10	2.94
41 - 60	1.18	1.42	2.21
61 - 80	1.18	1.42	1.68
81 -100	1.18	1.42	1.42
101-900	0.69	0.69	0.71
over 900	0.42	0.42	0.42

guage Location-ID) code, point-of-presence (POP) rate center V and H coordinates, POP's wire-center V and H coordinates, and availability vector for the various services such as T-1, DDS (Dataphone Digital Service), and VG (Voice-grade) applicable to each common carrier. Furthermore, one will need more than one tariff for computing and comparing the monthly costs of leasing a circuit. AT&T employs three tariffs, 9, 10, and 11, to compute the monthly costs of the various services listed, as follows:

1. AT&T Tariff 9 defines the cost of the Inter-Office Channel (IOC) that connects the two POPs, the CO connection between the POP and the CPE CO, CO functions such as the Common-Control Switching Arrangement (CCSA), and Enhanced Private Switched Communications Service (EPSCS), M-24/M-44 multiplexing, customer controlled reconfiguring, and transfer and conferencing arrangements. See Tables 2.4 and 2.5 for the IOC mileage charges and CO connections rate elements, respectively.
2. AT&T Tariff 10 contains no rate information, but defines the data bases necessary to evaluate the V and H coordinates of POPs and the availability of various services. Such a tariff requires more than five hundred pages to describe it.
3. AT&T Tariff 11 contains all of the rate and service data relative to local channels (LCs) provided by AT&T with Access Coordination Function (ACF). AT&T obtains such channels from the applicable local exchange carrier (LEC) within the two LATAs associated with the two ends of the circuit. The rate structure for AT&T's local channels (LCs) is very similar to that for the LEC's special access tariffs, and consequently it varies by each associated state. Such a tariff requires about a thousand pages to describe it.

Two important points need to be made here. First, AT&T does not guarantee the minimization of the private line cost for the user who can, in some cases, find alternate POPs and lower the cost by separately leasing the appropriate IOC and special access circuits. The user can also purchase the ACF from AT&T to guarantee the end-to-end performance. Second, whereas the special access tariffs used by LECs employ the mileage between the rate centers, tariffs for IOCs require circuit mileage between wire centers. See Figure 2.12 for an illustration of this feature.

See Figures 2.13, 2.14, and 2.15 for the illustration of IOCs with intrarate center and interrate center LCs, and the LEC bridged multipoint circuit configurations, respectively. One can summarize the various steps required to compute the monthly cost of a leased private line as follows:

1. Determine the NPA-NXX (area code-CO codes) for the two applicable CPs (Customer Premises).
2. Determine the V and H coordinates of the wire centers that serve CPs.
3. Determine the LATA numbers for the two CPs.
4. Obtain a list of the V and H coordinates of the wire centers that serve POPs in the two LATAs.
5. For each POP that is in the same LATA as that for the CP, compute the distances between the wire centers serving the two CPs and the POPs within each LATA.

6. Determine the two POPs that provide the lowest cost for the LEC tariff.
7. Compute the inter-LATA distance between the two selected POPs.
8. Compute the monthly cost for the IOC considering the appropriate service (e.g., VG).
9. Add the LEC and IOC monthly costs.
10. Add the costs of ACF and other special access surcharge, inside wiring charge, and channel options, as listed in Table 2.5.

The above steps must be repeated for each circuit connection and for each service option. Some of these options require different data bases, as defined by Tariffs 9, 10, and 11. In case the costs of each individual local channel separately leased from a LEC

Table 2.4 AT&T's Inter Office Channel (IOC) Tariff 9 (3/13/87)

Voice-Grade Mileage Band	Fixed Monthly Cost	Per Mile
1 - 50	$ 58.74	$2.33
51 - 100	122.24	1.06
101- 500	164.24	0.64
Over 500	324.24	0.32
DDS (2.4- 9.6kbps) Mileage Band		
1 - 50	$ 79.34	$2.33
51 - 100	142.34	1.06
101-500	184.84	0.64
Over 500	344.84	0.32
DDS(56kbps) Mileage Band		
1 - 50	$ 302.00	$10.07
51 - 100	566.50	4.78
101-500	742.50	3.02
Over 500	1407.50	1.69
Accunet T1.5 Mileage Band		
1 - Over 1000	$2600.00	$15.50(1-2 Years)
1 - Over 1000	$2600.00	$14.85(3 Years)
1 - Over 1000	$2600.00	$14.25(5 Years)

Table 2.5 CO Connection Rate Elements and Other Functions (Tariff 9)

Element/Function	Recurring (Per month)	Non-Recurring (One Time)
Voice Grade (CO connection)		
DDS 2.4-4.8 Kbps-Type 1 or 2 (CO Connection)		
DDS 9.6 Kbps-Type 1 or 2 (CO Connection)		
DDS 56 Kbps-Type 1 or 2 (CO Connection)		
Accunet T1.5 (CO Connection)		
CCSA, EPSCS, SCAN (CO connection etc.)		
M-24, M-44 Multiplexing (hardware, software etc.)		
Customer Controlled Reconfiguration (software etc.)		
Transfer Arrangement (software etc.)		
Conferencing Arrangement (software etc.)		
Per Automatic Protection Capability		
Fiber Only Routing (software etc.)		
Signaling-Signal Range Extension		
Conditioning(e.g. C1,C2,..,Echo Control)		
DDS: 56 Kbps Diversity		
Accunet T1.5 -Special Conditioning		
Cusomer Directed Testing		
Service Discontinuation, etc. etc.		

Note: The reader is advised to check the need for these elements/functions and their latest costs.

must be computed, V and H coordinates of the rate centers that serve all possible CPE locations and POPs will also be needed. The pricing process is both time- and computer-storage consuming. The succeeding chapters will show some simplified methods that are adequate for synthesizing optimum network topologies. Some of the commercially available line pricing packages can be used to compute the monthly cost of a final network, if desired.

Cost may not be the only criterion for selecting a particular LEC or OCC or a tariff. The network reliability should also play an important part. If a node corresponds to a

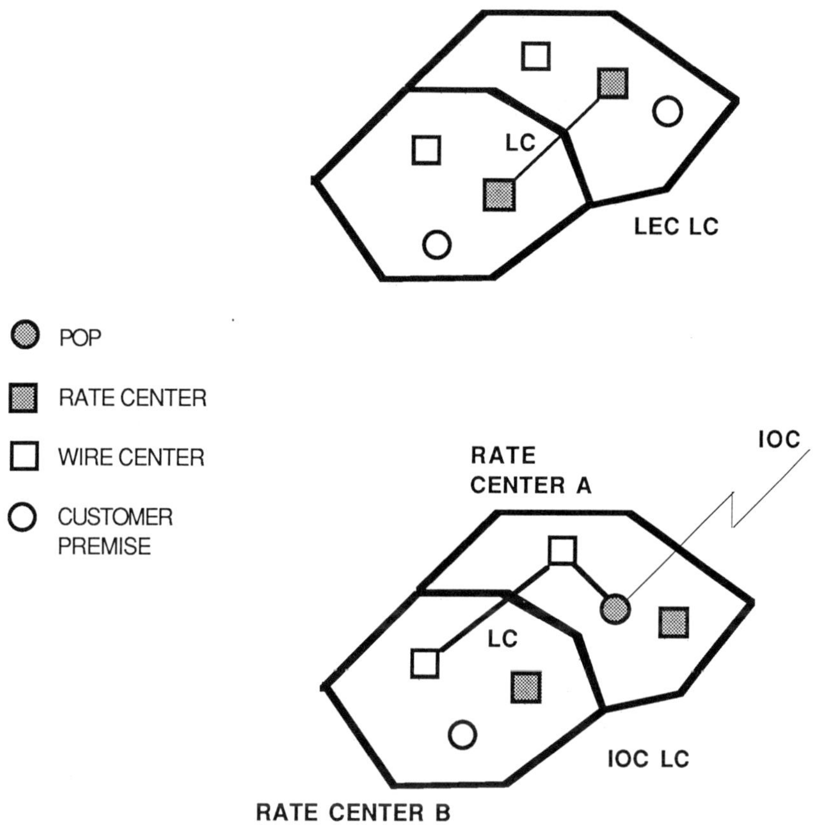

Figure 2.12. Local channel routing using AT&T's Tariff 11 or Exchange Carrier Special Access.

major corporate location, one should consider multiple access lines to that location by selecting alternate OCCs, leasing additional dial-up facilities, or even some *bypass* facilities. This will guarantee some communications with the outside world in case the LEC's CO/WC or OCC's POP fails. In most cases, bypass will also reduce the transmission costs. The amount of savings can be computed as shown in Chapters 7 and 8. In order to achieve the reliability objectives, the network designer must be familiar with the actual layout of each major switch and transmission facility and take the necessary steps to guarantee the flow of (at least) the high-priority traffic from every major location. These steps should influence the final network topology.

2.4.6 TRAFFIC ENGINEERING TECHNIQUES

2.4.6.1 Some Introductory Remarks

Traffic plays an important factor in the network design process. Its basic unit is the "erlang," as defined earlier. If a network system already exists, one can measure the

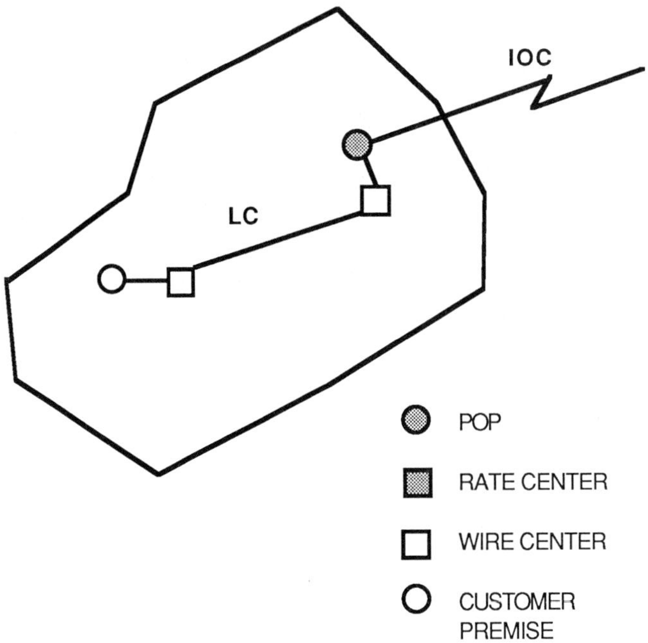

Figure 2.13. Topology of an intrarate center local channel.

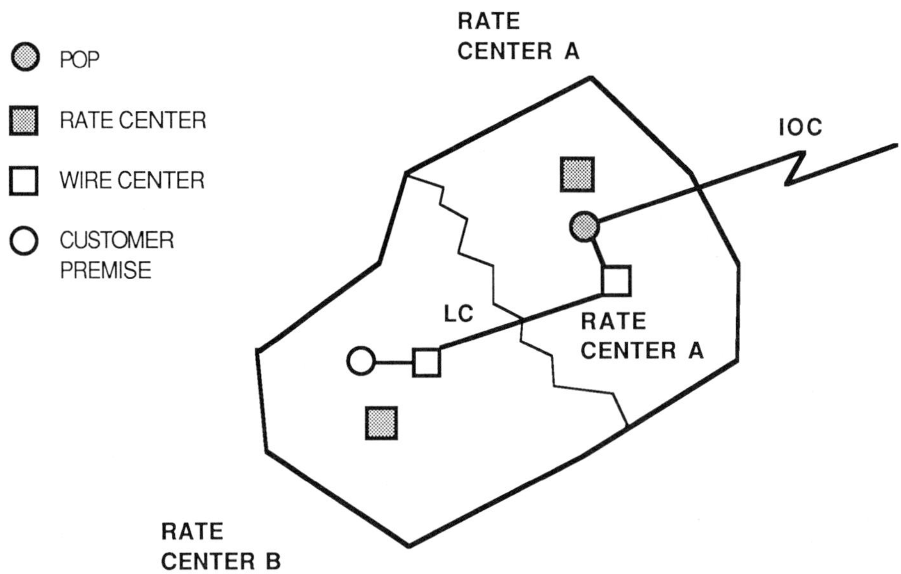

Figure 2.14. Topology of an interrate center local channel.

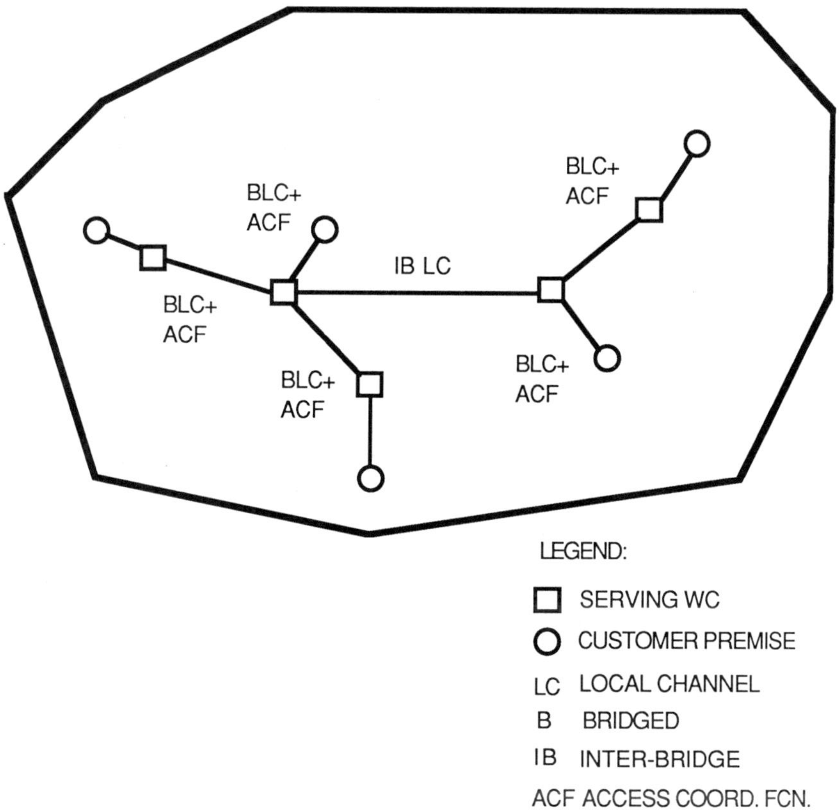

Figure 2.15. LEC bridged multipoint circuit topology.

traffic intensity for each network link and then derive the system grade-of-service (GOS) in any desired form. In case the GOS is better than that required, the designer may achieve some savings through better network dimensioning. If the designer has the resources, additional savings may be computed through the synthesizing of a new optimum network design.

Unfortunately, most networks are designed either from scratch or for environs that are quite different from the existing ones. To illustrate, one corporation has just merged with another large corporation. Although, the two existing networks maintained very good traffic data, each network operated in different areas with some overlaps, and was based on different GOS requirements. Furthermore, experience has shown that the combined network will generally create entirely new traffic loads on its resources as a result of different traffic flow patterns. Simple network tuning may not be adequate. For that reason, it may be necessary to redesign the network based on the new traffic patterns.

Even when the designer obtains the busy hour (BHR) traffic statistics based on actual observations or engineering assumptions, it will be found that a network design based on peak hour traffic loads is too expensive. Alternatively, the GOS of a new system based on peak traffic intensities will exceed those required by the new system. It should, therefore, be desirable to present a method for computing the traffic intensities that are finally used to dimension the network. Such values are generally called the time-consistent-averages (TCAs).

The following material presents some proven methodologies for obtaining BHR traffic intensities and the TCAs required for designing new networks or for redesigning given networks.

2.4.6.2 Estimating Traffic Loads for New Systems

The International Telecommunications Union (ITU) provides a wealth of data regarding the erlangs associated with various types of subscriber lines or terminals. Such data are based on the study of public networks of the world. Andrews (1967), AT&T (1966), and CCITT (1964 thru 1968) provide average values of busy hour (BHR) traffic intensities (incoming and outgoing calls) as .058, .04, and .037 for a typical telephone subscriber line of an AT&T CO, AT&T rural CO, and CCITT CO. These numbers must be modified if dealing with a doctor's or lawyer's office or a business PABX line. Such a line generally exhibits a busy hour traffic intensity of 0.25 erlangs. The use of personal computers (PCs) by many professional offices as a means of accessing data from host computers that manage commercial data bases (DBs) is becoming quite commonplace. Since these PCs use the public telephone network (PTN), their effect on the traffic intensities must be evaluated. The values of BHR traffic intensities on such lines may vary from 0.2 to 0.9 per each CO business line. The serious reader is advised to consult Eldin (1975) for an excellent discussion of CCITT recommendations on traffic engineering and network planning.

For data applications, there are no such things as the average traffic intensity figures for each terminal or port. Each application and each customer determines its own traffic intensities. In case there are no existing traffic models, one may try out several assumptions to establish a BHR. For example, one way to compute the BHR traffic intensity is to assume that each terminal produces a given (minimum) fraction of its capacity. Then, by using a traffic growth factor (TGF), one can easily vary the total traffic handled by the network.

The software design tool described in Chapter 6 allows the following techniques for specifying the BHR traffic intensities of each location:

1. Define the BHR traffic intensities (millierlangs for voice, and bits per second for data) for each location in the VHD file;
2. Define the types of data terminals located at each location and compute the total BHR traffic loads in bits per second using the specified nodal traffic weighting factors; and
3. Using a predefined BHR from-to traffic file, compute the traffic intensities for each location and for each AL/trunk facility for each selected network topology.

These techniques have been found to be quite useful for most network design situations, and their use will be illustrated in Chapters 7, 8, and 9.

2.4.6.3 Estimating Traffic Loads for Existing Systems

Most network systems employing SPC switching nodes have the capability of providing a magnetic tape file with a recording of each voice or data call. There are two types of traffic tapes for voice networks. The SMDR (Station Message Detail Recording) tape is produced by the cognizant PABX on a daily or monthly basis. The AMA (Automatic Message Accounting) tape is generally produced by the network management node on a monthly basis. Whereas the SMDR tape records the details of all telephone calls switched by the PABX, the AMA tape records details of calls handled by switches. Similarly, the data switching nodes have also the capability of creating traffic files with detailed call records. Each call record can be analyzed for the source-destination pair, time when the call was made and the call duration or the amount of data handled. Such an analysis can be used to create the From-To traffic tapes which in turn can be used to design a network. Detailed traffic flows for each nodal pair associated and for each business hour can be computed. In order to dimension the facilities of an affordable network, one also requires a methodology for deriving the meaningful traffic intensities (also called the time consistent averages or simply TCAs) from the From-To traffic files described above.

2.4.6.4 Traffic Engineering Practices for System Dimensioning

There are basically two approaches to obtain the TCAs required for network dimensioning: (1) North American practice, which requires the use of a typical month's traffic data, and (2) CCITT practice, which requires the use of 30 peak hours from yearly traffic data.

The major difference lies in the size of the data base employed. For the North American practice, one needs only a single month's traffic tape. The CCITT practice requires the use of traffic tapes for each of the months of a typical year.

The first approach requires the development of a point-to-point traffic flow chart as shown in Figure 2.16. All 30 days and all 24 hours of each day are displayed in two

Hours	Day1	Day2	Day 30	TCAs
H1	A1,1	A1,2	A1,30	TCA1
H2	A2,1	A2,2	A2,30	TCA2
H3	A3,1	A3,2	A3,30	TCA3
.	.	.	.	.
.	.	.	.	.
H24	A24,1	A24,2	A24,30	TCA24

(1) TCAh = AVERAGE (TCAh1,TCAh2,TCAh,3,...TCAh,30)
(2) TCA = MAXIMUM (TCA1,TCA2,TCA3,....TCA24)

Figure 2.16. North American methodology for computing TCAs for each network node pair.

HValue1	HValue2	Hvalue30	
A1	A2	A30	TCA

TCA = AVERAGE (A1,A2, A3,.....A30)

Figure 2.17. CCITT methodology for computing TCAs for each network node pair.

dimensions (horizontal and vertical columns of the display), as done for a public network. Only the valid business days and valid business hours should be considered for designing a private network for a corporation.

The CCITT practice for computing TCAs is illustrated in Figure 2.17. According to this practice, only the first 30 highest values of yearly traffic flows are considered for computing TCAs for any given point-to-point traffic flow. The earlier remarks dealing with the computational complexities encountered in computing TCAs are still valid.

It should be emphasized that the methodologies of Figures 2.16 and 2.17 are applicable to any point-to-point traffic flow. When there is no reference to a network topology, each point-to-point traffic flow refers only to the traffic flow between a given pair of subscribers. It is an enormous task to compute the TCA for each subscriber pair. Furthermore, this data is not fully useful for dimensioning the network without a topology. Therefore, TCAs for traffic flows or traffic intensities, are generally computed for either each network node (e.g., PABX) pair or each network node. Such TCAs are then used to dimension the resources of each network topology considered.

In order to compute the savings achieved through the use of time-of-day (TOD) routing schemes in the network, the designer must compute the TCAs for each hour and for each AL and trunk bundle of a given network topology. Such information helps the designer to dimension the circuit bundles as a function of the special TOD routing schemes. Experience has shown that significant savings can be achieved for large corporate and public networks that encompass several time zones of a large country, such as the U.S.A. and Canada.

One may choose to reduce the computational work associated with traffic engineering by choosing to compute the total traffic intensities (considering both input and output traffic) handled by each networking node based on the above methodology for TCAs for only the busy hour. These values can then be directly employed to dimension the AL bundles. To dimension the trunk bundles, one can use the procedure presented in subsection 2.4.7. It should be emphasized that such a procedure assumes symmetric traffic flows.

It would be instructive to show how the TCA values are combined with the analytical methods for dimensioning the AL and trunk bundles. Three methods may be used, as described in the following paragraphs.

1. The detailed method: (a) first define the network topology by choosing the concentrator and switch locations, (b) superimpose all of the From-To traffic values on each

related network resource for each business hour of interest, (c) develop the most economical TOD routing tables, including the alternate routing approach, and (d) using the analytical methods of Chapters 3 or 4, dimension the AL and trunk bundles for each TOD route while considering alternate routes.

2. The semidetailed method: (a) define the network topology consisting of all AL and trunk bundles, (b) develop TCAs for each From-To nodal pair only for the busy hour, (c) superimpose From-To TCA values on all affected network resources, and (d) using the analytical methods of Chapters 3 and 4, dimension the AL and backbone trunk bundles using the alternate routing approach, if applicable.

3. The simplified method: (a) define the network topology consisting of all AL and trunk bundles, (b) develop TCAs for each AL bundle, considering total traffic flowing in both directions only for the busy hour, (c) compute the total traffic flowing on each trunk bundle through simple addition and using the analytical methods of subsection 2.4.7, and (d) dimension the AL and trunk bundles using the analytical methods of Chapters 3 and 4.

The software design package described in Chapter 6 allows the designer to specify the traffic intensities handled by each location in the following ways:

1. Using the well-known engineering practices, assign meaningful traffic intensities to all locations of a brand new network for the busy hour (Method 3), and
2. Analyze the SMDR and AMA tapes on a separate system to create a From-To traffic file. Use this From-To file to compute the BHR traffic intensities handled by each location, and each AL/trunk bundle (Method 2).

2.4.7 ESTIMATION OF BACKBONE NETWORK TRAFFIC

Traffic intensities for the backbone network can be correctly derived only when the exact traffic flows are known between every nodal pair. However, if a From-To traffic file is not known, one can compute the traffic intensities on each network trunk by using an assumption that traffic from each source is uniformly distributed to all destinations as a function of their loadings. A symmetric traffic matrix can be defined for a backbone network as follows:

Step1: Compute the total traffic (TT_i) handled by the ith network node (or switch)
Step 2: Compute the grand total (GT) of all the traffic handled (both intra- and internode)
Step 3: Compute the traffic (T_{ij}) flowing on the trunk from the ith node to the jth node, as follows:

$$\begin{aligned} T_{ij} &= [TT_i * TT_j]/GT \\ &= T_{ji} \text{ (for a symmetric traffic matrix)} \end{aligned} \tag{2.26}$$

If the value i equals j, one obtains the value of intranodal traffic. The intranodal traffic implies that the source and the destination are terminated on the same backbone switch of the network node. As a result, this traffic does not influence the sizing of any trunk. One can vary the amount of internodal traffic by dividing the values, T_{ij}, by an arbitrary value of the backbone trunking factor (BBTF) and adding the difference to the intranodal traffic. The use of the BBTF is illustrated in Chapters 7 and 8.

After computing the internodal traffic intensities for each direction of flow, the designer should then apply the most appropriate traffic analysis formula to determine the size of each trunk bundle and its length in miles. One can also compute the cost of trunks by using the most appropriate tariff. In most instances, the resulting traffic flows on the backbone trunk bundles do not imply a minimum cost trunking network. There are two approaches to lowering the cost of a backbone network:

1. Introduce some hierarchy to network nodes. Using the concepts of high usage (HU) and final trunks, and the peaked behavior of overflow traffic, compute the optimum mix of HU and final traffic flow. The cost of the backbone network is usually

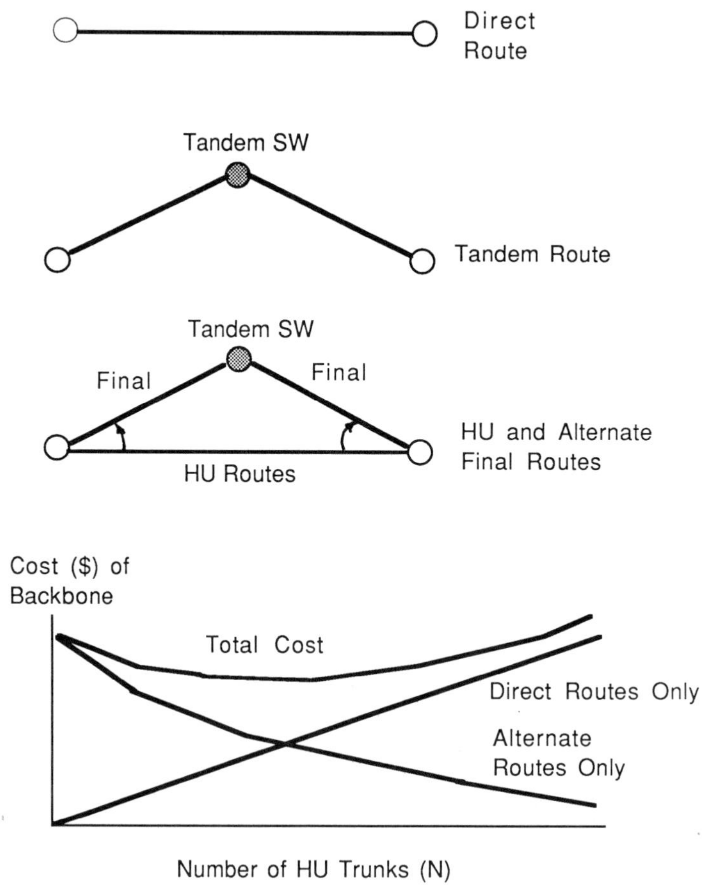

Figure 2.18. Illustration of the direct, HU, and mixed-hierarchical routing, and how these influence the cost of a backbone network.

reduced. This approach is employed in the design of traditional public telephone networks.

2. Remove the trunk with the lowest traffic and route the traffic on the shortest remaining path connecting the two nodes. Compute the backbone network cost. Repeat this process until the cost stops decreasing. This approach is usually employed in the design of private voice and data networks.

Traditional backbone networks for voice usually employ alternate routing schemes to reduce the transmission cost. Such schemes require the use of high usage (HU) and final trunk bundles, as shown in Figure 2.18. The alternate routing plan first attempts to find a direct route through the HU trunk bundle. If all trunks in the HU bundle are busy, an alternate path through the arbitrarily defined tandem node, and through the two final trunk bundles is found. If all trunks in the final trunk bundle are busy, the call is blocked. In some large public networks, several levels of HU trunk bundles may be employed before invoking a final trunk bundle.

The savings in total trunking costs are realized through two factors:

1. high utilization of HU trunks bundles by allowing large blocking, and
2. the peakedness of overflow traffic requires a lesser number of incremental trunks in the final trunk bundle than otherwise.

Sharma et al. (1982) discuss an analytical method for dimensioning the backbone network using the HU and alternate hierarchical routes. Such a technique is not useful for the emerging new systems that employ the dynamic, nonhierarchical routing (DNHR) scheme associated with ISDN-based systems. Neither is such a scheme useful in designing modern corporate networks that employ T1 multiplexers and T1 highways connecting major locations. For these systems, the second technique is quite appropriate. See Chapter 9 for several applications of this approach in the design of integrated voice and data networks for corporations.

3

Analysis and Design Tools for Voice Networks

3.1 TRAFFIC ANALYSIS TOOLS

Traffic analysis is required to compute the performance of a CS voice network in terms of internodal blocking and the number of internodal circuits as a function of the allowed blocking probability. There are generally two formulas employed to compute the blocking probability for a loss system in which calls are lost if service cannot be provided at the time of a call:

1. Poisson's formula
2. Erlang-B formula

Both of these formulas have been found useful. Using the traffic and system performance concepts of Chapters 1 and 2, useful relationships and charts can be developed to assist the systems analyst. Books by Syski (1960), Collins and Pederson (1973), and Bell Telephone Labs (1977) should be consulted for additional understanding.

3.1.1 POISSON'S LOSS FORMULA

The Poisson's formula assumes that (1) infinite sources are responsible for a steady rate of random call arrivals, and (2) lost calls are held for a duration equal to call holding times. If A represents the call intensity measured in erlangs (see Section 1.4 for a definition), the probability of n calls in the system can be expressed as follows:

$$P_n = A^n * e^{-A}/n! \qquad (3.1)$$

The probability of blocking on a system with N outlet circuits can be expressed as follows:

$$B = \sum_{n=N}^{\infty} P_n = 1 - \sum_{n=0}^{N-1} P_n \qquad (3.2)$$

This relationship (Equation 3.2) is commonly known as the Molina formula. Figures 3.1, 3.2, and 3.3 illustrate the relationship between A and N for various values of B.

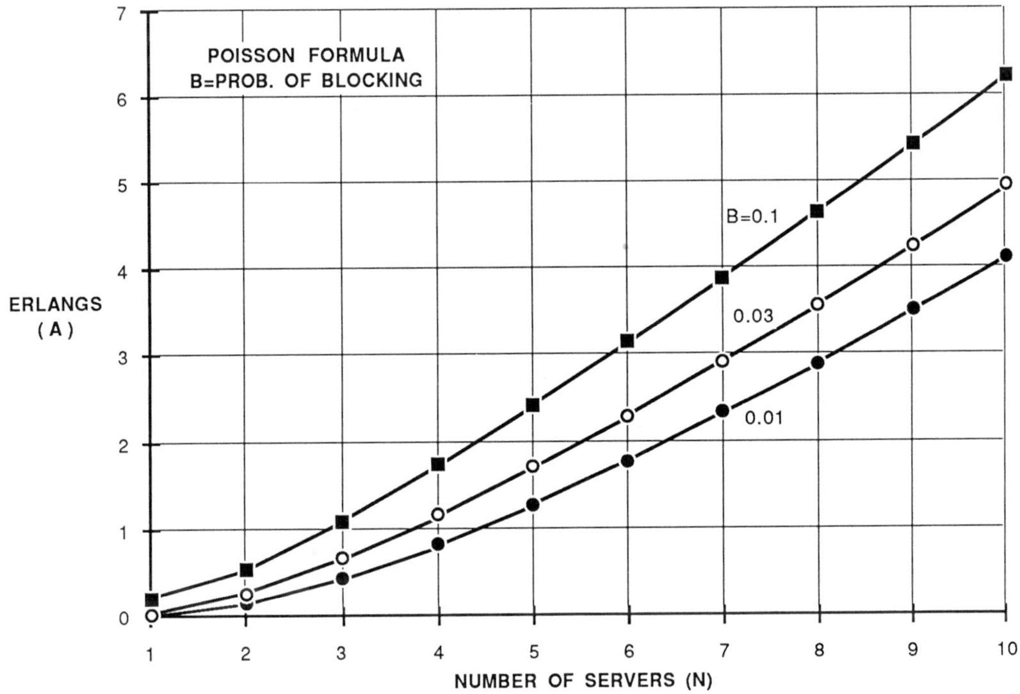

Figure 3.1. Erlangs (A) versus the number of servers (N) for some useful values of blocking (B): Poisson formula for the range $\{1 \leq N \leq 10\}$.

Such curves can be useful in computing the exact number of circuits required to handle A erlangs of traffic intensity for a specified blocking probability B.

3.1.2 ERLANG-B LOSS FORMULA

The Erlang-B formula assumes that (1) infinite sources are responsible for a steady rate of random call arrivals, and (2) the lost calls are cleared as soon as they are blocked, and these calls never return. If A represents the offered call intensity measured in erlangs and N is the number of servers, the blocking probability $B(N)$ is given by the Erlang-B loss formula:

$$B(N) = [A^N/N!] \Big/ \left[\sum_{i=0}^{N} A^i/i! \right] \tag{3.3}$$

The probability of blocking $B(N+1)$ can be expressed as a function of $B(N)$ in a recursive form, as shown in Equation 3.4.

$$B(N+1) = A * B(N)/[N+1+A * B(N)] \tag{3.4}$$

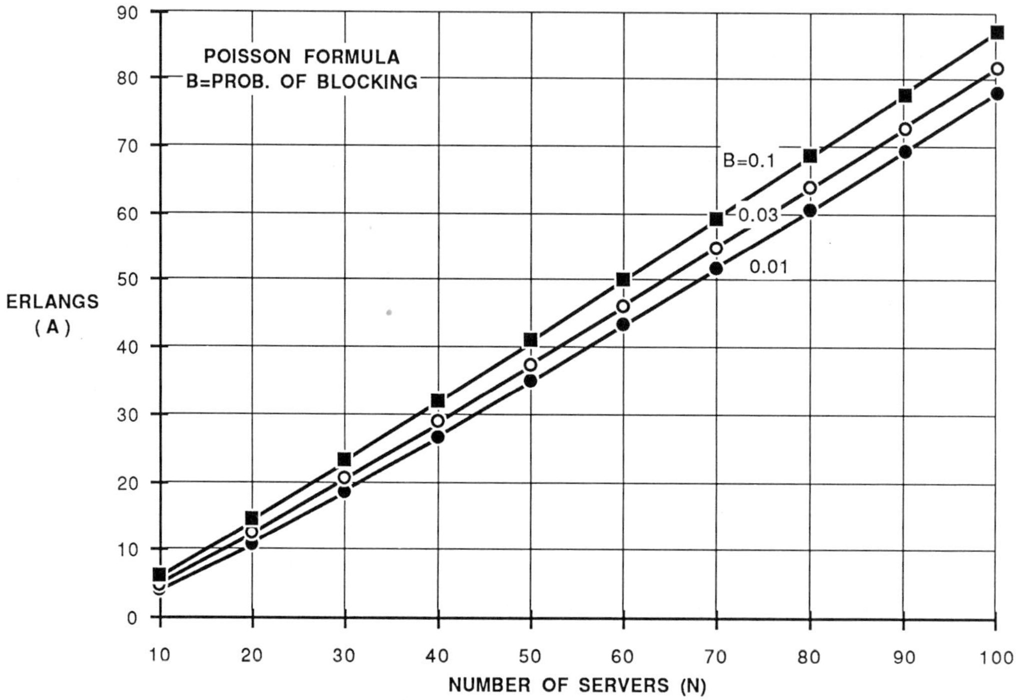

Figure 3.2. Erlangs (A) versus the number of servers (N) for some useful values of blocking (B): Poisson formula for the range $\{10 \leq N \leq 100\}$.

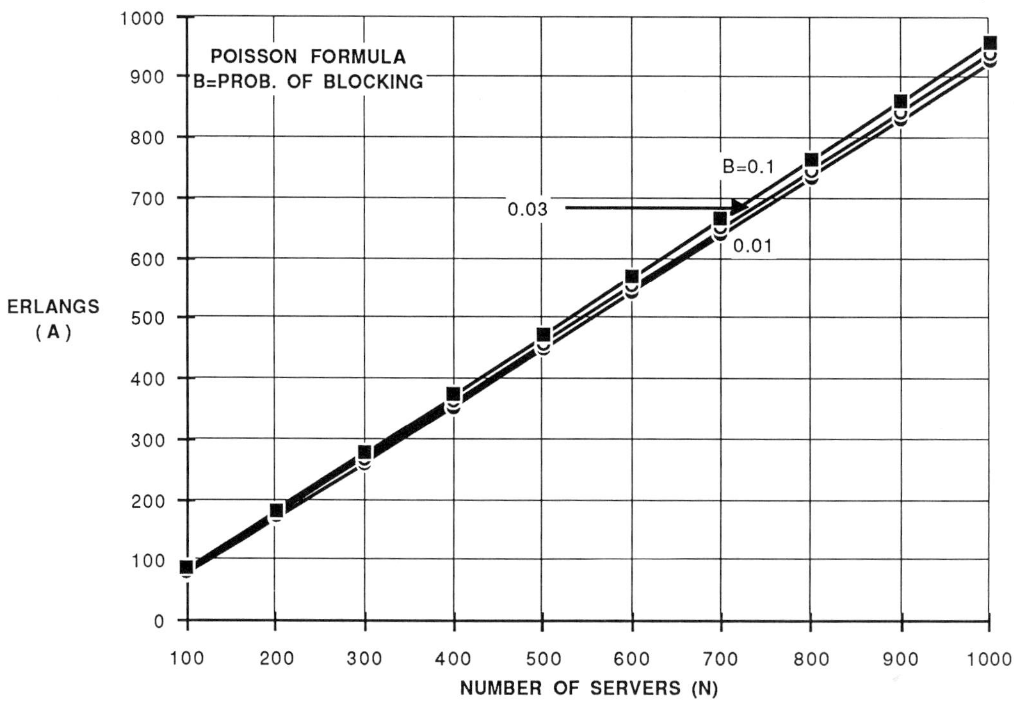

Figure 3.3. Erlangs (A) versus the number of servers (N) for some useful values of blocking (B): Poisson formula for the range $\{100 \leq N \leq 1000\}$.

Table 3.1 A Software Program Written in Basic for Computing the Blocking Probability (B) Given A Erlangs and N Servers

```
1000 REM ERLANG B FORMULA GIVEN A AND N
1010 INPUT "ENTER OFFERED ERLANGS  "; A
1020 IF A=0 THEN END
1030 INPUT "ENTER NUMBER OF SERVERS  "; N
1040 IF N=0 THEN END
1050 GOSUB 5000
1060 PRINT "ERLANGS (A) = "; A
1070 PRINT "N0. OF SERVERS (N) ="; N
1080 PRINT "PROB. OF BLOCKING ="; B
5000 REM SUBR FOR COMPUTING B GIVEN A AND N
5010 B = 1.0
5020 FOR I=1 TO N
5030   NUM = A*B
5040   DENOM = I + NUM
5050   B = NUM/DENOM
5060 NEXT I
5070 RETURN
```

This formula uses the fact that the probability of blocking B is equal to 1 if there are no servers. Thus $B(0) = 1$. Similarly, we can also get $B(1) = A/[1 + A]$, and so forth. This relationship (Equation 3.4) can be transformed into a short program written in BASIC language to compute B for any given values of A and N. See Table 3.1 for a listing of such a program. See Figures 3.4, 3.5, and 3.6 for the curves relating N to A for some useful values of B. Although these curves can help one to compute the exact number of outlet circuits required to handle A erlangs for the given values of B, one can also approximate the value of B for given values of A and N if enough values of B are used in the figures.

One can also compute the average utilization of the outlet trunk bundle that carries only $A(1 - B)$ erlangs. Such a quantity represents the efficiency of the network. The average utilization of the trunk bundle is therefore as follows:

$$\rho = A(1 - B)/N \tag{3.5}$$

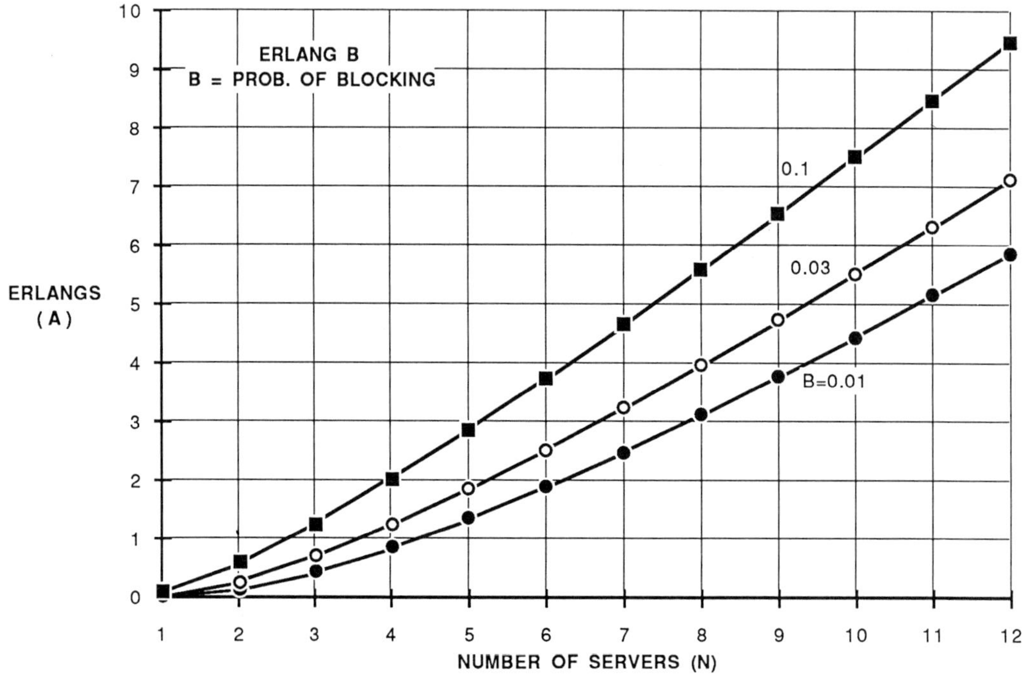

Figure 3.4. Erlangs (A) versus the number of servers (N) for some useful values of blocking (B): Erlang B formula for the range $\{1 \leq N \leq 12\}$.

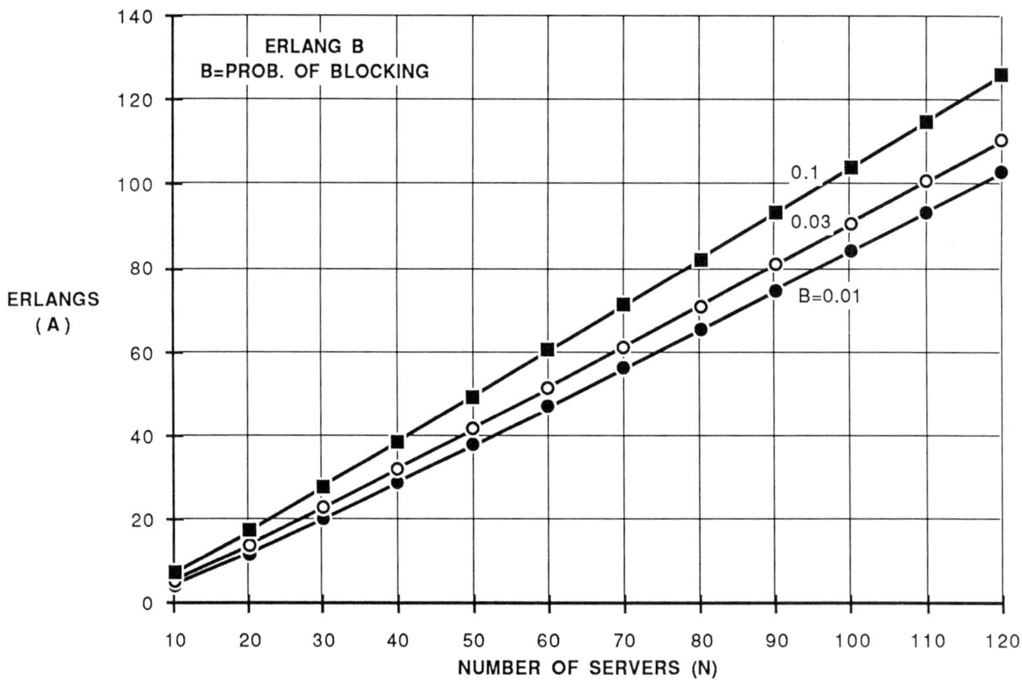

Figure 3.5. Erlangs (A) versus the number of servers (N) for some useful values of blocking (B): Erlang B formula for the range $\{10 \leq N \leq 120\}$.

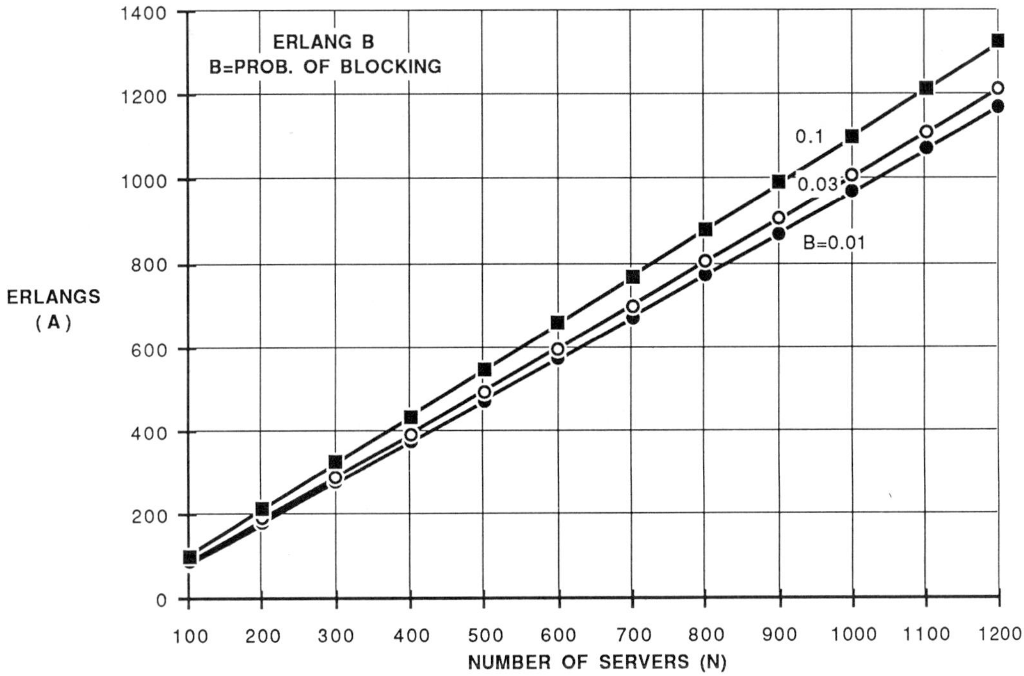

Figure 3.6. Erlangs (A) versus the number of servers (N) for some useful values of blocking (B): Erlang B formula for the range $\{100 \leq N \leq 1200\}$.

See Figure 3.7 for several curves relating average trunk utilization to the number of trunks in the bundle and B. To obtain the value of ρ, one must first use the curves of Figures 3.4, 3.5, and 3.6 to compute N for a given A and B, and then substitute these values in Equation 3.5.

3.1.3 ERLANG-B LOSS FORMULA WITH RETRIES

If the lost calls that were cleared from the system are allowed to return at random, the value of the enhanced offered traffic intensity A' can be expressed (see Siemens 1970) as follows:

$$A' = A(1 + E + E^2 + \ldots) = A/(1 - E) \tag{3.6}$$

where

E = blocking probability computed for A' erlangs

It follows that $A * E$ calls will be rejected on the first attempt. These calls will return with $A * E^2$ calls being rejected again and so forth. See Figure 3.8 for a comparison between the above three formulas for the case consisting of 10 trunks. The curves show that the difference between the Erlang-B formula with retries and the Poisson's formula

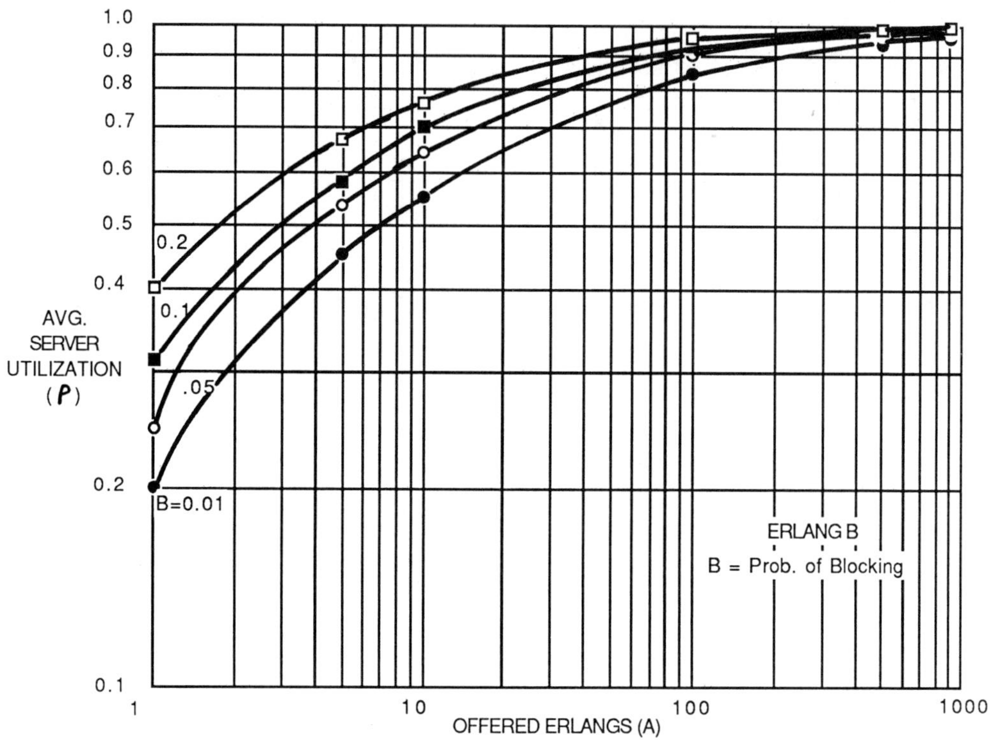

Figure 3.7. Average server utilization versus the number of servers (N) for some useful values of blocking (B): Erlang B formula for the range $\{1 \leq N \leq 1000\}$.

becomes small for the useful range of B. The curves also show that the three formulas provide similar results, for B smaller than 0.01.

3.1.4 ERLANG-B FORMULA WITH FINITE NUMBER OF SOURCES

The previous formulas assumed an infinite number of sources. It implies that there are occasions when a large number of active sources can result in a large number of call arrivals requiring a large number of servers. On the other hand, if the number of sources was very small, the worst-case situation would require a bounded number of servers. The value of N can be expressed for offered A erlangs, N servers, λ call arrival rate per idle source, $\overline{T}_s$ call service time (also equal to ECD), and M sources using the following relationships:

$$B = \left[\binom{M-1}{N}(\lambda\overline{T}_s)^N\right] \bigg/ \left[\sum_{i=0}^{N}\binom{M-1}{i}(\lambda\overline{T}_s)^i\right] \tag{3.7}$$

$$P_n = \left[\binom{M}{n}(\lambda\overline{T}_s)^N\right] \bigg/ \left[\sum_{k=0}^{N}\binom{M}{k}(\lambda\overline{T}_s)^k\right] \tag{3.8}$$

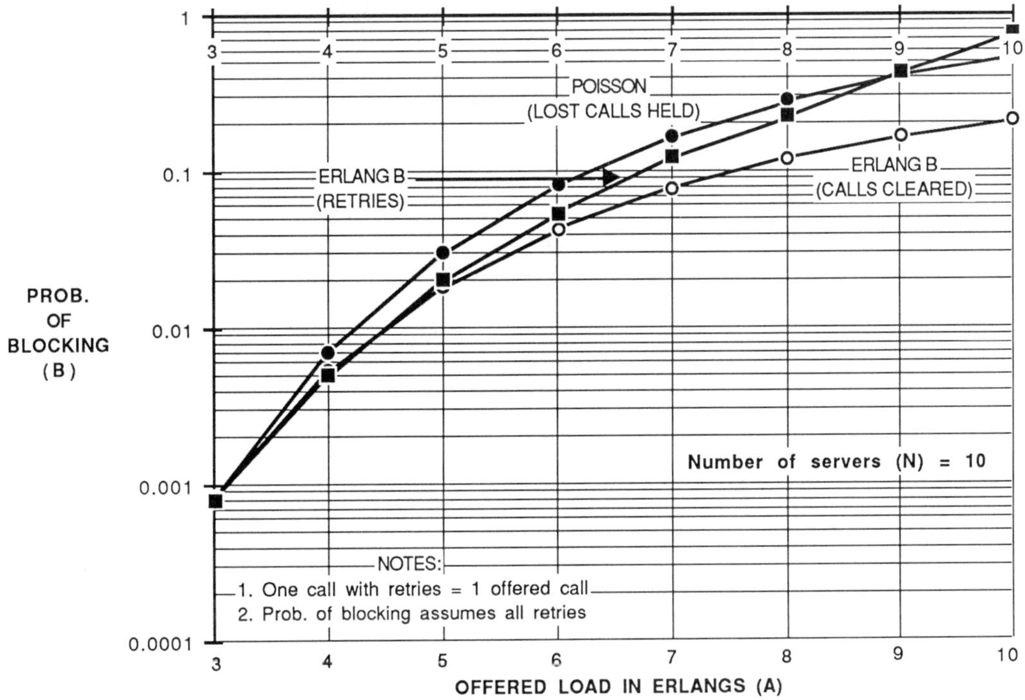

Figure 3.8. Probability of blocking (B) versus offered load in erlangs (A) and different teletraffic formulas.

$$A = \lambda \overline{T}_s (M - n) \tag{3.9}$$

$$n = \sum_{n=0}^{N} nP_n \tag{3.10}$$

Use Equation 3.7 to compute the value of $\lambda\overline{T}_s$ for given values of B, M, and N. Then compute the value of P_n using Equation 3.8. Compute the average value of n and A using Equations 3.9 and 3.10. The reader should remember that the expression $\binom{x}{y}$ is also equal to $C(x, y)$, as defined in Chapter 2.

Figure 3.9 shows plots of N for some useful values of A, B, and M. The results show that the infinite source case always requires more servers than the case of finite M for given values of erlangs and B. The results also show that, as M exceeds 500, the results for infinite sources apply.

3.1.5 ERLANG-C FORMULA FOR AUTOMATIC CALL DISTRIBUTION (ACD) SYSTEMS

Automatic call distribution (ACD) systems provide uniform distribution of incoming telephone calls to P available agent positions and detailed and timely reports dealing with traffic statistics, ACD resource utilizations, and agent performance levels. See Figure 3.10 for a block diagram of an ACD system. N incoming trunks allow the incoming

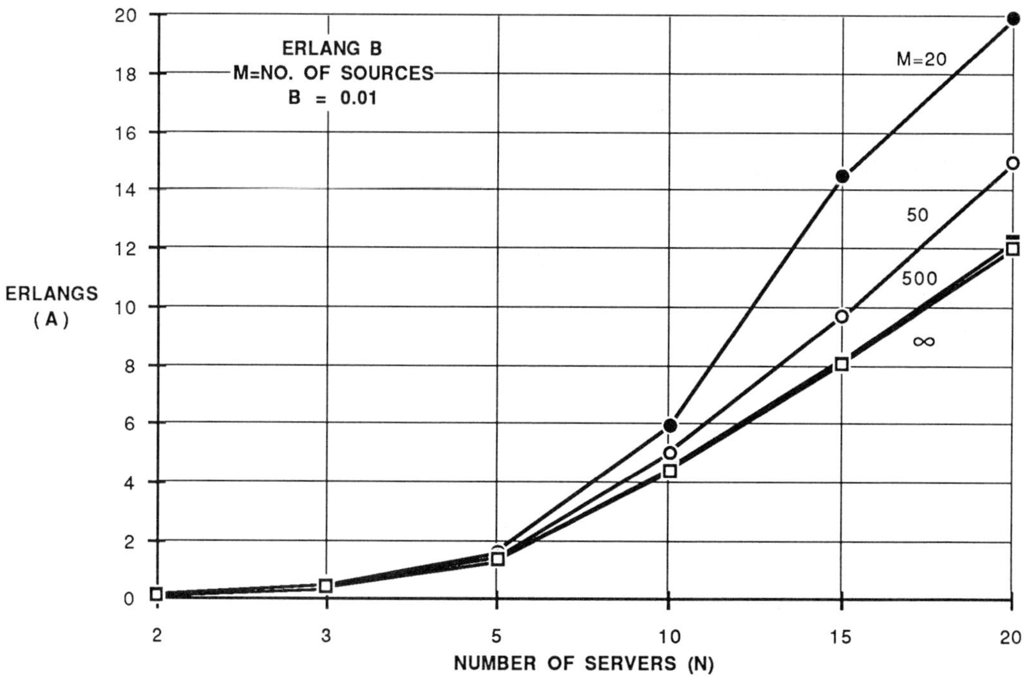

Figure 3.9. Erlangs (A) versus the number of servers (N) and some useful values of sources (M).

voice calls to arrive at the ACD node. As the call arrives, it is sent to the next available agent. If all of the agents are busy when the call arrives, it must wait for the first agent that becomes available. The calls are handled on a first-in-first-out (FIFO) basis. The caller may wait forever until the call is answered. However, the caller may leave the

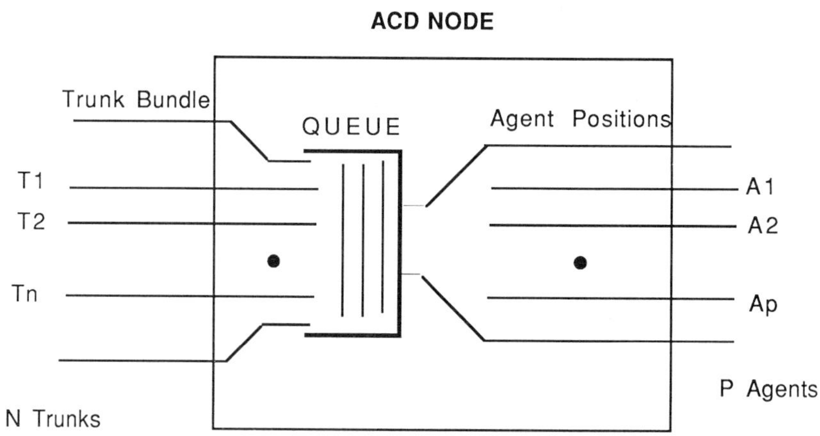

Figure 3.10. Block diagram for an ACD node.

system after waiting a certain amount of time. A well-designed ACD system minimizes the average wait time before the call is answered by scheduling a correct number of agent positions for each half-hourly period, as dictated by the observed daily traffic intensities. The probability of delay exceeding any given time t and an average delay are expressed by the following equations:

$$\text{Prob.}(\text{Delay} > t) = P_0 * \text{Exp}\left[-(P - A)t/\overline{\text{T}}_s\right] \tag{3.11a}$$

$$\text{Avg. Delay} = \left[\overline{\text{T}}_s/(P - A)\right] P_0 \cdots \text{ for all calls} \tag{3.11b}$$

where

A = offered traffic in erlangs = $\lambda * \overline{\text{T}}_s = \lambda * \text{ECD}$
P = the number of agent positions that serve the incoming calls
P_0 = Prob.(Delay > 0) = $(B * P)/[P - A(1 - B)]$
B = Probability of blocking, given A and P
$\overline{\text{T}}_s$ = expected call duration (equal to ECD)
Exp (x) = e^x

Some useful plots of computed agent positions P for useful values of A and $\overline{\text{T}}_s$ or ECD are shown in Figure 3.11 for three GOS values. A typical airline reservations system requires that the probability of exceeding 10 seconds should not exceed 0.15. See Equa-

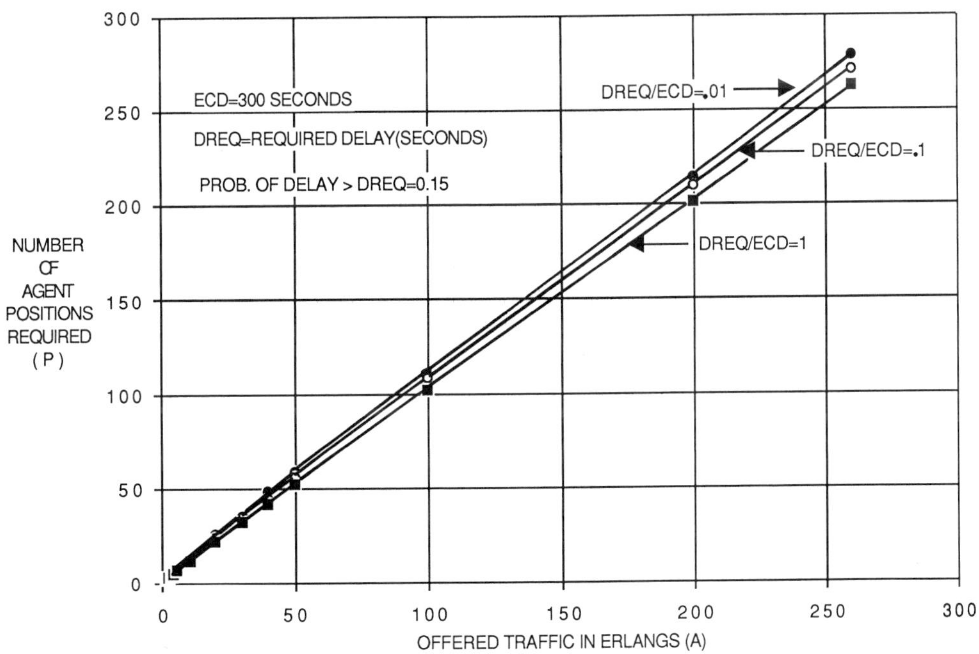

Figure 3.11. The number of agent positions (P) versus offered erlangs (A) for some values of required delay (DREQ) and probability of delays exceeding DREQ = 0.15.

tion 3.13 for computing the average delay per delayed call in case that becomes the GOS criterion. The Equation 3.11 can be useful for those ACD applications where the hourly traffic patterns for each day of the week are predictable. For applications where traffic is unpredictable (e.g., televised home merchandising), it may be impossible to provide a given GOS for all hours. The number of manned agent positions is generally determined by marketing considerations alone.

3.2 NODAL PERFORMANCE ANALYSIS TOOLS

There are many types of network nodes that handle voice calls. See Figures 3.12 and 3.13 for illustrations of a typical private/corporate network and a public telephone network (PTN). Whereas a typical private network consists of many PABXs and a few tandem switching nodes, a PTN generally consists of COs and four levels of tandem toll offices. Some exceptionally large corporations or organizations may require two levels of tandem nodes to reduce costs by using the principle of traffic concentration. Some levels of toll offices in a typical PTN may now be unnecessary if one uses the new ISDN architecture and the newly available large switching nodes. A typical voice network must perform the following applications:

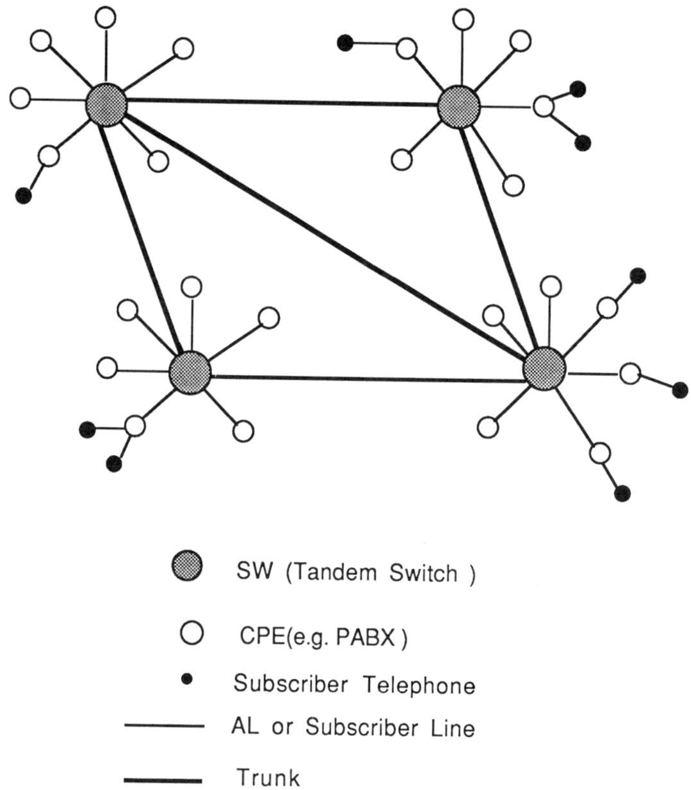

Figure 3.12. Topology of a typical corporate voice network.

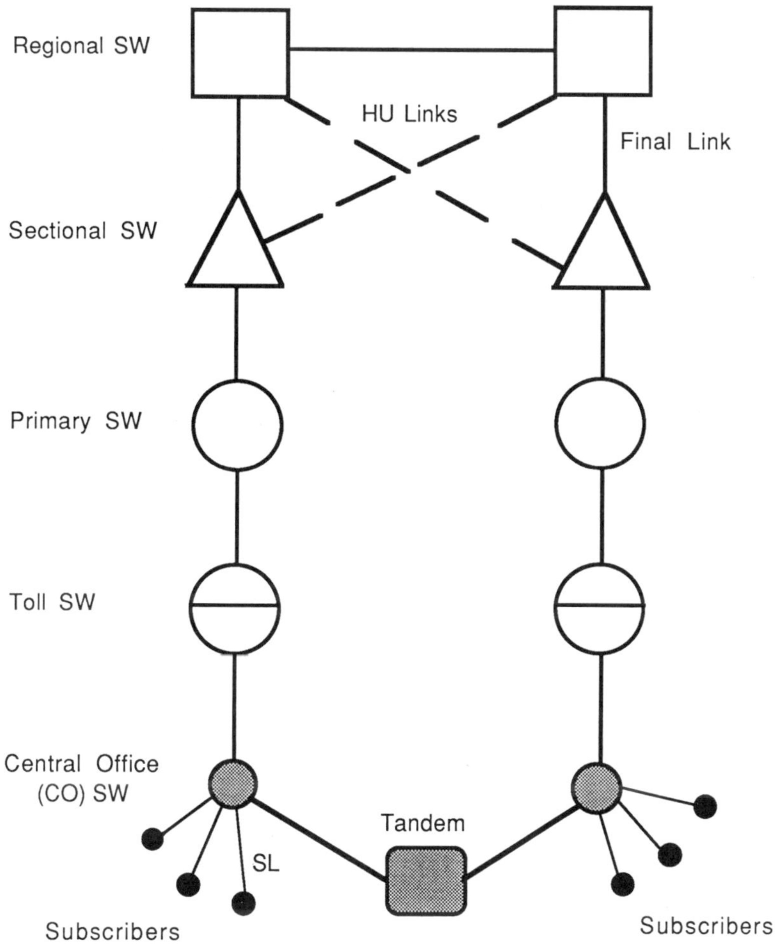

Figure 3.13. Topology of a conventional public telephone network.

1. request analysis (for class-of-service or COS, etc.) and provision of the dial tone
2. called-party address analysis
3. pathfinding and routing for single- and multi-addressed calls
4. circuit switching function
5. NMC functions

Some of the applications (e.g., NMC) may be distributed over a subset of network nodes. Other applications may be handled by several sub-nodes within a network node. For example, the subscriber-line scanning function may be handled by several line modules, each controlled by either a powerful microprocessor or a minicomputer. Each sub-node utilization should be modeled independently as a function of the applications handled. Each sub-node requires an operating system for allocating the time used by the central processing unit (CPU) or input/output device to the various functions handled. The time spent by the applicable operating system for each sub-node must be accounted for, along with the time spent on all the functions handled. Using this approach, find the sub-node with the highest utilization factor applicable for the required peak period traffic. If it is higher than 60 percent, then redesign the resource to satisfy the constraint. A simple design criterion to be followed is "the lower the utilization factor of the busiest sub-node, the higher the robustness of the network node."

3.2.1 THROUGHPUT ANALYSIS

A throughput analysis deals with computing the number of calls that can be handled by a network node. Find the sub-node with the highest utilization computed as a product of the number of instructions related to the call functions and the average instruction execution time. The number of instructions are computed either by simply adding all of the instructions actually executed by a typical call-processing environment or by measuring actual program executions. The average instruction execution time is measured by computing the execution time of each instruction and the probability of instruction use. An alternate way for computing the average instruction time is to actually measure the time spent on typical sets of instructions.

The throughput or capacity of a network node can then be computed as follows:

$$C = 3600/\mathrm{T}_c \tag{3.12}$$

where C = nodal capacity in calls per hour
T_c = time spent per call on the most active subnode of a network node

The throughput is commonly determined in terms of either call attempts or calls in progress. Each call attempt generally requires only the servicing of requests and finding a path through the network. Each call in progress also involves the handling of post-call functions such as billing, call-data recording, etc.

3.2.2 RESPONSE TIME ANALYSIS

There are two quantities of interest for voice networks: (1) the average time for receiving a dial tone after going off-hook, and (2) the average time required to make a connection. For a local call handled by a PABX, the average time required to make a connection is equal to the time it takes to make a connection between the caller and the called parties. For a call requiring the help of another network node, the average time required to make a connection is simply a sum of the times required to find the best route and make the hard-wired connection between the inlet and the outlet ports through the circuit switches (CS).

A CS network node employs *P* registers as part of common control equipment shared among all of the users. The network node scans all of the inlet ports for an off-hook condition. As soon as it discovers a request for service, it connects a common control (CC) register to the inlet port. The first function of the register is to provide a dial tone to the user. The second function of the register is to receive all of the dial digits. The third function is to hold the dial digits while the connection is made. The register is then released, and it is put back into the register pool. The average time of register occupancy will be called the *call connect time*.

The average time to receive a dial tone after going off-hook can be computed using the Erlang-C formula (Equation 3.11), and the facts that mean service time (equal to $\overline{T}_s$) is equal to the call connect time and that the *P* servers are equal to the number of dial-digit receiving registers. The formula provides the probability that delay will exceed any given time *t*. Since most designers employ the criterion that delay will exceed 1 second only 15 percent of the times that the user goes off-hook, Equation 3.11 is quite appropriate. There are times when the designer must know the average wait, defined as the average elapsed time between the moments that the user goes off-hook and the user hears a dial tone. The average wait for only delayed calls can be expressed as follows:

$$\overline{T}_w = \overline{T}_s/(P - A) \tag{3.13}$$

See Figure 3.14 for some useful curves relating $\overline{T}_w$ to *P* and *A*. The connection time ($\overline{T}_c$) is simply the sum of the routing and CS connection times. The routing time is the time spent in analyzing the routing tables for computing the best route. The CS connection time is the time spent in making the hard-wired connection through the circuit switch. This time is generally supplied by the vendor.

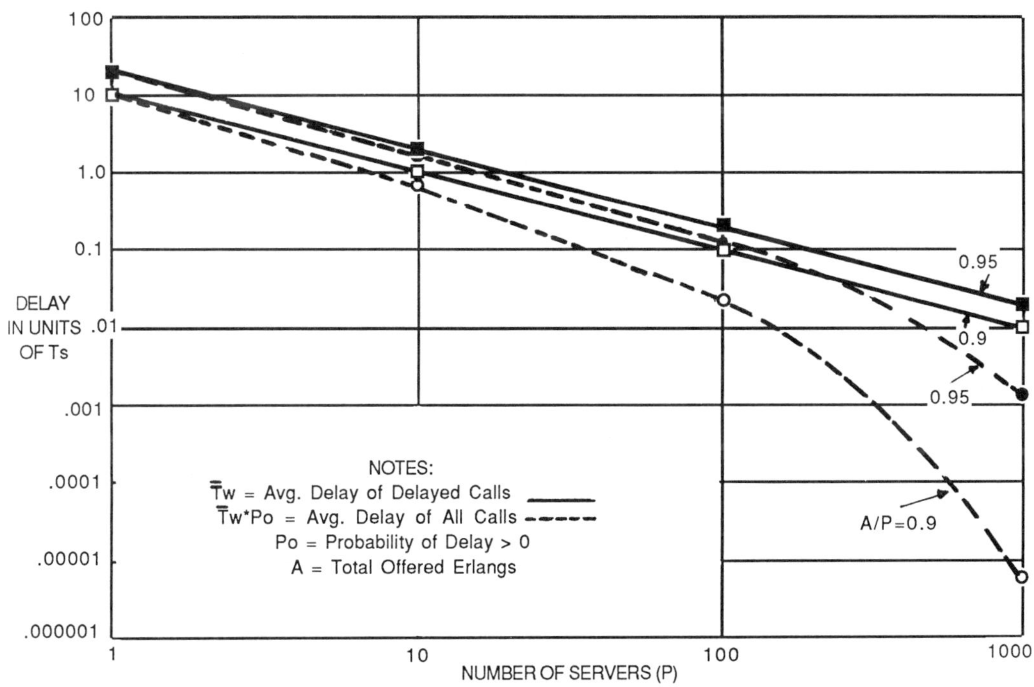

Figure 3.14. Average delay (in units of $\overline{T}_s$) versus offered erlangs (A) for some useful values of A/P ratios.

3.3 NETWORK PERFORMANCE ANALYSIS TOOLS

There are occasions when several CS network systems must be compared with one another in terms of their performance. This will require the computation of network throughput and the GOS parameters described as follows:

1. effective throughput of the entire CS network system
2. total end-to-end connection time measured as the elapsed time between the moment the last digit is dialed and the moment the destination terminal is made busy,
3. end-to-end blocking probability experienced by a call request.

3.3.1 SYSTEM THROUGHPUT ANALYSIS

One way to compare a CS network system with another is in terms of the total number of calls handled during a busy period. If each CO-type network node handled only local calls, the throughput of the entire system would be the sum of all CO nodal throughputs ($C_s = C_1 + C_2 + C_3 + \ldots$). This would also imply no need of tandem switches or trunks. In the other extreme, if every CO-type node handled only tandem calls, the total network throughput, C_s, must be divided by a factor, k, that represents the number of identical CS nodes in the path of the call. The actual throughput of a practical CS network system will be somewhere in between the two limits.

3.3.2 END-TO-END CONNECTION TIME

Another way to compare two CS network systems is to study their end-to-end connection time distribution. The total connect time can be expressed as the sum of (1) processing time delays experienced by each network node in the path of the call, and (2) the sum of the delays experienced by the signaling data on each internodal link. Since the delays encountered in each node and on each internodal link is a random variable characterized by an average and a variance, the end-to-end connection time will also exhibit randomness. Its average and variance can be expressed as follows:

$$\text{Avg}(\text{T}_c) = \sum_{i=1}^{N} \text{Avg}(\text{T}_{in}) + \sum_{j=1}^{L} \text{Avg}(\text{T}_{j1}) \tag{3.14}$$

$$\text{Var}(\text{T}_c) = \sum_{i=1}^{N} \text{Var}(\text{T}_{in}) + \sum_{j=1}^{L} \text{Var}(\text{T}_{j1}) \tag{3.15}$$

where

T_c = total connection time
Avg(T_c) and Var(T_c) are the average and variance of T_c
T_{in} = delay in the ith node
T_{j1} = delay on the jth internodal link
N, L = total number of nodes and links in the path of the voice call

In case no single node determines the final value of T_c, the Center Limit Theorem (CLT) should apply (consult Feller (1957) and Parzen (1965) for further study). Using the CLT, one can derive several useful expressions for 90 percentile, 95 percentile, 99 percentile, and the 99.9 percentile values of T_c, as follows:

$$T_c\,(90\%) = \text{Avg}(T_c) + 1.28\ \text{SQR}[\text{Var}(T_c)] \tag{3.16}$$

$$T_c\,(95\%) = \text{Avg}(T_c) + 1.65\ \text{SQR}[\text{Var}(T_c)] \tag{3.17}$$

$$T_c\,(99\%) = \text{Avg}(T_c) + 2.33\ \text{SQR}[\text{Var}(T_c)] \tag{3.18}$$

$$T_c\,(99.9\%) = \text{Avg}(T_c) + 3.09\ \text{SQR}[\text{Var}(T_c)] \tag{3.19}$$

The preceding methodology provides only approximate answers for a large network system during the planning phase. The exact solution is always unavailable, either due to the unpredictability of exact traffic intensities or due to a lack of closed-form mathematical solutions. Consequently, the approximate methods should be more than adequate for most cases.

3.3.3 END-TO-END BLOCKING PROBABILITY

Two CS network systems can also be compared in terms of their end-to-end blocking probability. The blocking probabilities experienced in the N network nodes in the path of a call can be summed to get the average and variance of the end-to-end network system blocking, B_s, as illustrated in the previous section. If the individual blocking probabilities are very small (i.e., $B_i <= 0.05$), the average system blocking becomes:

$$B_s \simeq B_1 + B_2 + \ldots B_N = \sum_{i=1}^{N} B_i \tag{3.20}$$

4

Analysis and Design Tools for Data Networks

4.1 TRAFFIC ANALYSIS TOOLS

Whereas traffic analysis tools for voice traffic are required to compute the values of nodal/system blocking or connection times, traffic analysis tools for data traffic are required to compute the nodal/system response times. In order to develop meaningful analysis tools for data, one must first understand the behavior of data traffic.

Data traffic differs from voice traffic in many significant ways. Whereas each voice call can be characterized by its well-behaved service time, T_s, data traffic is generally very bursty in nature. Data traffic generally involves the transmission of messages, both small and large, as required by the associated protocol defined as the set of rules for interchanging data between two communications devices. See Figure 4.1 for an illustration of the traffic flows between the terminal controller and a host computer for the asynchronous (or ASYNC) and Binary Synchronous Communication (BSC) protocols, respectively. In many cases, some of the messages have to be created by human operators while the connection (virtual or permanent) is being maintained. Some other messages may have to be retrieved by a host computer from its memory before transmission. In some applications, large files are transmitted continuously for a long time. Some voice networks allow the transmission of data through the use of circuit-switched data calls. Many corporations are experiencing a large growth in their circuit-switched data traffic as a result of the increase in PC-to-mainframe communications.

Almost all data applications require some form of store and forward techniques involving some delays. Consequently, data traffic analysis tools can only be expressed in terms of queuing theory. See Figure 4.2 for a representation of two queues within a PS / MS network node. One queue represents the input process and the other the output process.

Basically two types of queuing models are of interest here. Each model assumes a unique traffic model. The first type is characterized by random arrivals and negative exponential distributions of the service times, T_s and N servers. It is defined as the M / M / N type traffic, with either a finite or infinite number of sources. The second type of data traffic is characterized by random arrivals and constant service times with N servers. It is defined as the M / D / N type traffic with either finite or infinite sources. These two types of traffic determine delays in the queue within a PS / MS nodes and data networks. The works of Feller (1957), Parzen (1965), Martin (1972), Syski (1960), and

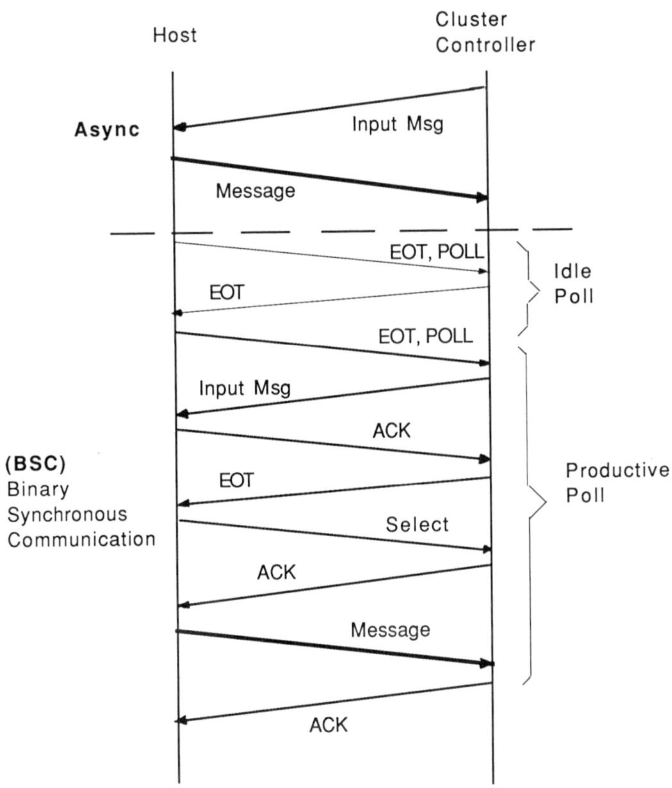

Figure 4.1. ASYNC and Binary synchronous communication protocols.

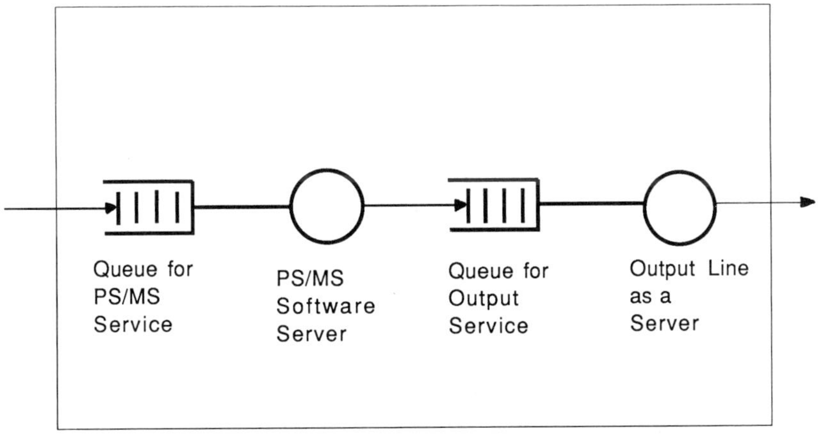

Figure 4.2. Queuing model of a typical PS/MS with two-stage service.

Schwartz (1987) should be consulted for an in-depth treatment of the concepts used in this chapter. We will first consider $N = 1$, since it provides a better understanding of the material. Only those cases with infinite sources will be considered herein.

4.1.1 THE M/M/1 QUEUE ANALYSIS

Considering that time in queue (T_q) is the sum of the waiting time and service time, and using ρ for server utilization (or loading) and $E(T_s)$ for the expected value of service time, one can express the expected value of T_q for a single server as follows:

$$E(T_q) = E(T_s)/(1 - \rho) = 1/(\mu - \lambda) \tag{4.1}$$

where

$\rho = \lambda * E(T_s) = \lambda * \text{EMD}$
$\lambda =$ message arrival rate
$\mu = 1/E(T_s)$

The standard deviation of T_q (equal to: Std.Dev. (T_q)) is also equal to $E(T_q)$, or

$$\text{Std.Dev. } (T_q) = E(T_s)/(1 - \rho) = 1/(\mu - \lambda) \tag{4.2}$$

The average number of items in the queue (L) and the average wait time (T_w) are also of interest:

$$E(L) = \rho/(1 - \rho) \tag{4.3}$$

$$E(T_w) = \rho/(\mu - \lambda) = (\rho * T_s)/(1 - \rho) \tag{4.4}$$

4.1.2 THE M/D/1 QUEUE ANALYSIS

The time spent in the queue is the sum of time in waiting and getting served. Using ρ for the server utilization and T_s for fixed service time, one can express the expected time in queue as follows:

$$E(T_q) = T_s(2 - \rho)/2(1 - \rho) \tag{4.5}$$

The standard deviation of T_q can be expressed as

$$\text{Std.Dev. } (T_q) = [T_s/(1 - \rho)] * \text{SQR}[\rho/3 - \rho^2/12] \tag{4.6}$$

where

$$\text{SQR}(x) = \text{Square root of } x$$

The average number of items in the queue and average wait can be expressed as follows:

$$E(L) = [\rho^2/2\,(1 - \rho)] + \rho \tag{4.7}$$

$$E(T_w) = (\rho * T_s)/[2\,(1 - \rho)] \tag{4.8}$$

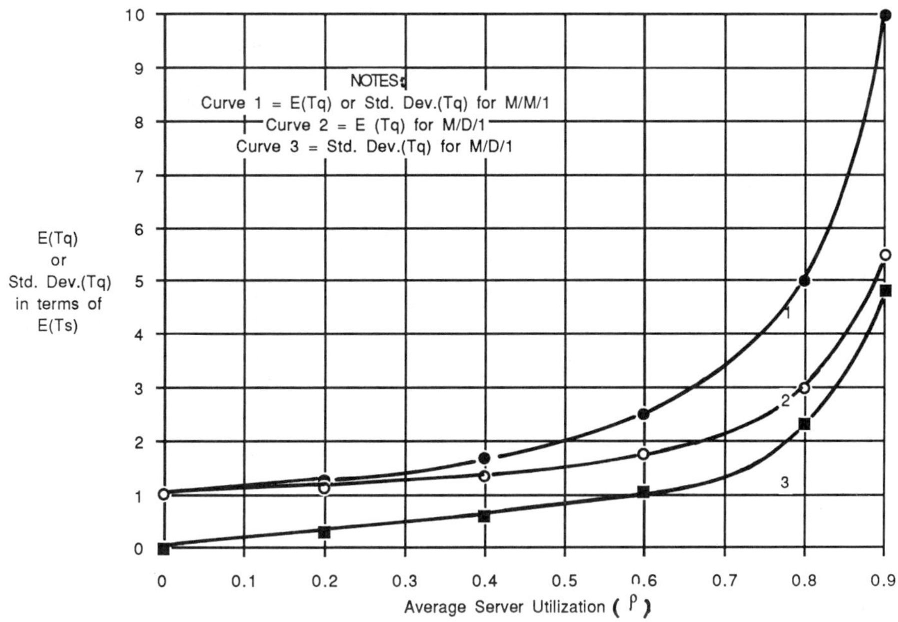

Figure 4.3. Useful curves for M/M/1 and M/D/1 queuing disciplines.

See Figure 4.3 for a plot of $E(T_q)$ and Std.Dev. (T_q) for both the M/M/1 and M/D/1 traffic and queuing models. It should be obvious that the M/D/1 yields the lower bound and M/M/1 yields the upper bound of delays for most practical systems.

4.1.3 THE M/M/N QUEUE ANALYSIS

Most queues in a large network system employ more than one server (i.e., $N > 1$) to handle a large number of requests for data transmission and minimize delays experienced within the network nodes. For the M/M/N type queue characterized by random arrivals, negative exponential service times, and N servers, one can define the performance by not allowing the probability of delay to exceed a given value. Denoting the traffic intensity by A erlangs, one can express several useful relations valid for infinite sources, as follows:

$$\text{Prob.}(\text{delay} > \text{T}) = (P_0) * \text{Exp}[-(N - A)\text{T}/\overline{\text{T}}_s] \quad (4.9\text{a})$$

$$\text{Avg }(\text{T}_w \text{ or delay}) = \overline{\text{T}}_s/(\text{N} - \text{A}) \cdots \text{ delayed calls only} \quad (4.9\text{b})$$

$$\text{Avg }(\text{T}_w \text{ or delay}) = \overline{\text{T}}_s * P_0/(\text{N} - \text{A}) \cdots \text{ all calls} \quad (4.9\text{c})$$

$$\text{Std. Dev. }(\text{T}_q) = [\overline{\text{T}}_s/(N - A)] * \text{SQR}[P_0(2 - P_0) + (N - A)^2] \quad (4.9\text{d})$$

where

P_0, Exp (x), and $\overline{\text{T}}_s$ have already been defined in Chapter 3.

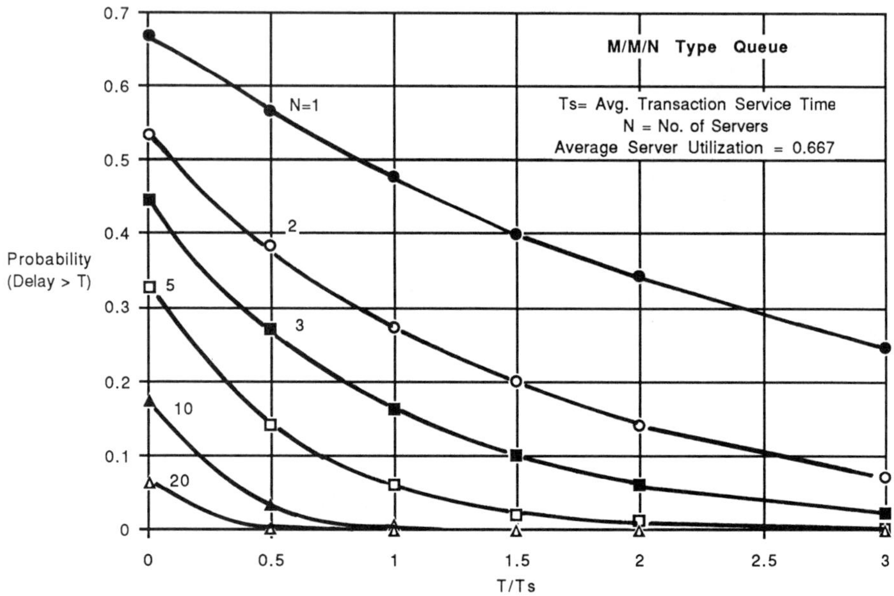

Figure 4.4. Probability of delay exceeding required delay (DREQ) versus DREQ for some useful values of the number of servers(N): M/M/N queuing discipline.

Equations 4.9a-d are the same as those used for an ACD system, where N agents handle the incoming traffic intensity, represented by A. Refer back to Figure 3.11, which illustrates some useful curves relating N (or P) to A for a typical design constraint of delay exceeding 10 seconds only 15 percent of the time. See Fig. 3.14 on page 73 for some curves relating Avg(delay) to A for some useful values of server utilization. Figure 4.4 illustrates another set of useful curves that relate Prob.(delay > T) for T expressed as a multiple of $\overline{T}_s$ and average server utilization (equal to A/N) of 0.667.

4.1.4 THE M/D/N QUEUE ANALYSIS

The analysis of a M/D/N type queuing system characterized by random arrivals, fixed service time, and N servers is quite complex. See Appendix A for a computer solution to derive the probability of delay exceeding any given T, for A erlangs and N servers and the first two moments of delay. Some useful curves relating Prob.(delay > T) are plotted in Figure 4.5 for $A/N = 0.667$. The curves of Figures 4.4 and 4.5 can be very useful for designing systems with the desired response time constraints. In case the traffic characteristics fall in between the worst case (i.e., M/M/N) and the best case (i.e., M/D/N), it may be adequate to interpolate the results for design guidance.

4.1.5 THE MULTIPRIORITY QUEUE ANALYSIS

No priorities were considered in the queuing formulations presented in the preceding subsections. All items that are serviced according to the first-in-first-out queuing discipline are assumed to have the same priority. There are applications that require the

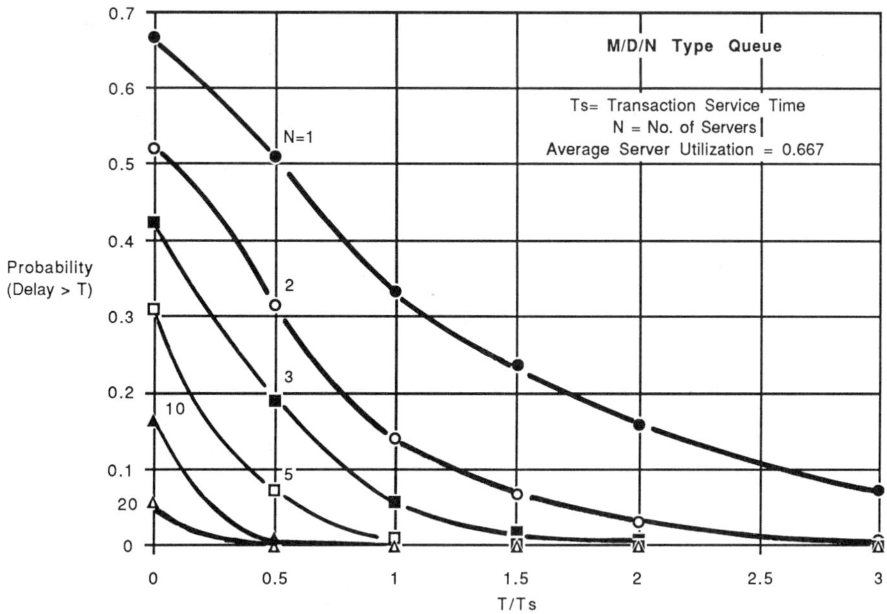

Figure 4.5. Probability of delay exceeding required delay (DTREQ) versus DREQ for some useful values of the number of servers(N): M/D/N queuing discipline.

computation of average wait times for items characterized with *R* (where *R* is greater than one) priorities. Generally, two cases are of interest:

1. No interrupts are allowed for lower priority items when a higher priority item arrives. Only after an item is fully serviced can an item of higher priority in the queue be serviced.
2. Interrupts are allowed for lower priority items when a higher priority item arrives. The so-called premptive-resume priority queuing discipline is employed.

The queuing analysis becomes quite complex for most practical situations. A limited number of solutions for the M/M/1 type queuing model have been enumerated by Martin (1972). The Multipriority Queue model on the ANALYSIS Menu of the software package discussed in Chapter 6 can also be used to study the average wait times in a queuing system characterized by up to five priorities. The model is valid only for the noninterrupt, first-in-first-out service discpline.

4.2 NODAL PERFORMANCE ANALYSIS TOOLS

There is a definite need to compare data communications systems. It can be done in terms of system performance. But that cannot be done until one computes the performance of each node. A network system consists of several types of network nodes. Some provide users with an access to the network services. Some others provide transport to the user and/or supervisory traffic. One can characterize any network node in terms of two types of performance parameters:

1. Throughput, in terms of packets or messages or bits handled per second by the node; and
2. Delays experienced by each packet or message while passing through the node.

4.2.1 NODAL THROUGHPUT ANALYSIS

The throughput of a network node can be described in terms of the bits or packets or messages handled per unit time. The contents of subsection 3.2.1 dealing with nodal throughput for a CS network will also apply to the PS or MS nodes. The values of nodal throughputs are generally provided by the vendor.

In order to compute the throughput of a network node when such data are not available, one must first model all of the relevant sub-nodes within a network node. Whereas some sub-nodes may only involve microcomputers or minicomputers, other sub-nodes may involve both microcomputers and peripheral devices. Each sub-node performs a set of selected tasks. The sub-node showing the highest utilization determines the nodal throughput for a peak hour. In the case where the most active sub-node reflects an excessive loading (>60 percent), such a sub-node must be redesigned by using a faster microcomputer and/or a peripheral device and/or a better software module. While computing the nodal throughput, one must never forget the response time requirements. Only the highest throughput that guarantees a given response time should be considered. The task of computing nodal throughput is generally tedious and time-consuming.

The nodal throughput for a PS node is generally dependent upon the CPU speed or the limited speed of the input/output links. It is also related to the availability of the large amount of random access memory (RAM) required for temporary storage of a large number of packets being handled by the node. There are many vendors that provide PS nodes with the capacity to switch 1000 to 100,000 packets per second. Assuming 1000 bits per packet, a throughput rate of 1Mbps to 100Mbps can be reached if and only if input/output links have an adequate transmission capacity.

The nodal throughput of a message switch is generally determined by the disk capacity or speed. Each message is divided into blocks. A header block is prepared for each message and chained to the first message block and other message header blocks. The remaining message blocks are also chained together. The address of each header block is generally preassigned when the last block of the message is received on a given link. Each disk read/write request can be made according to the FIFO scheme. This will result in a low nodal throughput, since each disk access will encounter full disk latency. The throughput of an MS node can be increased by batching disk read/write requests, and ordering these according to the rotational latency before servicing. While a previous batch is being served, a new batch is being created. A steady state is reached when the size of the previous batch and the next batch of disk requests become equal during a peak traffic period. See Figure 4.6 for an illustration of such a state. The larger the batch, the higher the throughput and the higher the delay through the node. Sharma (1968) has developed a complete analysis of the batching and ordering approach based on the Chapter 2 combinatorial models. This paper also considers the effects of the arm-

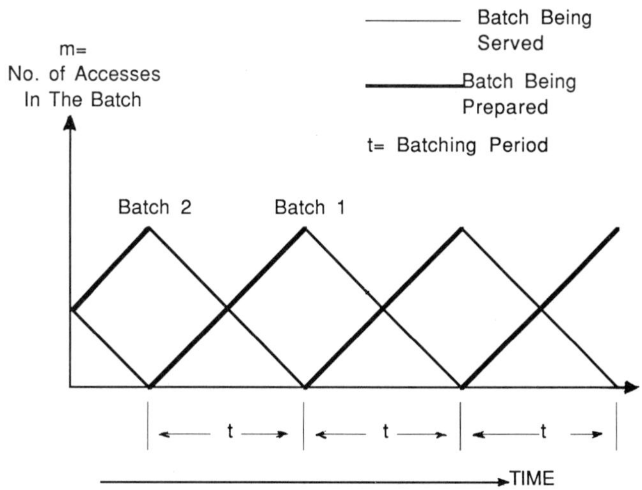

Figure 4.6. Batching, ordering, and servicing of disk accesses.

position mechanism of the disk. Many disks nowadays do not use the arm positioning mechanism. See Figure 4.7 for some useful curves relating the number of disk rotations to the batch size and some useful values of cells per disk track.

4.2.2 *NODAL RESPONSE TIME ANALYSIS*

One can compare the performance of two network nodes in terms of their response times. In order to analyze the average delay experienced in a network node, one must first model the queue, as shown in Figure 4.2. The number of servers and average service

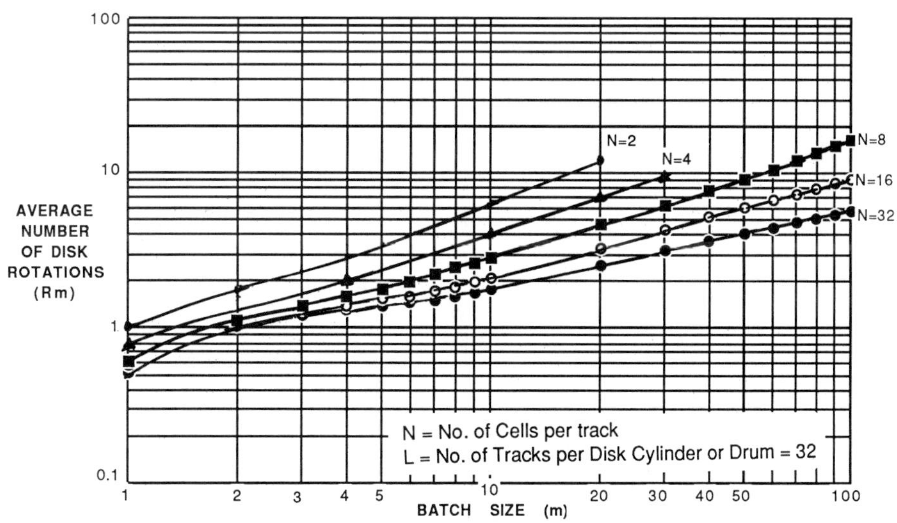

Figure 4.7. Average number of disk rotations versus batch size and some useful disk parameters.

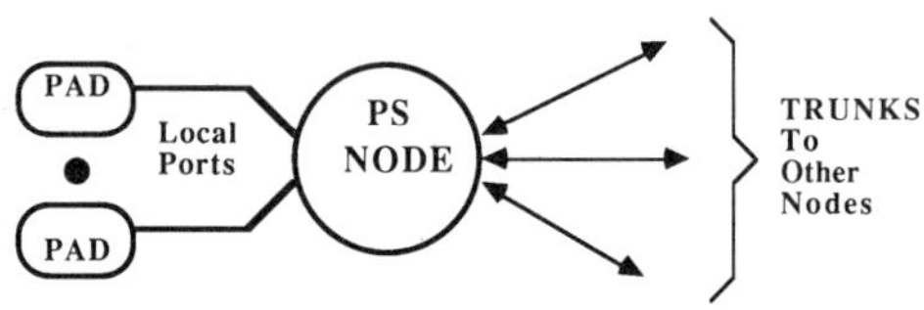

Figure 4.8. Simplified topology of a PS/MS node.

time of each message or packet must be known before computing the response times. There are basically four different techniques for merging locally generated traffic with the en route traffic passing through the node. These are described as follows:

1. Each and every packet is subject to the store and forward mechanism using the FIFO (first-in-first-out) queue discipline. The asynchronous time division multiplexing (ATDM) technique is used for the output process in all PS nodes. Many local ports and long-haul trunks may be served by the node using the star topology. For example, all X.25 nodes obey these rules. See Figure 4.8 for an illustration of the nodal topology.

2. Each node serves several local ports, access links, and long-haul trunks for information exchange, link/trunk sharing, and trunk concentration. The concepts of a shared (or virtual) packet and a shared contention bus are employed to reduce the nodal response time for interactive traffic. An example of this system is AT&T's ISN/Datakit system. The topology of its PS node is illustrated in Figure 4.9, involving several communication modules for serving CPUs, local ASYNC ports, access links to concentrators, and trunks to other PS nodes. Most modules transmit on a shared contention bus and receive on the broadcast bus. The switch listens on the contention bus and transmits on the broadcast bus. Fraser (1983) describes the operation of such a node in some detail, but without developing a generalized equation for computing the nodal response time.

3. Each node serves several local ports and one fully shared trunk for concentration and information exchange. See Figure 4.10 for an illustration of the nodal topology

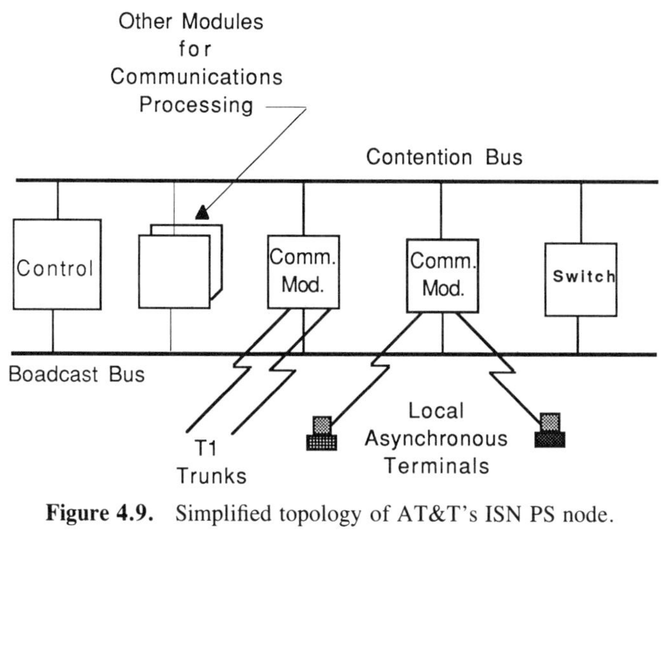

Figure 4.9. Simplified topology of AT&T's ISN PS node.

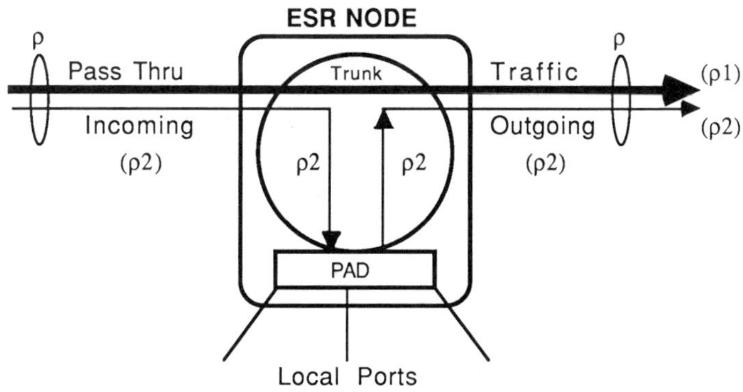

Figure 4.10. Simplified topology of Doelz's ESR PAD/PS node.

forming a basis for the Doelz network architecture, as described by Doelz and Sharma (1984). The time on the shared trunk is divided into very small slots. Each time slot equals a virtual packet transmit time. The PAD unit enclosed within each netlink node creates protocol-independent virtual packets from each active source. The PAD performs the reverse function for the virtual packets arriving for a local port. The en route traffic on the trunk has priority over the local port traffic queued for the trunk. The locally generated packets are queued and sent on empty slots. An en route packet not destined to a local port is not subject to the S/F mechanism, and therefore bypasses the node. This technique is also known as the Extended Slotted Ring (ESR).

4. A traditional multipoint network consists of a host, a front end (FE), and several control units (CUs) sharing a single transmission line. The host and the FE act as one store and forward unit. The network nodes employ the packet-switching technology while handling both short and long messages. Each point-to-point link connecting any two nodes must be analyzed separately for statistics applying to message lengths and arrivals. The nodal response time analysis is similar to that for a conventional PS node.

Now we will study the nodal response times for each of the above-described nodal designs.

4.2.2.1 Response Time Analysis for a Conventional PS Node

A study of Figure 4.8 shows that a typical PS node receives packets from several input channels and puts these packets in an input queue. The PS node then analyzes each packet for addresses and other functions. The processed packets are then put into an output queue. As soon as the output link(s) is available, these packets are transmitted. The type of queueing discipline and the number of servers should then be determined for each of the two queues. To compute the total delay through the network node, one must add the delays encountered in each of the input and output queues according to the analysis tools provided in Section 4.1 and Subsection 3.3.2. Simple addition assumes that these two queues are stochastically independant. Some twenty years ago, such independence was very hard to achieve due to the scarcity of cheap RAM storage and powerful microprocessors. See Equations 4.9a-d for expressions that enable the computation of the first two moments of delay for the M/M/N type queue. Also see Appendix A, which provides a methodology for computing the first two moments (i.e., average delay and Std. Dev. of delay) valid for the M/D/N type queue. By computing the correct values of delays in the two nodal queues and assuming independence, it is possible to estimate the probability of total nodal delay exceeding any given value T. One can express some useful results as follows:

$$E(T_q) = E(T_{q1}) + E(T_{q2}) \text{ or} \tag{4.10a}$$

$$[P(T_{q1}) + P(T_{q2})] \text{ is } > \text{ a given value T, and can be computed using Figures 4.4 and 4.5} \tag{4.10b}$$

where

$E(x)$ = Expected value of x
$P(x)$ = Probability of x
(T_{q1}) = Delay in the input nodal processing queue
(T_{q2}) = Delay for the output transmission queue

4.2.2.2 Response Time Analysis for the ISN Node

According to Fraser (1983), each packet assembled for the output trunk may be composed of several segments of interactive data streams from local ports or data from other trunks. This technique is used to minimize the delay associated with assembling a packet/message for slow interactive traffic. Furthermore, the high capacity shared-contention bus and the switch provide very-high-speed packet switching.

No generalized equations for computing the nodal response times on each VC can be derived due to their dependence on the exact LAN topology (e.g., number and mix of local line speeds). Fraser (1983) showed that the switching time should be about 250 microseconds per packet. For a 40-byte output packet size, the average delay for the output transmission is about 10 ms and 0.5 ms for a 60 percent loaded 56Kbps and 1.35Mbps trunk, respectively.

4.2.2.3 Response Time Analysis for the ESR Node

Response time of an ESR node can be analyzed for two designs: (1) asynchronous point-to-point, whereby each user queues the inquiry message for transmission by pressing the 'send' button, and (2) virtual multipoint (VMPT), which permits a controlled sharing of the transmission facility among several virtual multipoint circuits through the mechanism of polling. One must emphasize that the conventional multipoint (MPT) allows a controlled sharing of the transmission facility among the terminals or stations associated with only a single MPT circuit.

Asynchronous Point-to-Point (or Simply ASYNC). Following the nomenclature of Figure 4.10, one can compute the nodal delay as follows:

$$\left.\begin{aligned} E(T_q) &= 0 \quad \text{[pass-through packet]} \\ E(T_q) &= [\rho_2(1 + \rho_1)/2(1 - \rho)] * T_s \quad \text{[local packet]} \end{aligned}\right\} \tag{4.11}$$

where

$\rho = \rho_1 + \rho_2$ = total input/output loading of ESR
T_s = slot time on the ESR
ρ_1, ρ_2 are the pass-through ESR and locally generated loads, respectively.

It should be emphasized that the protocol for the ASYNC traffic requires a single inquiry message from a data terminal and the single output message from the host. This protocol is much simpler than that for the BSC protocol illustrated in Figure 4.1. The messages are removed from the shared media by the destination nodes. Equation 4.11 was derived jointly by Doelz and Sharma. The expression has some similarities to the expressions for queuing delay for a M/D/1 queue such as Equation 4.5. This expression was arrived at by studying the results of computer simulations and fitting the results to an analytical formula. The precedence of the incoming line prevented a quick intuitive solution.

Virtual Multipoint (VMPT). The delays for a virtual multipoint (VMPT) circuit with N drop nodes is also of interest. In fact, the ESR architecture allows the sharing of the transmission media among several virtual circuits, each based on the conventional MPT operation. For the VMPT operation, one must realize that the incoming load is not reduced by the amount of traffic destined for the local ports of a given drop node. Furthermore, the traffic, originated from the local ports of a given drop node ρ_2, adds to the incoming load. In other words, the output load is related to the input loads as follows:

$$\rho_o = \rho_i + \rho_2 \tag{4.12}$$

The relationship between the input traffic and output traffic loads are illustrated in Figure 4.11. The queuing delay in the jth node for VMPT operation has been expressed by Doelz (1985) as follows:

$$\overline{T}_{qj} = [1/2N][\rho_i \, \rho_o]\,[\text{NUM}_1/\text{DEN}_1] * T_s \tag{4.13}$$

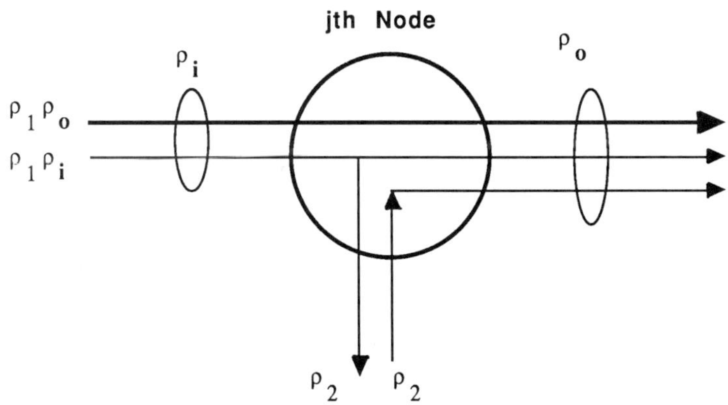

Figure 4.11. Illustration of a VMPT nodal traffic flow in a Doelz's ESR PAD/PS node.

where

ρ_i, ρ_o are as illustrated in Figure 4.11
T_s = slot time on ESR
N = number of drop nodes on the netlink
$NUM_1 = [\rho_i + \rho_o + j\,\rho_i/N]$
$DEN_1 = [1 - (j - 1)\,\rho_i/N]\,[j - \rho_o - j\,\rho_i/N] - \rho_i\,\rho_o/N$

4.2.2.4 Response Time Analysis for the Multipoint Node

A typical multipoint network consists of a host, a front-end (FE), and a network of modems, cluster controllers, and terminals. Most of the delays due to queuing generally take place at the host and the cluster controllers. The delays due to queuing can be modeled in a manner similar to that for a conventional PS node. The M/M/1 model generally applies for the nodal delays and the M/M/N model generally applies to the delays on output transmission lines. The effective value of transaction time for each links will be determined by the presence of many types of messages as shown in Figure 4.1.

4.3 NETWORK PERFORMANCE ANALYSIS TOOLS

There are times when several network systems must be compared with one another. This can be done by comparing their performance in terms of system throughputs and end-to-end response times as defined in Chapter 2.

4.3.1 NETWORK SYSTEM THROUGHPUT ANALYSIS

The nodes of a wide area network (WAN) are generally interconnected by nonshared links according to a well-defined topology. If every node of a WAN handled only the local traffic, the throughput of the system would be equal to the product of nodal throughput and the number of PS central offices. Such an extreme case will require no internodal trunks for user data. If each PS node handled only the tandem traffic, then the system throughput of a WAN can be approximately expressed as follows:

$$C_s = N * C_s/Lp \tag{4.14}$$

where

C_s = system capacity in packets or messages per second
N = number of identical PS/MS nodes in the network system
Lp = average number of nodes in the path of a packet or message

The reader should note that Equation 4.14 applies only to a data network with identical PS/MS nodes. The data wide-area networks (WANs) usually employ two types of nodes: (1) concentrator nodes (i.e., PADs) that serve subscriber lines only, and (2) large packet/message switching nodes. Equation 4.14 assumes that a sufficient number of concentrator nodes are available to handle the traffic intensities required to achieve the network capacity, C_s. The voice networks, on the other hand, employ two types of switching nodes: (1) PABXs or CO type nodes for handling the originating and destination traffic, and (2) tandem nodes for handling only the tandem traffic.

The nodes of a local area network (LAN) are generally connected by means of a shared media. In case LAN nodes are connected by separate links (e.g., Star topology), the LAN throughput can be computed using the same methodology as used for WANs. The throughput of a typical LAN depends upon the shared media capacity (R_o), the number (N) of stations, the interface delay T_{int}, and the type of access employed. Stuck (1983) describes a simplified methodology for computing the throughput of a LAN based on the three well-known access techniques, and for two cases of interest—Case A, when only one out of N stations are active, and Case B, when all stations are equally active. The throughputs of Token Ring, Token Bus, and CSMA/CD LANs can be expressed approximately for two special cases of interest, as follows:

1. Token ring: $$R_1 = L_m/[T_{msg} + T_p + N * T_{int}] \text{ bps} \cdots \text{Case A} \quad (4.15)$$

$$R_1 = L_m/[T_{msg} + T_{int} + T_p/N] \text{ bps} \cdots \text{Case B} \quad (4.16)$$

2. Token bus: $$R_1 = L_m/[T_{msg} + N * (T_{int} + T_p)] \text{ bps} \cdots \text{Case A} \quad (4.17)$$

$$R_1 = L_m/[T_{msg} + T_{int} + T_p] \text{ bps} \cdots \text{Case B} \quad (4.18)$$

3. CSMA/CD: $$R_1 = L_m/[T_{msg} + T_{ifg}] \text{ bps} \cdots \text{Case A} \quad (4.19)$$

$$R_1 = L_m/[T_{msg} + T_{ifg} + (2e - 1) * (T_s + T_j)] \text{ bps} \cdots \text{Case B} \quad (4.20)$$

where R_1 = maximum mean throughput rate in bps
T_s = slot/frame time = $(2T_p + T_{ifg})$
T_j = jam time for CSMA/CD = $48/R_o$
T_p = Propagation time on the shared media
T_{msg} = message transmit time over the shared media = $(Lm + 96)/R_o$
T_{ifg} = interframe gap time for CSMA/CD = 9.6 microseconds
T_{int} = interface delay time = 4 microseconds (token bus)
= $1/R_o$ (token ring)
L_m = bits per message
e = 2.71828
R_o = transmission capacity of the shared media

The reader should consult the works of Schwartz (1987), Stuck (1983), and Stallings (1984) for additional insight into LAN operations and related design parameters. See Figures 4.12 through 4.15 for the plots relating net output data rates to transmission capacity, R_o, in bps for all three LAN architectures, 100 stations, two message lengths L_m, and cases A, B of interest. The capacity of the shared medium (R_o) is varied between 0 and 24 Mbps. The value of $2e$ represents a crude approximation for the number of retransmissions required to achieve a correct transmission. The curves of Figures 4.12 to 4.15 were obtained by converting Equations 4.15 to 4.20 into a spreadsheet and varying some of the design parameters. The results show that a token ring is generally the most efficient of the cases considered. The reader should realize that LANs based

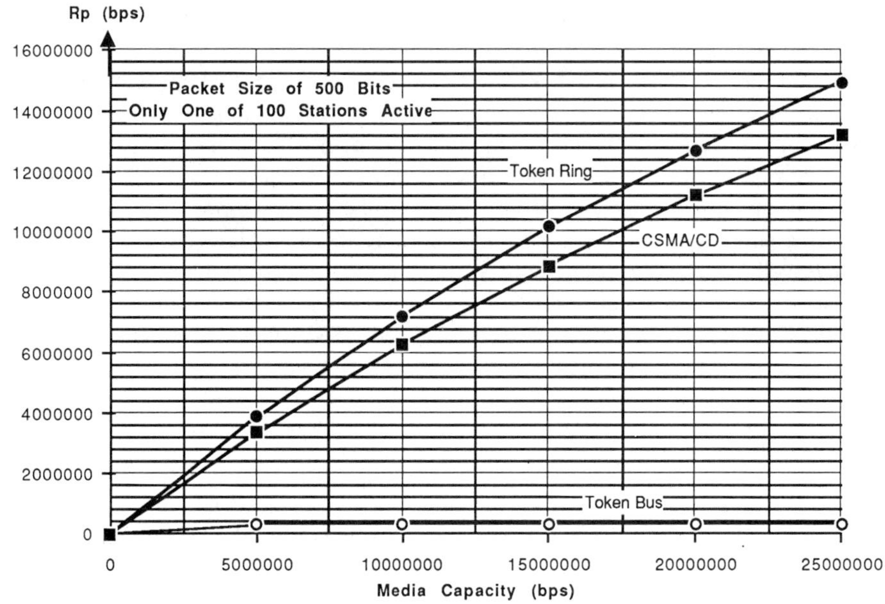

Figure 4.12. Potential throughput (R_p) versus media capacity (R_o) and some useful LAN design parameters: 500 bits per packet and only 1 of 100 stations active.

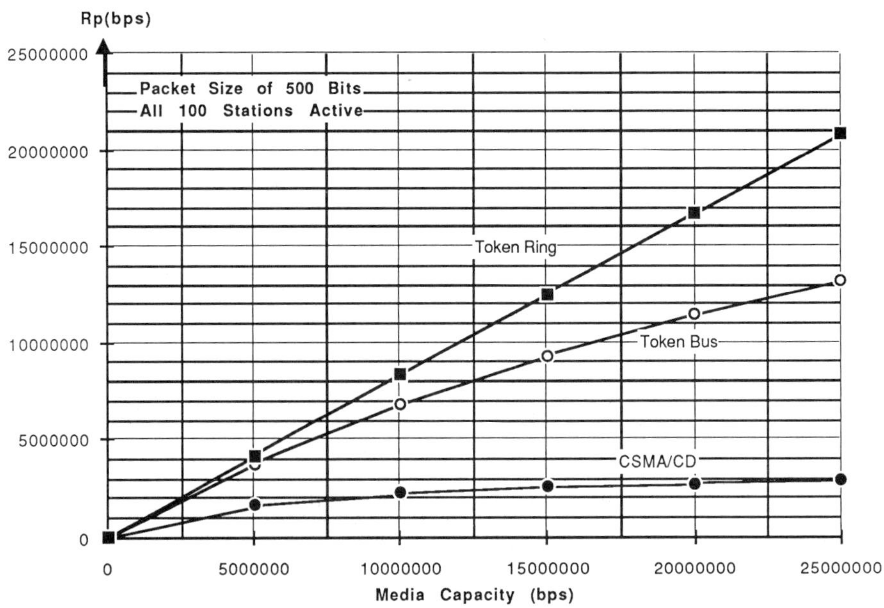

Figure 4.13. Potential throughput (R_p) versus media capacity (R_o) and some useful LAN design parameters: 500 bits per packet and all 100 stations active.

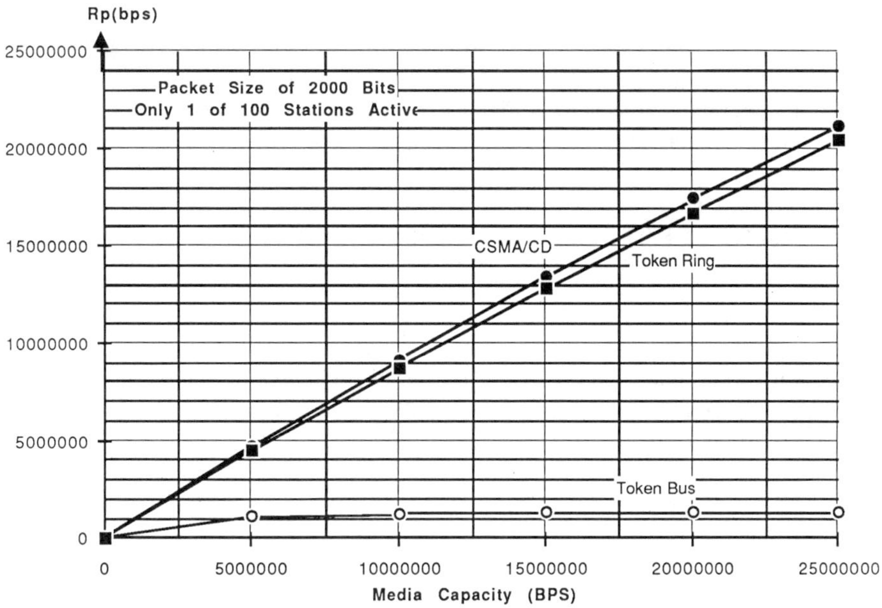

Figure 4.14. Potential throughput (R_p) versus media capacity (R_o) and some useful LAN design parameters: 2000 bits per packet and only 1 of 100 stations active.

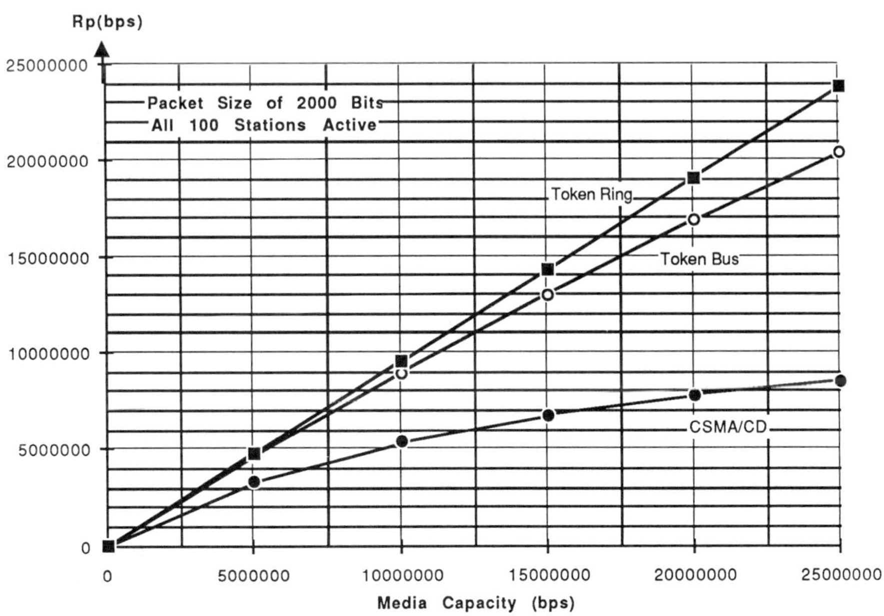

Figure 4.15. Potential throughput (R_p) versus media capacity (R_o) and some useful LAN design parameters: 500 bits per packet and all 100 stations active.

on the integrated voice and data PABX can also provide interstation communication at lower (<64Kbps) data rates than those obtained on the most popular LAN architectures. It may also be interesting to observe that LANs using fiber optical buses can provide interstation communications at very high (< =1000Mbps) data rates. For such environments, the relative merits of a token ring may not be important. In that case, other criteria such as the cost of interface devices may become important. The interface cards for a token ring/bus are generally more expensive than those for CSMA/CD. The cost of an interface card for the IVD PABX is generally hard to quantify for identical LAN operational environs.

4.3.2 NETWORK RESPONSE TIME ANALYSIS

On occasions, a designer must compare several data networks in terms of their end-to-end response times. There are four types of WANs of interest for performing end-to-end response time analyses:

1. WAN's employing PS nodes and PADs (cf. X.25);
2. private networks using AT&T's ISN/Star topology;
3. private networks using Doelz's ESR directed link topology; and
4. private networks based on IBM's SNA and conventional multidrop topology.

4.3.2.1 End-to-End Response Time Analysis for a Public PS Network (PSN)

A typical service on a PSN involves the establishment of a virtual circuit prior to the transmission of data packets associated with the protocol messages. For example, the ASYNC protocol will require an inquiry message and a reply from a host. Another example deals with the transmission of a data file from one host to another. For most applications, one can define an end-to-end response time as the elapsed time between the sending of the first character of the message or packet and receiving the first packet of that message by the destination. The end-to-end communication path is shown in Figure 4.16. Figure 4.17 shows the various processes/queues in each network node.

The end-to-end response time is the sum of end-to-end queuing delays and fixed system delays. The end-to-end queuing delay (T_q) can be expressed as follows:

$$\mathrm{Avg}(T_q) = \sum_{i=1}^{N} \mathrm{Avg}\,(T_{\mathrm{in}}) + \sum_{j=1}^{L} \mathrm{Avg}\,(T_{jl}) \tag{4.21}$$

$$\mathrm{Var}(T_q) = \sum_{i=1}^{N} \mathrm{Var}\,(T_{\mathrm{in}}) + \sum_{j=1}^{L} \mathrm{Var}\,(T_{jl}) \tag{4.22}$$

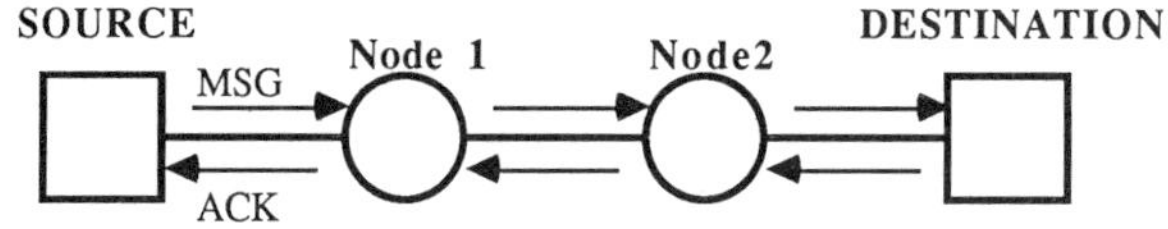

Figure 4.16. Model of a communication path in a PS network.

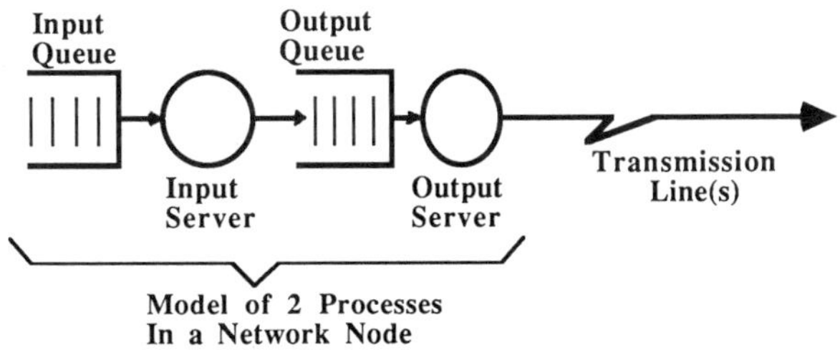

Figure 4.17. Queuing models within a PS node.

where

T_q = end-to-end queuing delay
Avg (T_q) and Var (T_q) are the average and variance of T_q
T_{in} = delay in the ith node (input process)
T_{jl} = delay on the jth internodal link (output process)
N = number of nodes in the transmission path
L = number of internodal links in the communication path

The fixed system delay, T_{fs}, consists of modem delays (if modems are used), propagation delay, and input/output port delays. End-to-end fixed system delays can be expressed as follows:

$$T_{fs} = T_{pg} + N * T_{pt} + N * T_m \tag{4.23}$$

where

T_{pg} = propagation delay for the entire path length
T_{pt} = processing delays (buffering/packetization/switching)
T_m = modem delay if modems are employed
N = number of PS nodes in the path of communication

For a PSN, user ports are generally served by a PAD. In that case, the PAD node should be added to both ends of the communication path. Equation 4.23 assumes identical PS nodes and user ports. If not, the processing delays must be computed for the input, the output, and the en route nodes independently. The end-to-end response time can be represented as the sum of Avg (T_q) and T_{fs}. Since there is no such thing as a typical communication path, and there are too many variables, it is quite difficult to present useful curves. The reader is advised to develop a meaningful model for each communication path and solve for end-to-end response times. In succeeding chapters dealing with actual case histories, comparisons between different systems will be made. The expression for end-to-end response time can be utilized to compute another useful response time defined as the elapsed time between the moment the user sends an inquiry to a host and the moment it receives the first character or packet of the response message. We can define it as the system response time, expressed as follows:

$$T_{sr} = 2(T_q + T_{fs}) + \text{HTT} = \text{system response time} \tag{4.24}$$

where

HTT = host think time spent in analyzing the user request and receiving the the response from its secondary storage

The quantity T_{sr} is sometimes called the bid-to-start (BST) time. Furthermore, the quantity $2(T_q + T_{fs})$ is also expressed as the block/packet turnaround time, T_{ta}.

4.3.2.2 End-to-End Response Time Analysis for an ISN-Based Private Network

The methodology for computing the end-to-end and system response times for a private network based on the AT&T's ISN is identical to that presented in the previous subsubsection. The only difficulty lies in computing the nodal delays. As mentioned earlier, no generalized equation for computing the delays in an ISN node can be employed to compute the delay through a node serving an arbitrary number of subscriber lines with varying capacities and loads. Consult subsubsection 4.2.2.2 for the recommended nodal delays for an ISN environment. These recommended values can then be added together in the manner shown in Equations 4.21, 4.22, 4.23, and 4.24 to compute the desired end-to-end response times.

4.3.2.3 End-to-End Response Time Analysis for an ESR-Based Private Network

Since it is quite an involved process to derive useful expressions for end-to-end response times for all types of ESR-based applications, we will express only a few, useful, expressions. These expressions have been derived from an extensive data base of actual experimental results. These expressions have been tested against simulation results, and excellent agreement exists. Turnaround queuing time, T_{aq}, and the average bid-to-start time, BST, have been expressed by Doelz (1985) for asynchronous point-to-point and virtual multipoint (VMPT) operations as follows:

Case 1. Asynchronous point-to-point operation (ASYNC):

$$T_{aq1} = [\rho_o/2(1 - \rho_o)] + (NUM_2/DEN_2] * T_s \tag{4.25}$$

$$BST = T_{ta} + T_{it} + T_{ir} + T_{rt} + T_{fc} + HTT + T_{sd} \tag{4.26}$$

where

ρ_i, ρ_o = the actual input/output loads at the node
NUM_2 = $[\rho_i/N][1 + (N - k)\,\rho_o/N + (k - 1)\,\rho_i/N]$
DEN_2 = $2[1 - (N - k)\,\rho_o/N - k(\rho_i/N)]$
N = number of drop nodes (not including the node with host port)
k = kth node from which the turnaround queuing delay is measured
T_{ta} = the turnaround time on the ESR, as defined in Equation 4.31
T_{it} = inquiry transmit time on user port
T_{ir} = inquiry receipt time on host port
T_{rt} = the effective time to transmit a response block on the host port
T_s = slot time on ESR
T_{fc} = the time to receive first response character on the user port

HTT = host think time
T_{sd} = the shipping delay

See Figure 4.18 for a simplified chart showing the events occurring on the transmit (T) and receive (R) sides of the user and host ports and the two associated ESR nodes for the ASYNC operation.

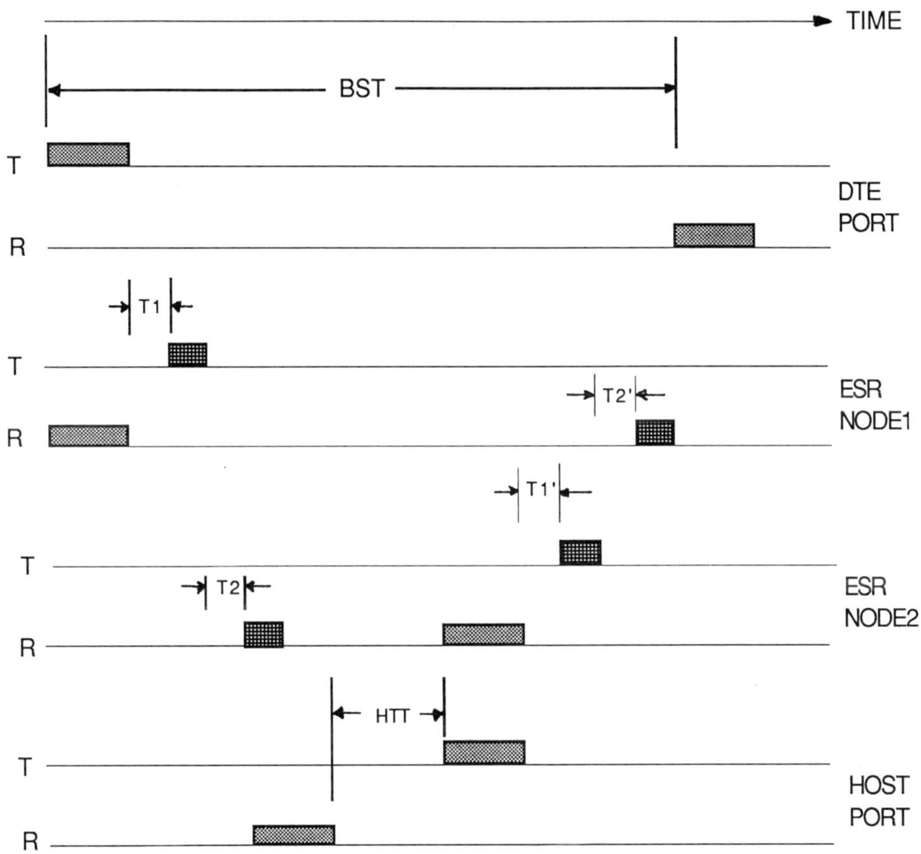

NOTES:
1. T1 AND T1' CONSIST OF SLOT LATENCY, PACKET PROCESSING AND QUEING DELAY
2. T2 AND T2' CONSIST OF BYPASS DELAY, PROPAGATION TIME AND ESR BLOCK TRANSMISSION TIME
3. BLOCK TURNAROUND TIME ON ESR = T1+T2+T1'+T2'

A BLOCK TRANSMITTED/RECEIVED ON DTE/HOST PORT

A BLOCK TRANSMITTED/RECEIVED ON ESR

Figure 4.18. A chart showing the major events occurring at the nodes served by an ESR for ASYNC operations.

Case 2. VMPT operation: This application allows the superimposition of several individual virtual multipoint circuits on the same physical media. Each virtual MPT circuit can be designed to handle N_{cu} logical control units. The total number of physical control units (N) must always be higher than N_{cu}, as expected. The value of turnaround queuing delay (T_{aq2}) and the bid-to-start (BST) delay for the case of VMPT can be expressed as follows:

$$T_{aq2} = \{[\rho_o/2(1 - \rho_o)] + [NUM_3/DEN_3]\} * T_s \tag{4.27}$$

$$BST = EPL + PMST \tag{4.28}$$

Here ρ_i, ρ_o are as illustrated in Figure 4.11; N = number of drop nodes on the VMPT netlink; $NUM_3 = [\rho_i/N][1 + \rho_o + (k - 1)\rho_i/N]$; $DEN_3 = 2[1 - \rho_o - k\rho_i/N]$; k = kth node from which the turnaround queuing delay is measured; EPL is the expected value of the polling latency; and PMST is equal to the poll-to-message-start time. The value of EPL can be determined in terms of the idle-poll-time (IPT) and PMST as follows:

$$EPL = 0.5(N_{cu})(IPT)[1 + \lambda(PMST + ERT)] \tag{4.29}$$

$$IPT \approx T_{pt} + T_{pr} + T_{ta} \tag{4.30}$$

$$PMST = HTT + 2T_{ta} + 147 + 168000(1/UPR + 1/HPR) \text{ milliseconds} \tag{4.31}$$

$$ERT = \max\{\text{response message receipt time on ESR or user port}\} \tag{4.32}$$

See Subsection 2.4.2 for computing the average value of ERT

$$T_{ta} = 4T_{np} + 2T_{sl} + T_{aq} + 4(N - 1)T_{hm} + 2(K_{pg})L_{esr} + T_{nbp}(N - 2) + 2T_{eb} \tag{4.33}$$

For these equations we employ the equivalences: N_{cu} = actual number of control units on the VMPT circuit; λ = actual message arrival rate on the particular VMPT circuit; T_{pt} and T_{pr} are the poll message transmit/receive times on the ports with data rates of HPR and UPR, respectively; T_{np} = nodal processing time; T_{sl} = slot latency time; T_{hm} = half modem time; K_{pg} = propagation constant; L_{esr} = ESR length; T_{nbp} = nodal bypass time; N = total number of drop nodes on the ESR; T_{eb} = effective ESR block transmit time; and T_{aq} is applicable end-to-end queuing delay.

The quantity ($2 * EPL = N_{cu} * IPT$), valid for no activity, is sometimes called the system walk time. It is equivalent to the minimum time required to poll all of the stations on a virtual circuit.

4.3.2.4 End-to-End Response Time Analysis for a Conventional Multipoint Network

In a conventional multipoint network of Figure 4.19, contention for the use of a multidrop link is resolved through the use of a roll-call type polling of N cluster controllers.

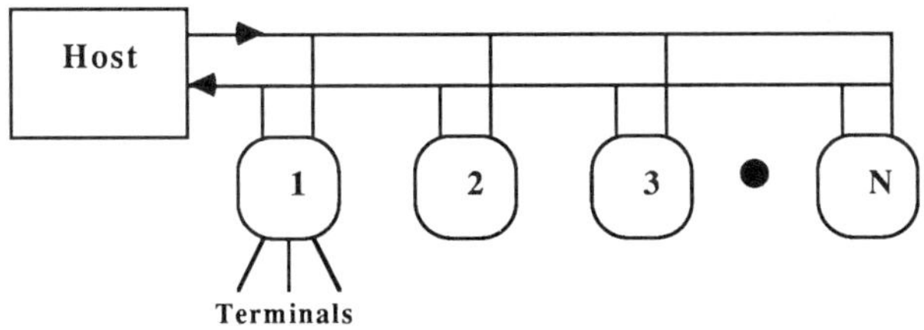

Figure 4.19. Model of multipoint polling for a network with N controllers.

The cluster controllers are polled either sequentially or according to a preset order. Several types of messages flow between the host and a cluster controller before the host sends the desired message. See Figure 4.1 for an illustration. One can study a useful quantity defined as the elapsed time between the moment a packet arrives at the controller from a local port and the moment that packet begins transmission towards the host. One should note that the elapsed time between the moment one presses the "send" button for the inquiry message and the moment one receives the first packet of the host message is equivalent to the expected delay. Schwartz (1987) has shown that the average access delay is a function of multidrop line loading (ρ), the number of drops (N), the average message arrival rate, λ, second moment of message duration, $\overline{m^2}$ and the system walk time (W), as shown below:

$$E(D) - (W/2)\,[1 - \rho/N]/(1 - \rho) + [N\,\lambda\,\overline{m^2}]\,/2(1 - \rho) \qquad (4.34)$$

where

$$W = 4N * T_{np} + 2N * T_{pm} + 2N * T_{m} + \tau(N + 1)/2 \qquad (4.35)$$

where T_{np} is the nodal processing time for each poll message; T_{pm} is a fixed poll message transmit time; T_m is modem time (which should be twice the half-modem time used for analyzing ESR netlinks); and τ is the total two-way round-trip propagation delay on the netlink. The factor of 2 implies the two directions of data flow, as shown in Figure 4.1. The factor of 4 implies four nodal processing functions occurring in the two nodes related to each idle poll. The propagation delay assumes that the N controllers are distributed equally on the netlink.

For the case of hub polling, where each controller passes the poll to the next controller after being serviced, the value of W can be expressed as follows:

$$W_h = 4N * T_{np} + 2N * T_{pm} + 2N * T_m + \tau \qquad (4.36)$$

By substituting the value of W_h in Equation 4.34, one can solve for the expected value of access delay for the case of hub polling.

It should be instructive to compare the equations for E(D) or BST for the ESR and the conventional multipoint network link. Using Equations 4.29, 4.30, and 4.33, the value of W for ESR can be defined as follows:

$$
\begin{aligned}
W_{esr} &= N_{cu} * \mathrm{IPT} \\
&\approx N_{cu} * (\mathrm{T}_{pt} + \mathrm{T}_{pr} + \mathrm{T}_{ta}) \\
&= N_{cu} * (\mathrm{T}_{pt} + \mathrm{T}_{pr} + 4\mathrm{T}_{np} + 4(N - 1)\,\mathrm{T}_{hm} + 2(K_{pg})\,L_{esr} + 2\mathrm{T}_{eb}) \\
&= N_{cu} * (\mathrm{T}_{pt} + \mathrm{T}_{pr} + 4\mathrm{T}_{np} + 4(N - 1)\mathrm{T}_{hm} + \tau + 2\mathrm{T}_{eb}) \qquad (4.37)
\end{aligned}
$$

where the value of τ has been substituted for $[2(K_{pg})\,L_{esr}]$.

A comparison between Equations 4.35 and 4.37 shows plenty of similarity. However, it does appear that the value of W_{esr} should be much higher than W for the conventional MPT networks for identical values of N and N_{cu}. The additional terms of T_{pt} and T_{pr} represent the poll message transmission times on the user/host ports, which were completely ignored in Equation 4.35. These values are generally negligible for most cluster controllers, which employ high-speed ports. The value of T_{eb} should be identical to T_{pm} for identical capacity of a shared media. The propagation time is about twice that in Equation 4.29 for identical values of N and N_{cu}. However, the major part of the difference is influenced by the term $[4N_{cu}(N - 1)\,\mathrm{T}_{hm}]$. The difference grows rapidly with increasing values of N_{cu} and N. Using the fact that ESR allows the superimposition of several virtual multipoint circuits on the same physical media, the value of N_{cu} is generally quite low per each VMPT circuit. In practice, the value of N_{cu} is rarely greater than 4 for most ESR applications. Since the value of N for a conventional multipoint circuit will generally be quite high, when compared to N for ESR applications, and since N is not generally allowed to exceed 6 for most ESR applications, W_{esr} should not be much higher than W for most practical cases. Furthermore, the ESR should always outperform the conventional MPT network at high traffic loads, due to its unique queuing mechanism. It was quite interesting to observe that Equation 4.28 provides almost identical results to those obtained from Equation 4.34 if one employs an identical value of walk times (W) and queuing delays. See Chapter 8 for some response time analyses of the two architectures for meaningful design parameters and assumptions.

We have now studied the nodal and end-to-end response times for WANs. We are now left with the end-to-end delay performance of LANs. Deriving simple expressions for the end-to-end response times for different LANs for identical operating environs is generally an impossible task. Since each WAN can be implemented by bridging several LANs together, the task becomes even more difficult. An attempt to derive a simple qualitative comparison between the end-to-end response times for LANs and WANs is discussed in the next section.

4.4 END-TO-END RESPONSE TIME COMPARISONS FOR LANs AND WANs

Since the emphasis of this book is on network topologies and their synthesis, we will only present a qualitative discussion of the end-to-end response times for LANs and some WANs. The reader should consult the works of Schwartz (1987), Stuck (1983), and Stallings (1984) for an insight into the operations of LANs and their performance. The curves of Figure 4.20 were derived from their works to illustrate the normalized response times to media (shared or otherwise) loading.

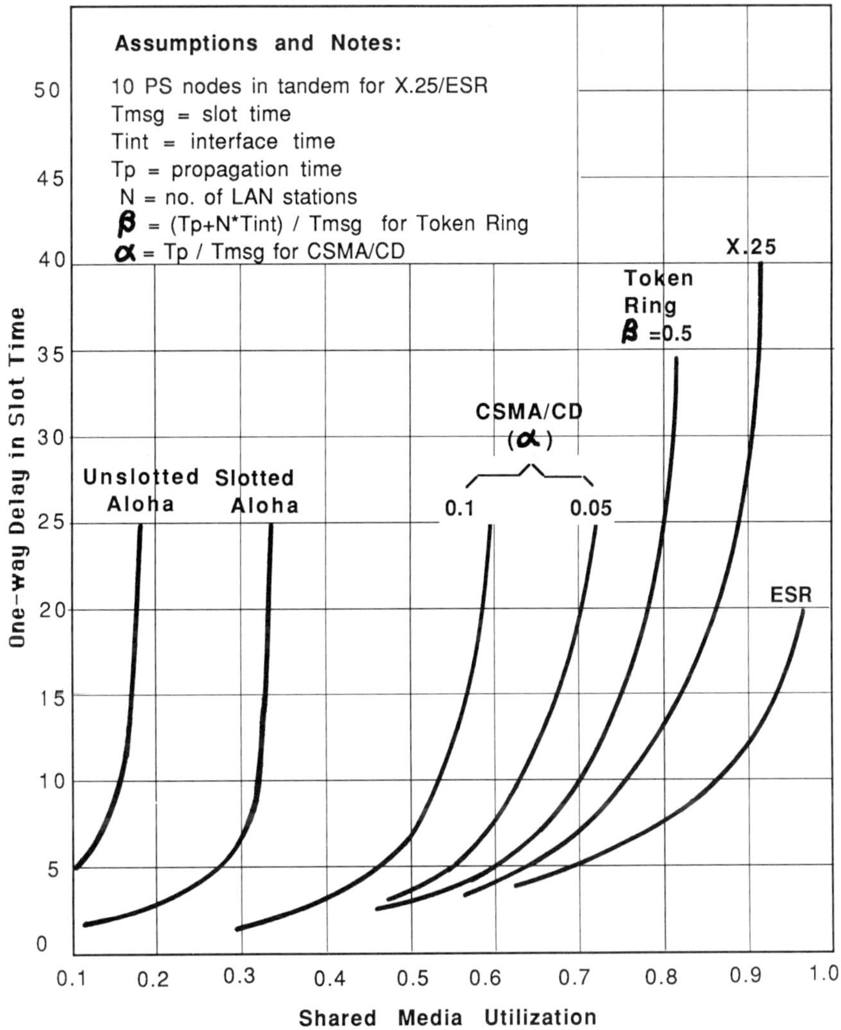

Figure 4.20. One-way delay versus shared media utilization in LANs and WANs.

The end-to-end response time is defined here as the elapsed time between the moment a packet begins transmission at the source node and the moment the packet starts to arrive at the destination node. Only token ring and CSMA/CD LANs are shown, along with the WANs based on the Aloha, Slotted Aloha, X.25 PS, and ESR access techniques. The netlinks based on the X.25 and ESR architectures were assumed to serve 10 nodes in tandem for each path of communication. The number of nodes served by a LAN indirectly affects certain design parameters, such as α and β, for LANs based on the concepts of CSMA/CD and token ring. It must be emphasized that the curves are idealized to illustrate the limits of their performance. Each one of the systems illustrated has an upper bound for media utilization. According to the results plotted, the ESR-based system is the best, and the Aloha-based WAN is the worst, of all considered in terms of media utilization. Other performance considerations, such as the cost of transmission media and hardware/software interface modules, should also be examined.

Detailed expressions for the response times for each type of network can only be derived for the exact environments under which each network operates. Furthermore, such design possibilities are never the same for any two practical systems. There are an infinite number of valid possibilities. It is for that reason that most references cited above shy away from discussing such expressions in detail. Even those papers that derive complex equations for a single system do so for a very limited set of idealized situations (almost useless). The actual set of design variables will generally fall outside of the limits assumed. A successful methodology employed by the author is as follows:

1. Measure the desired response times for several useful values of media loading.
2. Devleop an approximate analytical model for the end-to-end response time.
3. Produce some useful curve relating response times to various design parameters, such as media utilization.
4. Develop spreadsheets that will give answers as a function of design parameters on a ''what if'' basis.

Again, it should be emphasized that the above methodology is repeated for each network architecture and set of operational environments.

5

Algorithms for the Synthesis of Network Topologies

5.1 INTRODUCTORY REMARKS

There is no such thing as a closed-form algorithm for optimizing a network topology for most voice or data applications. The designer must employ a combination of heuristic algorithms and "what-if" types of analyses in an iterative manner until the best solution is obtained. See Chapter 2 for a discussion of such a methodology. Basically, the solution consists of the following steps:

1. Computing the lower bound on the number of concentrators/switches, as dictated by their maximum capacity to handle calls/messages/terminations;
2. Selecting the locations of a required number of concentrators/switches;
3. Assignment of customer premise equipment (CPE) locations to concentrators, and the assignment of concentrators to switches;
4. Connecting the CPE locations to concentrators and connecting the concentrators to switches with a correct number of links, according to the desired topology;
5. Computing the total cost of the network for the allowable sets of design parameters; and
6. Selecting the optimum solution by varying a set of design parameters.

By plotting the total network costs as a function of the number of concentrators/switches, as shown in Figures 5.1(a) and (b), the designer should be able to choose the most optimum network topology consisting of the correct number of concentrators/switches and links. It should be stressed that the word "optimum" used in this book refers only to the best solution found among those considered (and not from among all of the possible solutions). For those who consider themselves purists, this revelation may be disappointing. For those who consider themselves pragmatists, such a revelation only adds beauty to the art and science of networking. We will now discuss each of the above steps that can be applied to voice, data, and integrated voice/data applications.

5.2 COMPUTING THE LOWER BOUND ON THE NUMBER OF CONCENTRATORS AND SWITCHES

Using the traffic engineering techniques from Chapter 2, it is an easy matter to compute the total number of calls/messages/bits handled during the busy hour (BHR). By study-

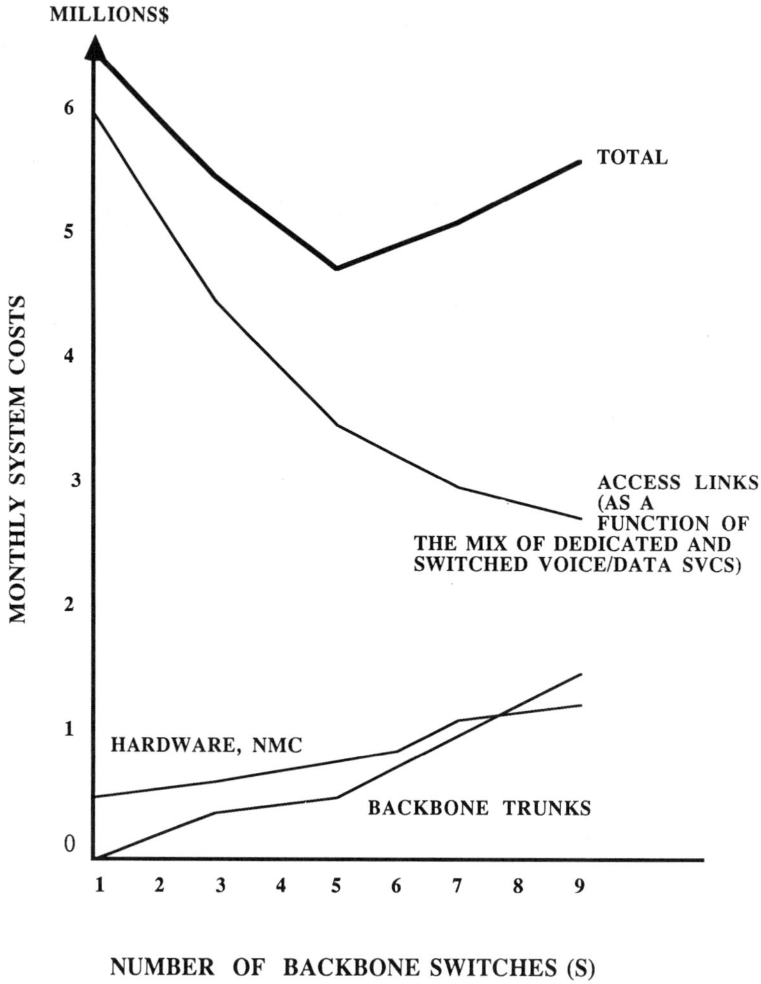

(a)

Figure 5.1(a, b). Distribution of costs in (a) a typical voice/data network, and (b) a typical ACD network.

ing the vendor data dealing with the capacity of the concentrators/switches to handle maximum traffic and/or number of lines and/or trunks, the designer can quickly determine the minimum number of concentrators and/or switches required to design the network.

For voice applications, the CPE location is generally served either by a PABX or a CO-Centrix. An unusual requirement may involve knowing only the number of telephones at each customer location. In that case, the first task may be to compute the optimum number of PABXs and/or CO-Centrexes. These days, the number of locations or PABXs served by the switches is determined by pure economics. One can now buy a PABX or a tandem switch (and even a combined PABX/tandem switch) with termi-

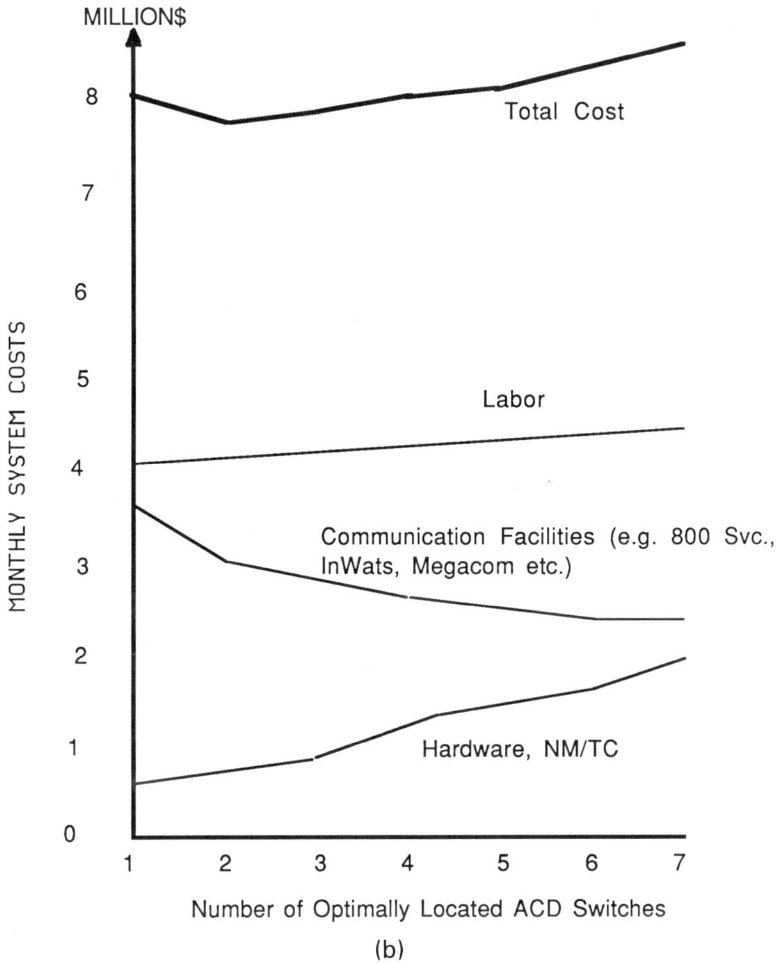

Figure 5.1(a, b). *(Continued)*

nations ranging from a small number to something above 20,000 telephones. Such a range of terminations does not, therefore, prevent a location from being served by a PABX. The design of an optimum network is then completed by computing the number and locations of tandem switches according to a least-cost criterion repeated for each selected number of locations (or PABXs) served.

For data applications, the number and type of data terminals at each location is generally given. The first task is to compute the minimum number of data concentrators needed by studying the capacity of a single concentrator in terms of messages or ports or both. A criteria for vendor selection may be a concentrator product that can handle a specified maximum number of ports (similar to that for a PABX). The next task is to vary the number of switches (starting from the minimum number of switches required) for serving a given number of concentrators until the least-cost criterion is met. This process may then be repeated by increasing the number of concentrators employed. This will result in several sets of plots, as illustrated in Figure 5.1, before the designer obtains an optimum network topology.

Cost may not be the only criterion for choosing the correct number of switches, as suggested by the curves of Figure 5.1. An additional criterion should be based on system availability or reliability. If the choice is between two or three switches, and the cost is not much different, one should choose a network topology with three switches. For one thing, the network system will be able to handle future growth better. Furthermore, with three switches, the network will be able to maintain a majority of trunk traffic flows in case a switch, or some trunk transmission facilities, fail. With two switches, the major nodes will fail to communicate in case of a switch or a trunk bundle failure. In general, the selection of an additional switch will always enhance the reliability of the system. The reliability criterion becomes even more significant when working with an ACD system (see Chapter 7 for a detailed design methodology) that employs virtual transmission facilities. Even the addition of a second switch will enhance the reliability of the one-node ACD system by rerouting the incoming traffic to the functioning ACD location. Such an option is provided by most common carriers for a small fee.

5.3 ALGORITHMS FOR LOCATING CONCENTRATORS/SWITCHES

For any given number of concentrators and/or switches, the designer must locate the concentrators and/or switches at existing customer locations (and not in a cornfield) to achieve the least cost. For voice applications, the CPE locations to be served by switches are specified. The task is to locate a given number of switches at the proper subset of CPE locations to be served. For data applications, the first task is to choose the optimum locations for placing the chosen number of concentrators. For that number/locations of concentrators, the next task is to optimally locate a given number of switches.

The author has developed a simple algorithm for locating a given number of concentrators/switches at a subset of the CPE locations as an initial solution. This algorithm employs the concepts of a center-of-gravity (COG) and successive decompositions, and it can be executed as follows:

Step 1. Define the V- and H-coordinates for each of the N CPE locations.

Step 2. Find the V- and H-coordinates of the mathematical COG for all N locations, using the following expressions:

$$\mathrm{V(COG)} = \left[\sum_{i=1}^{N} V_i * W_i\right] \Big/ \left[\sum_{i=1}^{N} W_i\right] \text{ and} \tag{5.1}$$

$$\mathrm{H(COG)} = \left[\sum_{i=1}^{N} H_i * W_i\right] \Big/ \left[\sum_{i=1}^{N} W_i\right] \tag{5.2}$$

where

V_i and H_i are the V- and H-coordinates of the ith locations, and W_i is the time-consistent average of the two-way traffic handled by the ith location.

Step 3. Find the closest CPE location to the above mathematical COG. This location is suggested as the optimum location for a single concentrator or a switch.

Step 4. To find the locations of two COGs, one must first draw a vertical line passing through the mathematical COG. Then enumerate the locations on the left and the right sides of the vertical line and find the mathematical COG for each of the two sets of locations, using the procedure of Step 2. Again find the closest locations to these two new COGs.

Step 5. To find the locations of four COGs, one must first draw two horizontal lines through the two COGs and obtain four sets of CPE locations falling in the newly decomposed regions. Again, find the four mathematical COGs and the closest CPE locations to these four COGs.

Step 6. Repeat the Steps 4 and 5 by decomposing additional regions and computing the COG locations and the corresponding closest CPE locations. The alternation of vertical and horizontal decompositions is necessary to eliminate any dependence on geographical boundaries. The above scheme of starting with a vertical decomposition (e.g., Step 4) is recommended for countries such as Canada or the U.S.A. In some cases, starting with a horizontal decomposition is preferred. In other cases, both types of initial decomposition must be tried to find the best solution.

By naming these COGs and corresponding CPE locations in a consecutive manner, one can illustrate the entire solution space, as illustrated in Figure 5.2. Once the solution space is known, it is a simple matter to select 1, 2, 3, . . . , M locations for concentrators or switches. As illustrated in Figure 5.2, the choice of 1, 2, 3, 4, 5, 6, and 7 concentrator/switch locations are obvious when 1, 2, and 4 COGs are computed successively, using the above methodology.

The network designer will find that the above algorithm provides a rather good solution for most cases. However, there may be some pathological cases where the initial solution can be improved. One method is to study each of the M clusters and relocate the corresponding network center/switch such that the sum of the costs for all access lines of that cluster is minimized, as follows:

$$\text{Cost} = \min_i \left[\sum_j d_{ij}\, w_j \right] \ldots \qquad (5.3)$$

where

d_{ij} = distance or cost between the ith center and jth location/node
w_j = weight of jth location/node in terms of millierlangs or the number of bits per second

An algorithm based on an exhaustive search can yield an optimal solution, but it will be extremely time consuming. Many other approximate algorithms for locating concentrators and switches have been described by Sharma, De Sousa, and Ingle (1982). Since none of these algorithms can yield an optimal solution, it is difficult to compare their efficiency with that for the above-mentioned simplified algorithm. Some day, it may be

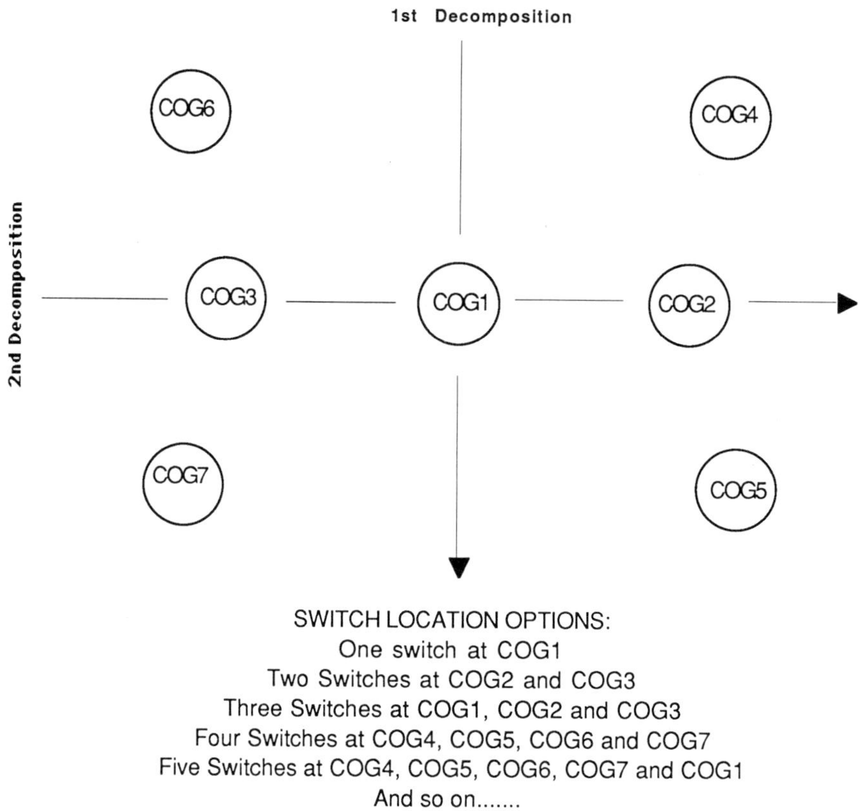

Figure 5.2. Decomposition algorithm for locating concentrator/switching nodes.

possible to compare all known heuristic algorithms in a scientific manner. At this time, no single algorithm can be proven as the best for all applications.

5.4 ALGORITHMS FOR CONNECTING CPE LOCATIONS TO CONCENTRATORS AND SWITCHES

The designer must employ the bottom-up approach to connect all of the nodes to one another. One must always start with the lowest level (e.g., CPE nodes) of unconnected nodes, and connect these to the concentrator nodes belonging to the next hierarchical level, according to an applicable topology. These concentrator nodes now take up the role of unconnected nodes, which must be connected to the concentrator nodes of the next level. This process is repeated until all nodes of a network hierarchy are connected together. There are several housekeeping chores that must be accomplished as the network evolves. These tasks can be listed as follows:

- Task 1. Accumulate the traffic intensity at each concentrator;
- Task 2. Accumulate the cost of network facilities (nodal hardware and transmission line) at each concentrator and level of network hierarchy; and
- Task 3. Eliminate the lower-level nodes from the nodes yet to be connected.

There are several applicable topologies for synthesizing voice and data networks at any stage of the above-mentioned process. The star topology is useful for designing almost all voice networks. The star topology is also useful for designing some packet-switched networks based on AT&T's ISN and CCITT's X.25 architecture. Such topologies are generally quite efficient for LAN applications. The multidrop (MD) topology is useful for designing WAN data networks based on the SNA architecture. Some PS networks, which employ PADs to handle SNA's SDLC and BSC protocols, will require a mixture of the multidrop and star topologies. The directed link (DL) topology is useful for designing LAN/WAN data networks based on Doelz's extended slotted ring (ESR) architecture, as described by Doelz and Sharma (1984).

The bus and the ring topologies for LANs should not generally require any design effort, since only simple network connectivity is employed. The cost of the entire network is generally influenced by the media interface modules and not by the network topology. The LAN links are generally owned and not leased from the local exchange carrier (LEC). In some cases, one may reduce the network cost through an optimal placement of telephone closets and main frame distribution (MFD) units in a large building, a multilevel ship, or a campus. Another technique for reducing the network cost is through replacing the expensive vendor-recommended transmission media with the existing twisted copper wires through the use of Baluns (see Chapter 9 for a proven methodology) or other special devices. The design of a premise wiring system (PWS) generally requires a lower-level design of voice/data network based on the star topology.

5.4.1 A GENERALIZED ALGORITHM FOR DESIGNING ALL NETWORK TOPOLOGIES

For most voice and data applications, the algorithm must perform the following tasks:

- Task 1. Select the number (equal to *C*) of concentrators/switches;
- Task 2. Find the optimum locations for *C* concentrators/switches;
- Task 3. Assign the unconnected nodes at the lower level to the various concentrator/switch locations based on the closest distance resulting in *C* sets of locations. In some cases, it may be advised to employ a more complex assignment criterion based on the from–to traffic intensity and the cost of all of the paths. In the discussions that follow, only the simplified assignment scheme will be considered;
- Task 4. Execute the steps of an algorithm required for designing any topology (see the subsequent paragraphs for the steps required for each of the Star, DL, or MD topologies);
- Task 5. Compute the total traffic (erlangs/bps/etc.) handled by each concentrator/switch by adding the traffic between each nodal pair (or traffic on each access line, AL);
- Task 6. Compute the number of circuits necessary to handle the required traffic on each AL bundle, while obeying the design constraint of, for example, the allowable data rate or blocking factor;
- Task 7. Compute the cost of each AL bundle as a function of the distance, number of circuits in the bundle, and the applicable tariff. Chapter 6 proposes a very flexible and modular approach of entering several tariffs into a tariff file and

choosing the applicable one for computing the cost of any AL or a trunk bundle;

Task 8. Compute the cost of each region and the entire system;

Task 9. Repeat Tasks 1 through 8 for each level of concentrator until only the backbone network remains to be designed;

Task 10. Synthesize the backbone network, connecting the *S* switches using the techniques presented in Chapters 2, 3, and 4;

Task 11. Compute the size of each trunk bundle as a function of the traffic handled, while obeying the design constraint of, for example, the allowable data rate or blocking factor;

Task 12. Compute the cost of each trunk bundle as a function of the distance, number of trunks in each bundle, and the applicable tariff. Also compute the cost of the entire trunk network;

Task 13. Repeat Tasks 1 through 12 for each combination of *C* concentrators and *S* switches, and for various combinations of transmission and nodal hardware technologies;

Task 14. Make a chart similar to that in Figure 5.1 plotting network-related costs (e.g., cost for the total AL, trunks, hardware, and NMC/TC). Select the most optimum network topology consisting of the correct number of concentrators and switches for a correct combination of nodal/transmission technologies; and

Task 15. Display all of the resulting network topologies in the form of graphics with a summary of network costs.

The above process will generally require several human interventions, especially when judgements are to be made regarding the chosen sites for concentrators and switches and the applicable tariffs (one for each transmission technology). But, in any case, most of the other steps must be executed on a computer to save human labor. Chapter 6 presents a desktop computer technology to handle most of the above steps. Several attempts are being made by researchers to develop a technique based on the so-called expert system (ES) or artificial intelligence (AI). However, ES or AI are useful only when cost-effective algorithms are unavailable, and the entire intelligence or experience is available only in the form of a large relational data base (e.g., in the area of human illness diagnosis). The approach taken in this book mainly relies on a rich source of cost-effective algorithms. Only a few human interventions are needed at critical points. Actually, the entire process, as illustrated above, can be accomplished within a few hours after the required input files (as defined in Chapter 6) are available to run the network design program. Since most networks have only three levels of nodes (i.e., CPE locations, concentrators, and switches) to be interconnected, only two passes are required to complete the first nine tasks of the earlier-described design process. During the first pass, the CPE locations are connected to the selected number of concentrators, and during the second pass, the concentrators are connected to switches before designing the backbone network.

5.4.2 ALGORITHM FOR DESIGNING A STAR TOPOLOGY

Only Task 4 in the generalized design process of Subsection 5.4.1 needs to be altered for connecting the nodes to associated concentrators/switches locations according to the star topology.

Step 1. Connect the locations in each set to the associated concentrator/switch location by an access line bundle. This step yields a multicenter, multistar topology network, as illustrated in Figure 1.8.

5.4.3 ALGORITHM FOR DESIGNING A DIRECTED LINK (DL) TOPOLOGY

Only Task 4 in the generalized design process described in Subsection 5.4.1 needs to be altered for connecting the given nodes to concentrator/switch locations according to the DL topology. This algorithm applies to the design of the first feasible network as part of the total network design process of U.S. Patent 3,703,006. It is fully described by Sharma and El-Bardai (1970), and summarized as follows:

Step 1. Divide the nodal traffic load W_i by the allowable maximum link weight W_m and compute the integer quotient Q_i for all i. Connect node N_i to the concentrator by Q_i special circuits to handle excess traffic for each node. Replace the original W_i by the value $(W_i - Q_i * W_m)$.

Step 2. Find the node (N_i) among the unconnected nodes that is farthest from the given concentrator node C_j.

Step 3. Find an unconnected mode (N_j) closest to N_i.

Step 4. Add the traffic weights of these two nodes to get a running total as $W_t = W_i + W_j$.

Step 5. If $W_t <=$ the allowable maximum, W_m, then connect N_j to N_i and remove these from the unconnected list of nodes. Call N_j as N_i. Repeat Steps 3 and 4 until $W_t > W_m$.

Step 6. If $W_t > W_m$, then connect N_i to the concentrator node.

Step 7. Repeat Steps 2 through 6 until all nodes have been considered and connected.

The above algorithm is illustrated in Figure 5.3. If one ignores the special circuits created in Step 1, the network now consists of several netlinks, each of which may be shared by several nodes. The concept of a multicircuit access line bundle for the star topology does not apply to a shared netlink based on the DL or a multidrop topology.

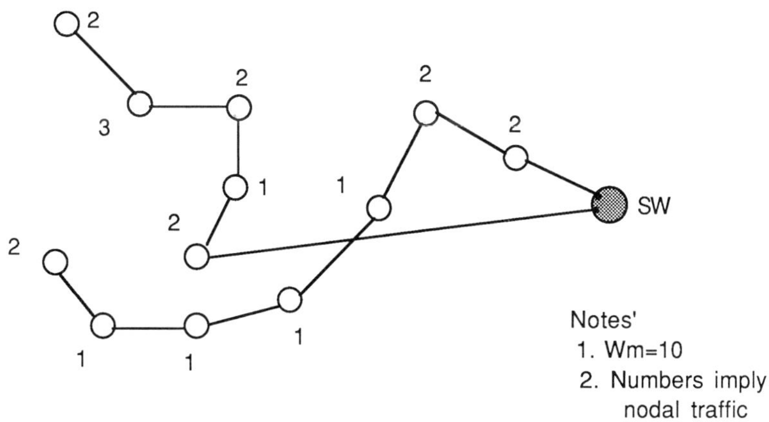

Figure 5.3. Sharma's algorithm for optimizing a DL topology.

The remaining steps of the network design process of Subsection 5.4.1 remain unchanged.

5.4.4 *ALGORITHM FOR DESIGNING A MULTIDROP (MD) TOPOLOGY*

Kruskal (1956) and Prim (1957) are credited with the development of algorithms for connecting a set of nodes to one another according to a minimal spanning tree (MST) topology. Such a topology guarantees a minimal sum of length associated with all links or connections employed to connect the nodes. An MST topology is analogous to a true multidrop topology. The original MST topologies were synthesized with no regard to capacity constraints. Therefore, the algorithms of Kruskal and Prim require modifications for designing practical networks.

The first practical algorithm for designing an MST with capacity constraints was developed by Esau and Williams (1966). See Esau and Williams (1966) for a description of their algorithm. Several practical algorithms were also developed by the author. Consult Sharma and El-Bardai (1972) for a description of three algorithms. Two of these algorithms always produced more expensive CMST (capacitated minimal spanning tree) networks when compared to the third algorithm, which later became the basis of U.S. Patent 3,703,006. Unfortunately, most of the published papers refer only to one of the first two algorithms, discarded by the author. Several other authors have developed additional algorithms since 1970 for synthesizing MD topologies with capacity constraints. See Frank, et al. (1971), Frank and Chu (1972), and Kershenbaum and Chu (1974) for a summary and comparison of several algorithms, including the ones discarded by the author.

Kershenbaum and Chu (1974) also present a unified, parametrized algorithm for designing an MD topology with capacity constraints. Since no single algorithm based on heuristics can be guaranteed to always yield the best results, the designer must try several parametrized approaches to obtain an optimum solution. As a result, their algorithm becomes quite cumbersome. Furthermore, their work does not document the cost reductions for some practical networks. Ideally, it should be possible to compare all known good algorithms against a fixed set of input data and design criterion and then choose the best. But that has been impossible until now. For the first time, the author will attempt to compare some of the well-known algorithms in Chapters 7 and 8 for identical inputs and constraints imposed by practical networks only. The reader is encouraged to compare these algorithms with others by using the input data presented in Chapters 7 and 8.

5.4.4.1 Algorithm for Designing a Minimal Spanning Tree (MST) Network with No Capacity Constraints

We will describe Prim's basic algorithm (see Prim 1957), considering no capacity constraints. Kruskal's algorithm (see Kruskal 1956) will yield identical results for no capacity constraints. The major steps for designing an MST network with no capacity constraints, using Prim's algorithm, can be described as follows:

Step 1. Start out with all unconnected (or isolated) nodes and no clusters.
Step 2. Consider the closest node to an unconnected mode. If the closest node is also

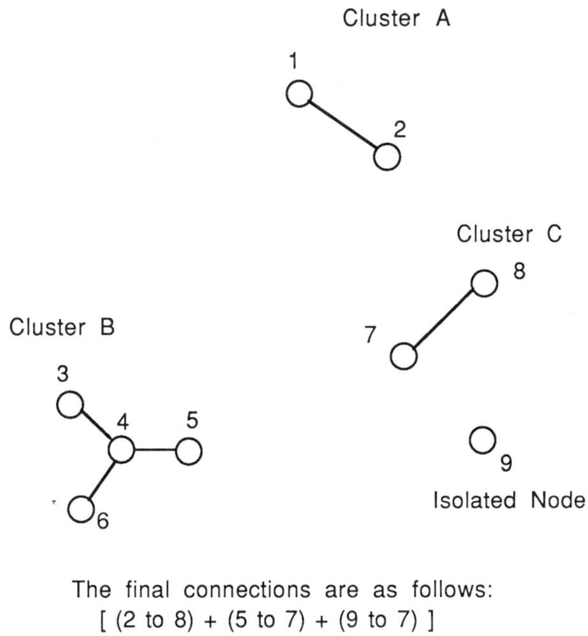

Figure 5.4. Prim's algorithm for optimizing an MST topology.

unconnected, make a connection between the two nodes. This step, when repeated for all isolated nodes, will yield either isolated nodes or clusters of connected nodes.

Step 3. Connect every isolated node to the closest cluster by the shortest connection.

Step 4. Connect the two closest clusters by the shortest connection.

Step 5. Repeat the above steps until only a single cluster is formed.

The above steps are illustrated in Figure 5.4 for a case consisting of three clusters and one isolated node. This algorithm will be applied to several cases of interest in later chapters, and the resulting networks will be compared to those obtained by using other algorithms while considering no capacity constraints. In the discussions that follow, Prim's algorithm will be consistently used to synthesize each netlink with an MD topology when the nodes have already been assigned to that netlink by one of the two algorithms described next. In other words, the selected algorithm will always be used first to assign a set of nodes to a netlink, and then Prim's algorithm will be used to derive the actual MD or MST topology. It should be stressed that the computer time required by Prim's algorithm becomes excessive for a very large (>40) number of nodes. Fortunately, the number of nodes per netlink allowed by practical implementations does not generally exceed 10.

Kershenbaum and Chu (1974) have employed the algorithms of Prim and Kruskal for designing CMSTs. However, the network costs always exceeded those synthesized by using the Esau-Williams or Sharma algorithm.

Prim's algorithm should not be confused with another closed-form algorithm that scientists have been trying to derive for several centuries for minimizing the total length of connectorized wiring. Such an algorithm allows the addition of junction nodes for providing connection to regular network nodes. Unfortunately, such an algorithm is not useful for computing the link costs for most of the modern telecommunication networks. It should be useful in problems dealing with the design of ultra-fast very large scale integrated (VLSI) circuits.

5.4.4.2 The Use of the Esau-Williams (EW) Algorithm for Designing MD Topologies With Capacity Constraints

While the original algorithm was explicitly described by Esau and Williams (1966), the author is aware of many variations of their algorithm. Each worker who developed a unique variation of this famous algorithm claims to have done so for improving the original algorithm. Since the Esau-Williams (EW) algorithm is based on heuristics, it is impossible to always guarantee better results than those achieved from the original algorithm. We will attempt to describe the original algorithm in the language we have developed thus far. Before describing the various steps of this algorithm, one must first define a function T(X, Z) that is the basis of Esau-Williams algorithm. One can define the T(X, Z) function in terms of two nodes X and Z for the starting network topology of Figure 5.5, as follows:

$$
\begin{aligned}
T(X, Z) &= [(DX + DZ)] - [\mathrm{Min}\,(DX, DZ) + XZ] \\
&= DX - XZ \ldots \text{ applicable for Figure 5.5} \qquad (5.4)
\end{aligned}
$$

where

$$
\begin{aligned}
DX &= \text{distance or actual cost of line DX} \\
XZ &= \text{distance or actual cost of line XZ} \\
DZ &= \text{distance or actual cost of line DZ} \\
\mathrm{Min}(DX, DZ) &= \text{minimum of the two quantities DX, DZ}
\end{aligned}
$$

If a single tariff applies to the network and the costs are proportional to distance, the distance should be adequate for computing the T function or the network. The objective of this algorithm is to synthesize netlinks consisting of more than two nodes sharing a link and satisfying the maximum link capacity, W_m. The algorithm can be described as follows:

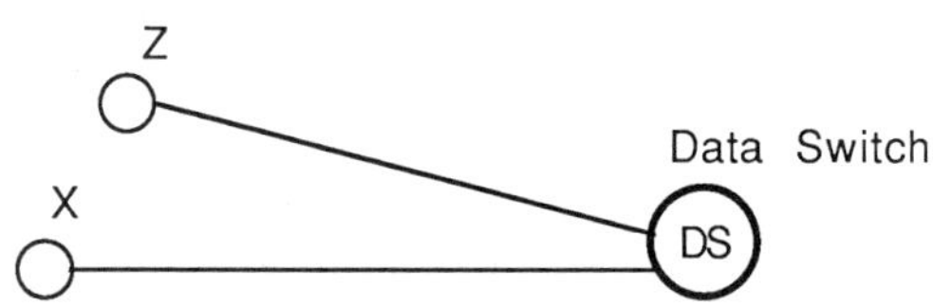

Figure 5.5. A starting topology for computing the T (X:Z) function involving three nodes, using the Esau-Williams algorithm.

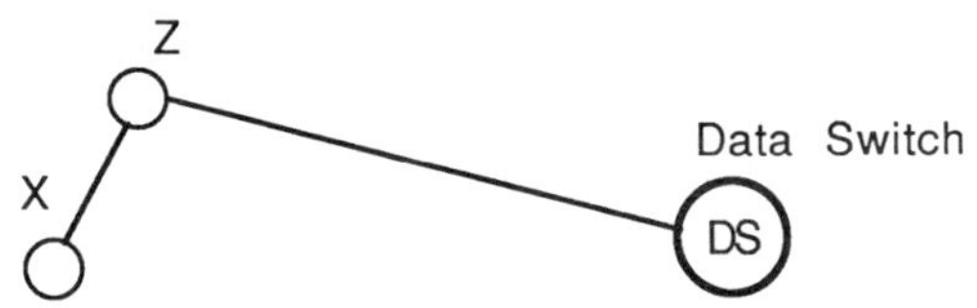

Figure 5.6. A three-nodes MD network topology using the Esau-Williams algorithm.

Step 1. Compute the maximum T function [equal to max (T)] for all but the switch node. This means the computation of the best secondary node, Z, for each primary X node.

Step 2. Form $(N - 1)$ components (or subsets or netlinks) with only two nodes in each component or netlink, one primary node and the data switch. See Figure 5.5 for a starting network with only three nodes.

Step 3. Search for the unselected primary node, X, with the max (T) function and consider the associated component (or subset or netlink).

Step 4. Consider the secondary node, Z, and the associated component.

Step 5. If the max (T) is =< 0, then remove the primary node, X, from the table. Go to Step 3.

Step 6. If the sum of component traffic loads exceeds the maximum capacity, W_m, due to prior changes, then recompute max (T) function and compute a new Z. Repeat Steps 4 through 6. If the sum of component traffic loads is <= W_m, then merge the two components or netlinks with only one connection to the data switch. Delete this X from the table.

Step 7. Repeat Steps 3 through 6 for all primary nodes until all primary nodes have been considered and deleted from the table.

Step 8. Apply Prim's MST algorithm to each set of components to obtain M netlinks based on the MD or CMST topology. See Figures 5.6 and 5.7 for an illustration of these steps for three and four nodes to be connected to the data switch in the network.

Since this process may be difficult to fathom in the written form, a flow chart, Figure 5.8, has been prepared to illustrate the EW algorithm. This flow chart replaces Task 4 of the generalized network design process of Subsection 5.4.1. In the succeeding chapters, several networks based on three variations of the EW algorithm will be synthesized and their costs compared to other MD solutions based on different algorithms and topologies.

The author believes that a large number of variations of the original EW algorithm

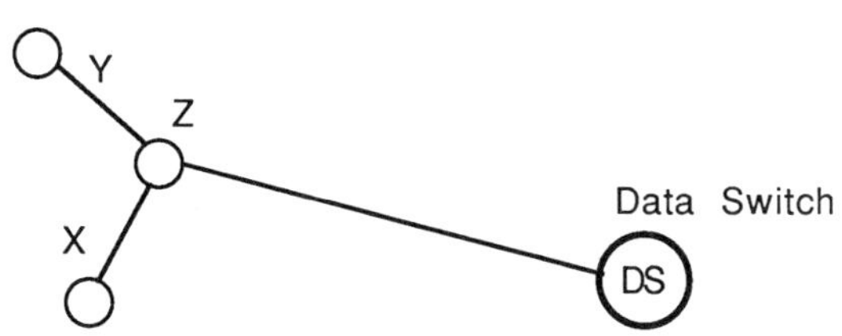

Figure 5.7. A four-nodes MD network topology using the Esau-Williams algorithm.

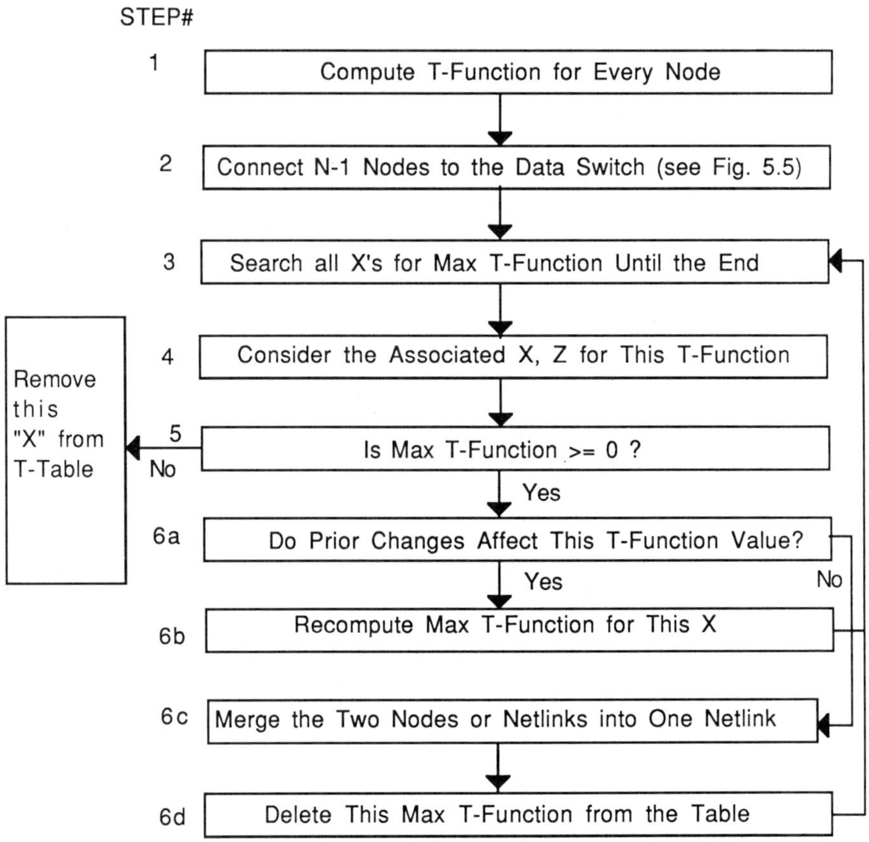

Figure 5.8. A flow chart defining the Esau-Williams algorithm for optimizing a MD topology.

exist in the world. It is not very difficult to develop a variation of the original algorithm that reduces the network cost slightly for a particular application, and declare it be the ultimate solution. But it is another matter to create a new variation of the EW algorithm and prove that it is always better than the original version. Such a goal may be an impossible task, considering the heuristical nature of the algorithm. A better alternative may be to use the original algorithm and its variation in rapid succession for a large number of applications, and examine the overall trend. Only in this way can we choose the better of the two algorithms. One can also apply the nodal exchange algorithm, as described in the following paragraphs on any EW network, and achieve additional sizeable savings, in most cases. This approach will be analyzed, along with two popular variations of the original EW algorithm, in the following paragraphs, for cost savings in Chapter 8.

5.4.4.3 Sharma's Algorithm for Designing MD Topologies With Capacity Constraints

This is a two-pass algorithm, as described by Sharma and El-Bardai (1970). During the first pass, a feasible network is realized by applying the MST algorithm to each netlink that was obtained through the use of the DL algorithm, as described in Subsection 5.4.2. The second pass employs a nodal exchange algorithm for reducing the cost of the first feasible network. Experience shows that cost savings of up to 8 percent can be realized for only an reasonably small amount of additional computer time. Each run for the second pass requires about the same time as that required for a normal EW algorithm. This process is stopped when no additional cost reductions are achieved. Kershenbaum and Chu (1974) wrongly reported that cost savings of only 1 percent could be realized at the expense of enormous additional computation time. This algorithm (the steps of this algorithm should replace Task 4 of Subsection 5.4.1) can be defined as follows:

Pass one (feasible network design algorithm).

Step 1. Using the algorithm of Subsection 5.4.2, define M unique sets of nodes for each netlink.

Step 2. Using the MST algorithm of Subsection 5.4.3.1, obtain the first feasible network consisting of M netlinks, each obeying the multidrop or minimal spanning topology.

Step 3. Compute the cost of each netlink, C_i ($i = 1$ to M), and the entire network, C_f.

Pass two (exchange algorithm).

Step 1. Find the farthest node, N_i, from the switch (SW) among the unconsidered nodes belonging to a link, L_i.

Step 2. Find a node, N_j, belonging to a link, L_j ($i \neq j$), and closest to N_i. Find the cost of two links, $C_0 = C_i + C_j$.

Step 3. Add node N_i to link L_j. If the weight of new link $W_{j'}$ exceeds the maximum allowable W_m, add one or more nodes (closest to the SW) of L_j to L_i, until the two link weights are $\leq W_m$. Now compute the cost of the newly created links, $C_1 = C_i' + C_j'$.

Step 4. Repeat Step 3 by adding N_j to L_i. Consider the cost of the two newly created links, $C_2 = C_i'' + C_j''$.

Step 5. Replace the original two links with the set of two newly created links that yield the lowest cost, $\text{Min}(C_1, C_2)$.

Step 6. Repeat Steps 1 through 5 for each of the other nodes of L_i.

Step 7. Repeat Steps 1 through 6 for all other links of the network. Note the new network cost, C_n.

Step 8. If $C_n < C_f$, then repeat Steps 1 through 7. Otherwise stop the exchange process.

It should be emphasized that the costs of each newly created link is computed only after applying the MST algorithm to the set of nodes belonging to the links. For some cases, additional cost reduction was observed if, after each successful nodal exchange, the nodes of each netlink were ordered according to their rank (or their distance from the SW). This algorithm has been described in precise language by Sharma and El-Bardai (1970). Since the exchange algorithm may be hard to fathom in the written form, a

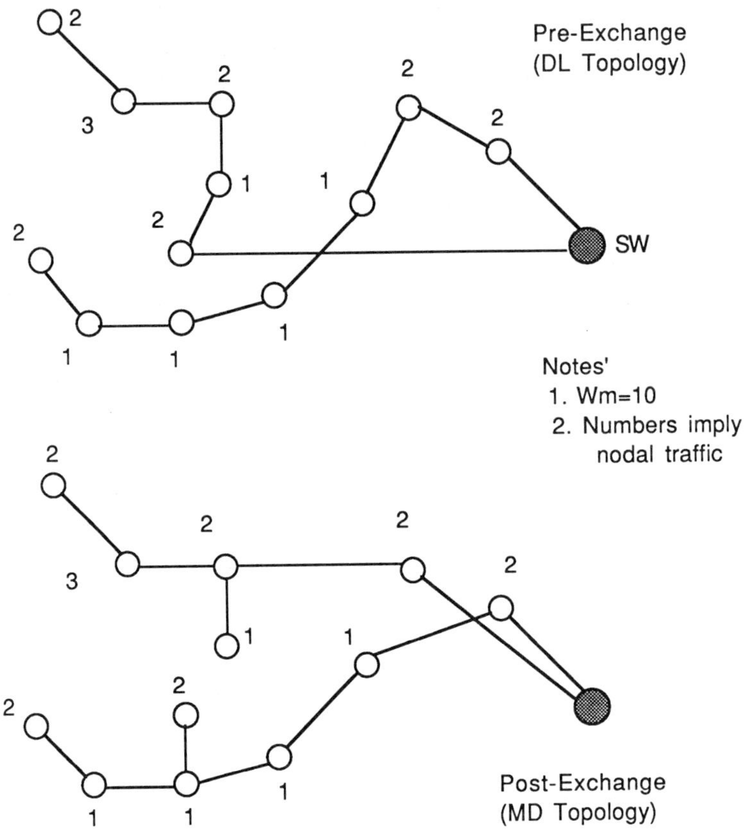

Figure 5.9. Sharma's exchange algorithm for optimizing a network with two netlinks.

transformation of a 2-link feasible network is shown graphically in Figure 5.9, through nodal exchange.

It is interesting to observe that the exchange algorithm can also be applied to any network synthesized through the use of the EW algorithm. The author has experienced a sizeable (1 to 4 percent) cost reduction with only a nominal increase in computational time. Such a cost reduction is significantly higher than what has been achieved through the use of other so-called greedy algorithms as developed by Kershenbaum and Chu (1974). The writer has also successfully used the exchange algorithm to reduce the cost of networks based on the DL topology. It was observed that only large but sparsely populated networks produced significant cost savings.

It must be stressed again that the steps required to apply each useful algorithm for an applicable topology (e.g., Star, DL, or MD) should replace the Task 4 to obtain a generalized, multilevel, network design process in Subsection 5.4.1. Prim's MST algorithm is only employed to synthesize each netlink of the EW or Sharma' network.

Several meaningful network design-related data bases will be developed, and many networks will be designed for the Star, DL, and MD topologies, using the above algorithms, and studied in the chapters that follow. Only a careful study of those results can help the reader appreciate the usefulness of each algorithm when considering the applications for voice, data, or integrated voice and data (IVD).

However, before studying the resulting topologies, one should consider a user-friendly software package that executes the above-mentioned steps for each applicable network topology. To execute these steps by hand will take nearly forever. The next chapter describes a software package that enables the designer to optimize the network topology quickly and compare the associated network costs for all applicable algorithms described in this chapter. Our literature does not provide examples of such a comparison for many of the well-known algorithms. Most of the published articles usually consider only one heuristic algorithm, that favored by the writer.

6

A Description of a Software Package Used for Network Topology Optimization

6.1 INTRODUCTORY REMARKS

The previous chapters have described the (1) basic telecommunications concepts, (2) meaningful analysis/design tools, and (3) appropriate algorithms required for the synthesis of voice, data, and integrated voice/data networks. By now, it should be clear that the job of designing even a small network is not a trivial one. The large number of inputs and the huge amount of computations required for designing a network should discourage even the experienced analyst. If one adds the complexities of statistical traffic theory, the task of networking appears unsurmountable. Consequently, the network design process has always required the use of mainframes and, as a result, the process has been a mysterious one, even for a majority of telecommunications managers.

The above difficulties can be overcome through the use of a user-friendly software package. Such a package should enable the designer to enter a large number of design parameters and to change some key design parameters effortlessly, for the sole purpose of optimizing a network topology through the process of iterative design, as described in Chapter 2 and illustrated in Figure 2.2. This chapter will describe a software package that has taken several years to develop and has already proven to be extremely flexible for optimizing and displaying network topologies as these are synthesized. Although the contents of this chapter may seem out of place when compared with that of earlier chapters, the philosophy behind, and the operation of, such a software package are essential aspects of network topology optimization. This will become clear when this software package is actually employed to synthesize optimum multilevel voice, data, and IVD network topologies and derive some important lessons as described in the next three chapters. For discussion purposes, this software package has been named as ECONETS which is short for economical networks.

6.2 THE PHILOSOPHY BEHIND THE ECONETS PACKAGE

In the past, only large mainframes were used for network design. The reasons behind this fact were many. Powerful and affordable desktop workstations were not available. Most network design programs required the use of a large data base related to the tariffs.

Consequently, the time necessary for computing the cost of a network topology was too high even on a mainframe. Furthermore, the MIS department prevented the distribution of computing power to workers and thus helped perpetuate the mystery of centralized data processing (DP) and network design. Some vendors are now allowing their customers to access their mainframe-based network design program through desktop PCs. But the cost of telecommunications far exceeds that for licensing the network design program.

The availability of reasonably priced fast desktop computers for network design is finally a reality now. One no longer needs to fight the MIS department for the scheduling of RJE terminals and batch processing times on the expensive mainframes. The availability of friendly man-machine interfaces on some of the desktop computers is reducing the training period required to master their operations and use. The availability of bit-mapped graphics on some desktop computers has already eliminated the dependency on mainframes for topology/map printouts during the "what-if-analysis" mode. The easy availability of a laser printer is also enhancing the quality of textual and graphical presentations. The entire network design cycle can now can be reduced to a few hours or days instead of months. But several other developments have also been instrumental in bringing the network design capability to the desktop.

Since the divestiture of AT&T in 1984, the forces of deregulation have been steadily destroying the stability of tariffs. Nowadays, one can lease DDS and T1 circuits from a bulk carrier for a song compared to the FCC regulated tariffs. Large customers can also lease private lines from AT&T at a discount (e.g., Tariff 12) or special rates (e.g., Tariff 15). Then there are the special tariffs from OCCs such as MCI and U.S. Sprint. Furthermore, the use of bypasses of BOC (and in some cases OCC) facilities creates additional opportunities to ignore the published tariffs. Thus, the stability of tariffs as they have been known for years has been eroded permanently. It has, instead, created the need for a two-stage network planning process: performance of the initial network topology optimization, using simplified but meaningful models of tariffs and, when the designer is satisfied with the network topology as obtained through the process of iterative optimization, use of one of the many commercially available PC-based line pricer packages to compute the network transmission cost to the "*n*th" degree of accuracy, if desired. This approach is especially attractive if one considers the fact (as proven in the next three chapters) that the network topology is influenced primarily by the network architectural considerations or network design parameters algorithms, and never by perturbations in the tariffs selected. This approach, if followed to the letter, can completely eliminate the mainframe from a network designer's life and make the process of network planning a joyful one.

The rapidly changing tariffs and networking technologies creates the need for ongoing planning and management of your own network as if it were your only strategic resource. One can reduce the cost of transmission facilities significantly through the process of negotiations with OCCs, bypasses, and in-house topological optimizations. The network design process must be a continuous one to keep up with the changing tariffs, evolving switching hardware, bypass technologies, and corporate goals/strategies. Mainframes, by their very nature and complexity, prevented this real-time planning process. The availability of a powerful desktop workstation, on the other hand, provides the designer with a totally flexible and productive tool for network topology optimizations in an ongoing fashion.

The above considerations were employed to create a very user-friendly software package for the modeling, design, and analysis of multilevel networks based on several topologies to serve a mix of applications, such as voice, data, and video. The name, ECONETS, was selected to represent its capability to design economical networks to a high degree of optimization. It consists of many subroutines that have been assembled to produce a holistic package for the modeling, design, and analysis of either a simple 2-level network, or a multilevel, integrated network for both voice and data. This software package was created using the ZBasic™ Compiler for the BASIC source code language. Portions of this code are © Copyrighted, 1985, 1986, 1987, and 1988 by Zedcor Inc. This package runs on the Macintosh II desktop computer. We will describe first the organization of the software package and then its detailed operations dealing with the preparation of input data files, viewing/updating of input data files, viewing of output (and help) files, network topologies, and the use of a dozen proven analytical tools.

6.3 ORGANIZATION OF THE ECONETS SOFTWARE PACKAGE

The application software package consists of many compiled, structured subroutines written in the ZBasic source language. It employs the principle of keeping the input files separate from the solution kernel of the algorithms and analytical tools used to synthesize a network topology. The designer has the option of using any set of applicable disk-resident input files. The user also has the option of storing the output files and the resulting graphs of the network topologies on the hard disk. These output files can be opened later for any desired editing and final printouts. The overview of the organization of an ECONETS package and its capabilities are shown in Figure 6.1. According to this illustration, there are three major subsystems:

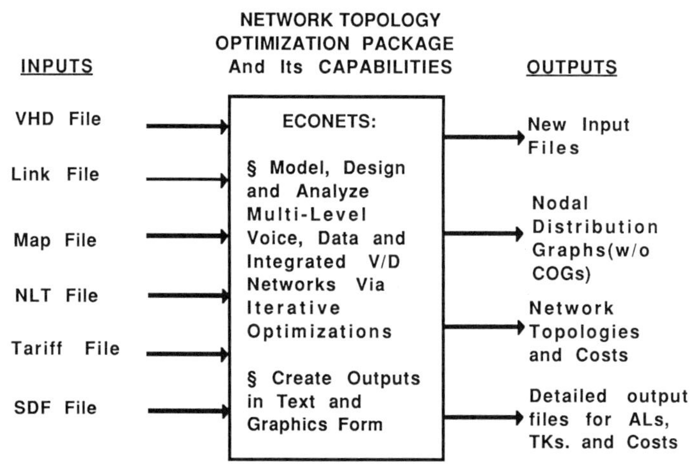

Figure 6.1. An overview of the organization and the capabilities of a software package used for network topology optimization.

1. Input Files, which define all of the inputs to the software system
2. The ECONETS package, with all of its operational capabilities
3. Output Files, which define the results of the modeling, design, and analysis

The input data are read from the set of input files, and these determine the type and quality of the outputs. If one is not careful, one can easily get meaningless output data. The designer must alway test the validity of the output data. The famous GIGO (garbage-in-garbage-out) adage still applies here in all its glory. On the other hand, a basic knowledge of the key telecommunications concepts, as elaborated in Chapter 1, will help the designer to create a meaningful network topology and interpret all of the output data correctly. Just as a hammer does not help a carpenter to create a beautiful house every time, the availability of a good network design program can never always guarantee a fully optimized network. The designer will always remain as the key to a good network. Also, he must always remain vigilant and above the task.

The ECONETS package provides all of the capabilities, in terms of creating new input files and creating output files, that define the results obtained from modeling, design, and analysis of telecommunications networks. The package also provides additional capabilities related to viewing input/output files, in addition to many utility/analysis tools. The package requires a minimum hardware system consisting of a Macintosh II computer equipped with 2 megabytes of RAM and an internal 40 Megabyte hard disk, a high-resolution monochrome monitor, a keyboard, and a mouse. Additional software applications, such as the EDIT or MacWrite, and MacPaint, may also be needed to edit/print the output results in text or graphics form on an imagewriter or a laserwriter (representing additional hardware).

The outputs, if desired, are also stored on the hard disk under selected titles. Each output text file contains a detailed description of the network topology and costs in terms of each link or connection, as well as other aspects of the analysis. The output graphic file describes the overall network topology in an object-oriented form. This network graph also contains summary costs of ALs, trunks, total network transmission, a voice call, or 1-megabits of data. In most cases, one does not need to store every output file and/or network graph. Also in most cases, a quick study of the summary costs is more than sufficient to tell the designer if additional network optimizations are needed or not.

We will now discuss each of the above three ECONETS subsystems in some detail. This discussion should familiarize the designer with the structure/meaning of input files, the capabilities of ECONETS and the structure/meaning of output files. This knowledge, coupled with the understanding obtained from reading the earlier chapters, should equip the designer to prepare all of the inputs required to model, design and analyze any meaningful network.

6.4 A DESCRIPTION OF THE INPUT FILES

There are at least six input files that are needed for network modeling, design, and analysis. These are as follows:

1. The VHD file which defines the V- and H-coordinates, and peak period traffic loads, for each location;

2. The LINK file, which defines the link type, for connecting the location to the network nodes.
3. The MAP file, which defines the map boundaries, for illustrating the resulting network topology;
4. The NLT file, which defines up to five node-link-tariff relationships for each link type;
5. The TARIFF file, which defines up to five tariffs that can be used for computing network costs; and
6. The SDF file, which defines up to 60 system design parameters required for computing system performance in terms of response times, cost, and presentations.

The FTD file may be required to define the peak period internodal traffic flows and deduce the actual nodal traffic loads from such flows. The LATA file may also be required, to compute the costs of local channels (LCs) if such channels are leased separately from a Regional Bell Holding Co. (RBHC) for an intra-LATA or an intrastate network. The NAME file defines the names of locations according to a 3-symbol city code and 2-symbol state code, represented as (CCCST). The NAME file is used only to create a data base (DB) with nodal names. Some specialized files, such as CSATBL, URTBL, WUTBL, MUTBL, and DTP (Daily Traffic Profiles), are required to compute the costs of virtual network facilities required to implement private voice and ACD networks.

We will now describe each file in terms of its structure and its meaning.

1. The VHD file is a sequential file with 4-element vectors chained together. Each 4-element vector defines the assigned site number, the associated vertical (V) coordinate, the associated horizontal (H) coordinate, and the peak period traffic load. The V and H coordinates are obtained from special tables that define such map-related coordinates to the area code-CO number (NPA-NXX) part of the telephone number associated with each site. The traffic load can be represented as millierlangs for corporate voice and ACD networks, or bits per second (bps) for data applications. The units used for representing peak period traffic loads must be remembered when interpreting the output files and SDF design parameters.

2. The LINK file is a sequential file consisting of an N-element vector, where N is the number of sites in the VHD file. Each element value defines the link type that will connect the associated site to the network nodes, such as concentrators and switches. The attributes of each link type are defined in the NLT file, as described later. A provision exists to ignore the LINK file. In that case, the link type of "0" is assigned to all sites. In some versions of the ECONETS package, a VCL (Virtual Circuit Link) file is employed to represent the traffic from up to seven types of virtual circuit types at a node and, of course, the associated LINK type.

3. The MAP file is a sequential file with 3-element vectors chained together. There may be as many vectors needed as to represent the map and state boundaries. A coarse map boundary for the U.S.A. requires about 90 vectors. The first element is either a 1 or 2, representing whether the map point is a starting point of a new boundary or a continuing point of the previously defined boundary, and the last two values of each vector represent the V and H coordinates.

4. The NLT file is a sequential file with 4-element vectors chained together for up to five link types. The first element defines the raw link capacity (LC) in bits per second. The second element defines the maximum allowable link rate (LR or W_m). The third element defines the potential nodal weight (PNW) or data rate from each piece of site equipment. Such potential data rates may be used to compute the site traffic loads if a design parameter (F_{nw}) in the SDF is defined as 1. The fourth element defines one of the five tariffs, as defined in the selected TARIFF file. The NLT file must always define the attributes of a ''0'' type link, since a provision exists to ignore the use of a LINK file.

5. The TARIFF file is a sequential file with 16-element vectors chained together for each of five tariffs. Each tariff is defined by sixteen values; the first value defines the average monthly cost of the two local channels on both ends of an Inter-Office Channel (IOC), the next five values define the distance limits employed to compute the monthly lease costs of a private line, the next five values define the fixed monthly lease costs associated with a private line whose length is between two distance limits, and the last five values define the per mile costs of that particular private line. Experience shows that five tariffs are more than sufficient to analyze any meaningful multilevel private network. One may construct several tariff files, each with a different combination of individual tariffs, including those for owned facilities employing digital coaxial cables, microwave, and fiber optical cables. A special design paramenter, F_{pf}, is available in the SDF file to account for the dependency on bandwidth or data rate of the owned facility.

6. The SDF file is a sequential file with a single 60-element vector. These elements define system design parameters used for computing system response times and network topologies/costs. These design parameters/elements can be described as follows:

The 0th element defines the analysis type (ATP) used for computing the response time in a data communication network. An ATP = 1 signifies an asynchronous (ASYNC) protocol, ATP = 2 signifies a virtual multipoint (VMPT) protocol as used by a Doelz network, and ATP = 3 represents a BSC or SDLC protocol employed in a polled, multidrop network.

The 1st element defines the user port rate (UPR) in bps for data networks.

The 2nd element defines the host port rate (HPR) in bps for data networks.

The 3rd element defines the input message length (IML) in bytes.

The 4th element defines the output response message length (RML) in bytes.

The 5th element defines the number of control units (N_{cu}) per virtual multipoint (VMPT) circuit of a data network used on the Doelz architecture.

The 6th element defines the average message arrival rate (R_{mph}) per hour on each VMPT circuit in a data network.

The 7th element defines the host think time (HTT) in seconds for a data network.

The 8th element defines the optimization factor (F_{opt}) that directs the program to reduce the cost of a feasible data network by using Sharma's exchange algorithm.

The 9th element defines the nodal processing time (T_{np}) per packet in milliseconds (ms).

The 10th element defines the half-modem time (T_{hm}) in ms.

The 11th element defines the propagation constant (K_{pg}) in ms/mile.

The 12th element defines the block (or packet) length (BLKL) in bytes.

The 13th element defines the information characters (ICPB) per packet/block.

The 14th element defines the traffic growth factor (TGF) to study the network costs or performance as a function of traffic growth.

The 15th element defines the factor (F_{lk}) that directs the program to read the LINK file.

The 16th element defines the factor (F_{nn}) that directs the program to read the NAME file.

The 17th element defines the factor (F_{lt}) that directs the program to read the LATA file.

The 18th element defines the FTD-use factor (F_{ftd}) (equal to 1 for using an FTD file).

The 19th element defines the nodal weight factor (F_{nw}) directing the program to use potential nodal weights (PNWs) of the NLT file to compute the nodal traffic loads.

The 20th element defines the link type for the access link (ALT) for a voice network.

The 21st element defines the access link multiplexing factor (ALMF). It defines the number of voice circuits carried by the AL in a voice network.

The 22nd element defines the allowed blocking factor for VoiceNet access lines (B_{al}).

The 23rd element defines the economical call cost (ECC) for a virtual network offering (in dollars).

The 24th element defines the expected call duration (ECD) of a voice call in seconds.

The 25th element defines the delay required (DREQ) in seconds for ACD design/analysis.

The 26th element defines the probability of delay exceeding DREQ (PEXD). It must be less than 1.

The 27th element defines the cost of labor (C_{lbr}) in dollars per hour.

The 28th element defines the fixed multiplier for the Rate Step Table (RSTBL) used for costing an ACD (F_{rst}).

The 29th element defines the ACD Type (ACDT) to be designed (types range: 1–3).

The 30th element defines the link type associated with a trunk (TKLT).

The 31st element defines the trunk multiplexing factor (TKMF). It defines the number of voice channels carried by a trunk circuit in a voice network.

The 32nd element defines the allowed blocking factor for voice trunks (B_{tk}).

The 33rd element defines the full-duplex factor (F_{fdx}) for trunks (1 for an FDX trunk).

The 34th element defines the maximum data trunk utilization factor (MTKU) (<1).

The 35th element defines the backbone trunking (traffic) factor (BBTF) for influencing the amount of interswitch traffic.

The 36th element defines the minimum value of the V-Coordinate (V_{min}) for mapping.

The 37th element defines the maximum value of the V-Coordinate (V_{max}) for mapping.

The 38th element defines the minimum value of the H-Coordinate (H_{min}) for mapping.

The 39th element defines the maximum value of the H-Coordinate (H_{max}) for mapping.

The 40th element defines the VCL-use factor (F_{vcl}) ($= 1$ for using VCL file).

The 41st–47th elements define the seven VCL utilization factors for use in the VCL file and for computing traffic loads at each site independently (not implemented in some ECONET versions).

The 48th element defines the self-healing-use factor (F_{sh}) for designing a Doelz network.

The 49th element (F_{np}) directs the program to print the nodal number on the network map.

The 50th element (F_{ph}) defines the equivalent number of peak hours per typical day. It is used to compute the average costs of a voice call minute or for transmitting 1 Mbits.

The 51st element defines the distance factor (F_{dis}) for computing distances on arbitrary V and H maps. It is equal to 1 for the U.S.A. and real numbers for other countries.

The 52nd element defines the factor (F_{pf}) used to compute the cost of an owned private facility.

The SDF parameters 53–59 are available for future expansion.

7. The NAME file is a sequential text file consisting of *N* site names, where *N* is the number of sites in the VHD file. Each name obeys the structure CCCST, where the three bytes CCC denote an abbreviated city/location/site name, and ST is the standard two-letter state designation. For example, Los Angeles, California is designated as LAXCA, and Dallas, Texas is designated as DALTX. The NAME file can be used to complete the data base before it is created and printed.

8. The FTD file is a sequential file consisting of M 3-element vectors chained together. The first element defines the "from" node number, the second element defines the "to" node number, and the third element defines the peak period traffic intensity (millierlangs or bps). One can use up to M such vectors to define actual traffic flows. The value of M is dictated by the available memory. If the 18th SDF parameter (F_{ftd}) is a "1", then the program computes the load on all access lines (ALs) and trunks, and also computes the actual traffic loads handled by each node. The contents of an FTD file can be viewed and/or updated using the FUPDATE menu. See Tables 9.2 and 9.4 for some examples of from-to traffic files used for designing the backbone networks of Figures 9.6 and 9.7. In many cases, this data is unavailable. For such cases, a statistical model is developed for designing a fully connected backbone network topology, according to a methodology described in Chapter 2. Consequently, for these cases, an FTF (or FTD file) may be ignored.

9. The LATA file is a sequential file that consists of the Local Access and Transport Area (LATA) numbers for each of the N sites as listed in the VHD file. The numbers can be obtained from commercially available tariff data handbooks. Such books have very good maps of each state, along with the boundaries of associated LATAs. The LATAs are essential for computing the costs of leased circuits for an intra-LATA or intra-state network. For an ACD application, the LATA file lists the state/region number for each location, as defined in the CSABDS (Customer Service Area Bands) file.

The other input files required for an ACD network design (i.e, CSABDS, UTBL, WUTBL, MUTBL, RSTBL, DTP) are too specialized to be described here. Fortunately, the contents of these specialized files do not change rapidly. And if they do, they change by a fixed parameter (F_{rst}), as defined in the SDF file.

See Figure 6.2 for a view of the first six input files in their sequential form. Some

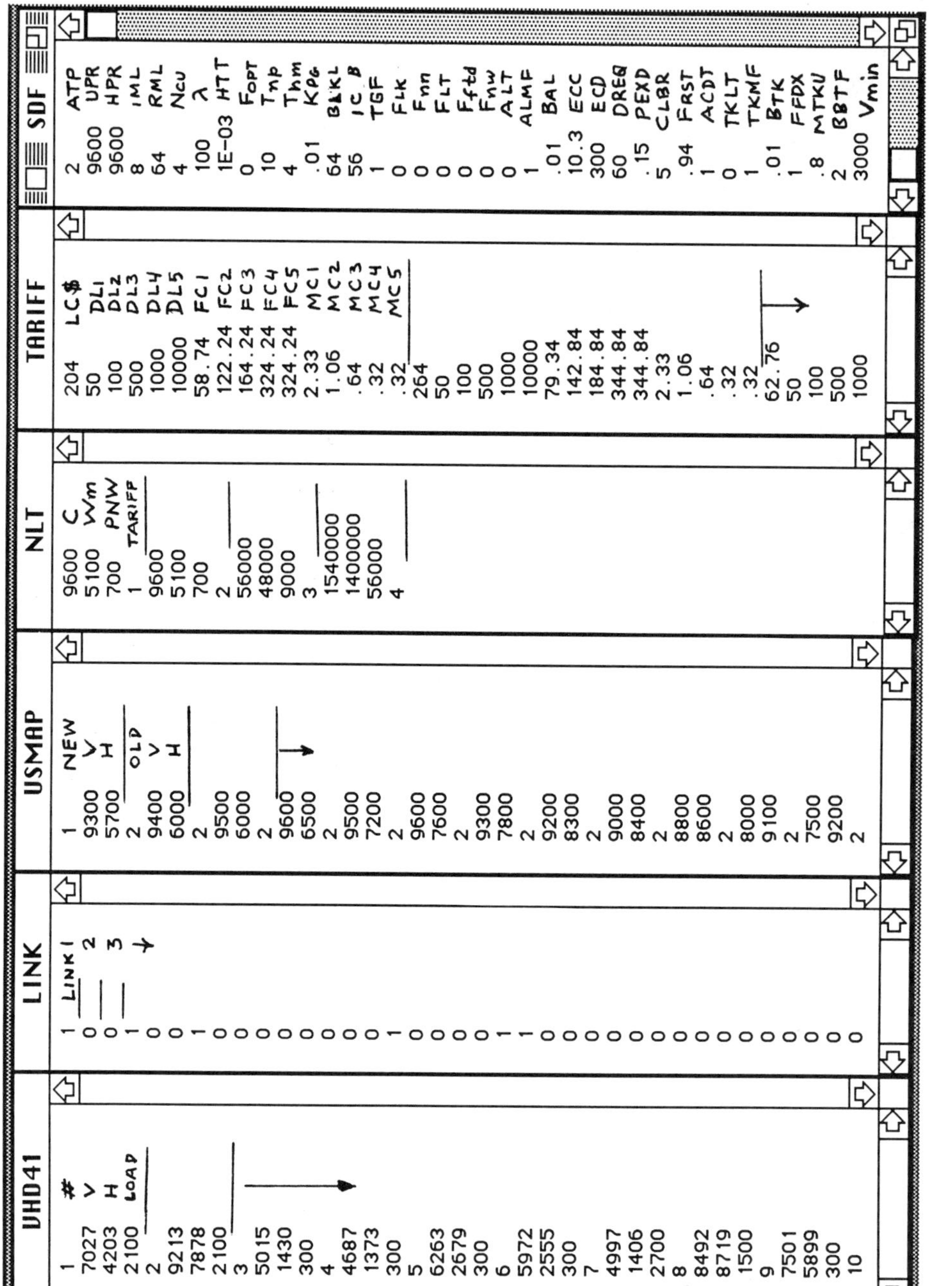

Figure 6.2. A display of the six most-used input files with data arranged sequentially.

handwritten remarks attempt to designate the boundaries of the various vectors and their elements.

By now, the reader should get an idea of how the input files are structured. In the next section, we will discuss the ways in which the contents of these input files can be created, viewed, and updated. The view/update capabilities are particularly important if one has to perform topological optimizations in quick successions.

6.5 OPERATIONAL CAPABILITIES OF THE ECONETS SOFTWARE PACKAGE

The ECONETS application performs like any other Macintosh application. A double click on the ECONETS icon with the mouse will open the application. The first window provides some important information about the ECONETS package. The first useful window that follows lists the names of up to sixteen input files that are read from an initialization file named FILES. The designer is given the chance to name all of the input files required for modeling, design, and analysis of a desired network. If these input files are not available, the designer has to reach the main MENU window, which can be done only if one starts with the names of six (6) sample input files included with ECONETS package. These are VHD41, LINK41, USMAP, NLT, TARIFF, and SDF. The icons of these six input files must be in the same folder in which the ECONETS application icon resides. After clicking on the DONE button on the screen, the ECONETS package goes through a system initialization process before presenting a clean window with pull-down menus with the titles NETWORKING, FCREATE, FUPDATE, UTILITY, and ANALYSIS. This set of menus was created especially to illustrate the capabilities of ECONETS. Other sets of menus can be easily constructed to satisfy user expectations. We will now discuss the various menus in a sequence dictated by the ECONETS structure, as illustrated by Figure 6.1.

6.5.1 FCREATE MENU

One should use the Menu FCREATE to create any of the required input files. Under the pull-down menu named FCREATE, one will find items called VHD, LINK, MAP, NLT, TARIFF, SDF, NAME, FROM-TO, and LATA. The nine items represent the type of input files to be created.

It is recommended that the designer have a source data sheet with all of the numbers already printed on the sheet, as suggested in Figure 6.3. Although such a table may take a long time to prepare, the effort is well worth the time spent on it. Many experienced telecommunications managers already require such tabulated data as part of the corporate inventory. Such a table will always be needed to check the accuracy of the data bases produced by ECONETS. Reading from such a table will also minimize errors during the creation mode. If an error is made during the file creation mode, one should not get too concerned. The designer should hope to detect such an error during one of the many test runs. The FUPDATE menu can be used later to quickly update the erroneous parameters of any input file.

When the FCREATE item (e.g., VHD, LINK, MAP, etc.) is selected, the program will first ask the name of the file to be created (e.g., VHD41, USMAP, NLT, SDF,

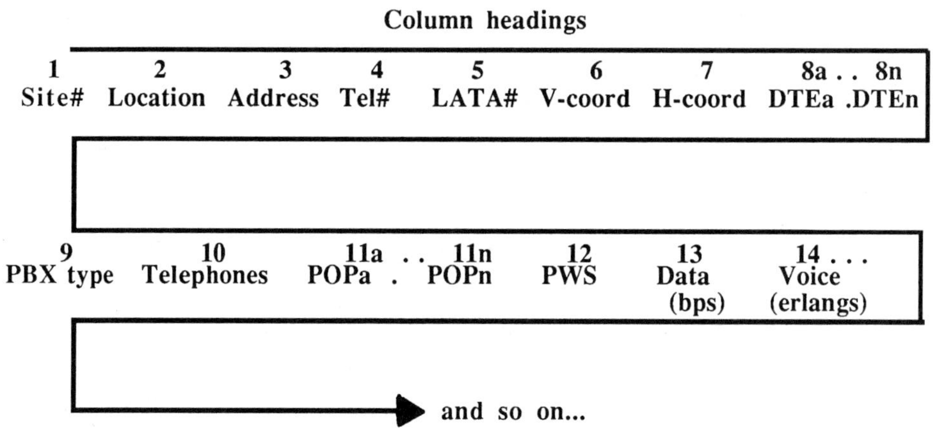

LEGEND:

1. Site# is a number assigned to each location or CPE site.
2. Name of a locaton or site in full (e.g. Dallas, Texas).
3. Address of the location or site (e.g. street number and ZIP code).
4. Telephone number including the NPA, NXX and 4-digit code.
5. LATA is the local access and transport area code for the CPE.
6. V-coordinate of the wire/rate center for the NPA-NXX code.
7. H-coordinate of the wire/rate center for the NPA-NXX code.
8. DTE-types and number of devices for each type at the location.
9. PBX type (vendor/model) and terminations (max, available).
10. Telephones, by type and number.
11. Closest POPs by OCC and their V&H coordinates.
12. Premise Wiring System (PWS) with type and size.
13. Data traffic in bits per second during a time consistent busy hour.
14. Voice traffic in erlangs during a time consistent busy hour.

Figure 6.3. A suggested source data sheet used for preparing input files.

etc.) Once that is done, the program begins providing a small spreadsheet with properly arranged data cells and vectors. The designer must fill these data cells and request for the next spreadsheet by clicking on the NEXT button. This process is repeated until there are no more vectors to be added to the input file. When this stage is reached, the designer should enter any negative value (e.g., -1) for the next requested element of a vector and click on the DONE button. This should terminate the process and an End-of-File (EOF) is added to the input file. The input file is then closed and stored away.

6.5.2 FUPDATE MENU

One should use the Menu FUPDATE to view and update any of the input files named in FILES. Under the pull-down menu named FUPDATE, one will find items called VHD, LINK, MAP, NLT, TARIFF, SDF, NAME, FROM-TO, LATA, and FILES. These ten items define the types of input files (as defined in FILES) to be viewed and updated. Once an input file has been viewed and/or updated, the program automatically goes to the main menu window after a short initialization process. The FUPDATE menu provides a very structured way of viewing/updating an input file. It is so user-friendly that the designer can now perform a large number of "what-if" type topological optimizations in quick succession, as illustrated in Figure 2.2 of Chapter 2. See Figures 6.4 through 6.11 for a structured view of the most often updated files, such as VHD, LINK, MAP, NLT, TARIFF, SDF, FROM-TO, and FILES. This structured view of input files is much easier to comprehend than the sequential view of these files as shown in Figure 6.2. The reader should note that these figures also illustrate the various menus employed by ECONETS.

To update any element of an input file, one simply brings the curser into the appropriate edit field and clicks on it with the mouse. The designer can then enter the new value. When updating is finished on the particular screen, one may ask for the next screen by simply clicking the mouse on the NEXT button. If the screen happens to be the last one, the designer should exit the FUPDATE menu by clicking on the DONE button. It should be pointed out that the designer can always add new vectors by simply filling out the empty cells in the current spreadsheet or in the NEXT one. Furthermore, the designer can also delete a vector by simply substituting the last vector in its place, after changing the nodal number. Again, the designer must never forget to add a negative number in the cell next to the last vector.

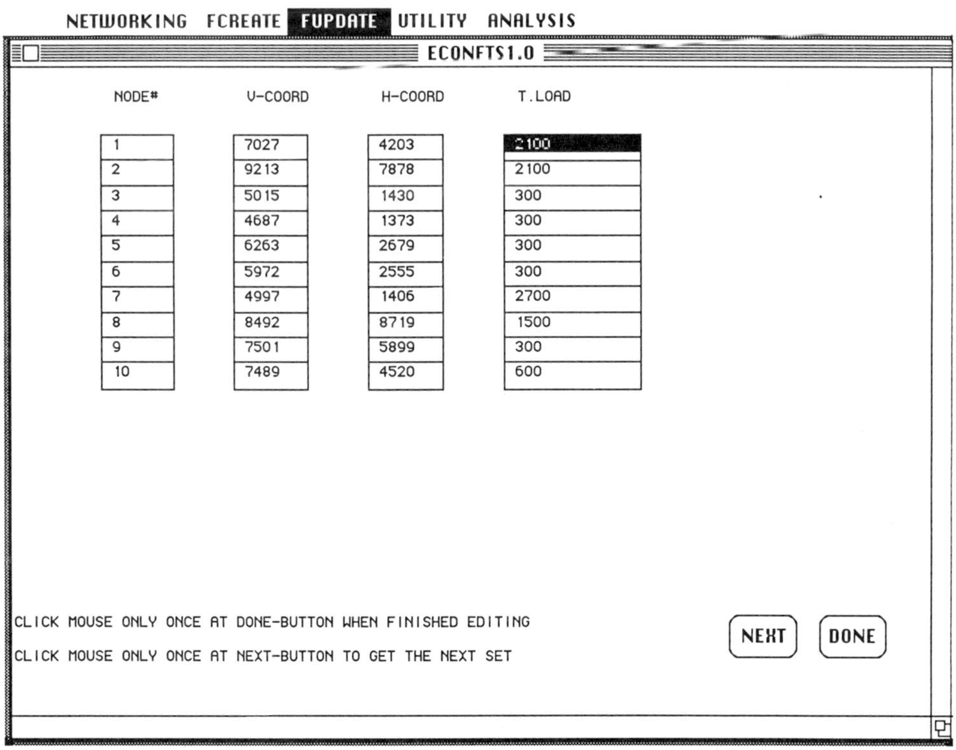

Figure 6.4. VHD file data in a structured form, as obtained from using the FUPDATE menu.

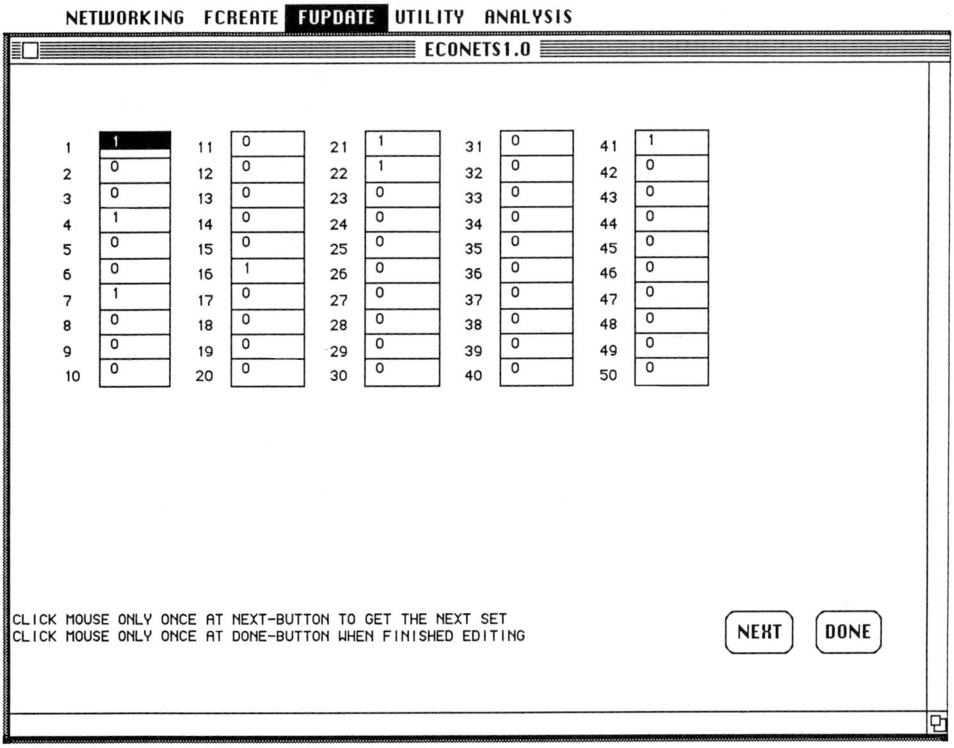

Figure 6.5. LINK file data in a structured form, as obtained from using the FUPDATE menu.

The designer can also view the contents of any input/output file by double clicking on the associated icon of an input file. This activates the ubiquitous EDIT application with Edit, Print, Search, and Format capabilities. But the designer will have trouble reading the contents (see Figure 6.2), since the associated site numbers are not included in most files. One has to count and mark certain values (by putting the blinking curser next to the element) before reaching the desired file-vector-element. The best way to view the input files is via the ECONETS's FUPDATE menu item for a particular input file. If the FUPDATE menu does not provide a view/update capability for a given input file (e.g., CSABDS, UTBL, WUTBL, MUTBL, RSTBL, DTP), the use of the EDIT application may be resorted to. If the EDIT application is also not available, one can alway use the MacWrite application to open the input file document. Before one can open the document, the MacWrite application will request input for a few choices for handling the carriage returns and paragraphs. The modified document can then be converted into a text file that can easily be saved or handled as a MacWrite document.

6.5.3 UTILITY MENU

The ECONETS package provides several utilities that can ease the task of changing input files and creating their backups without leaving the ECONETS application. Other utilities deal with capabilities of viewing HELP and Output Files, also without leaving the application. These utilities are described as follows:

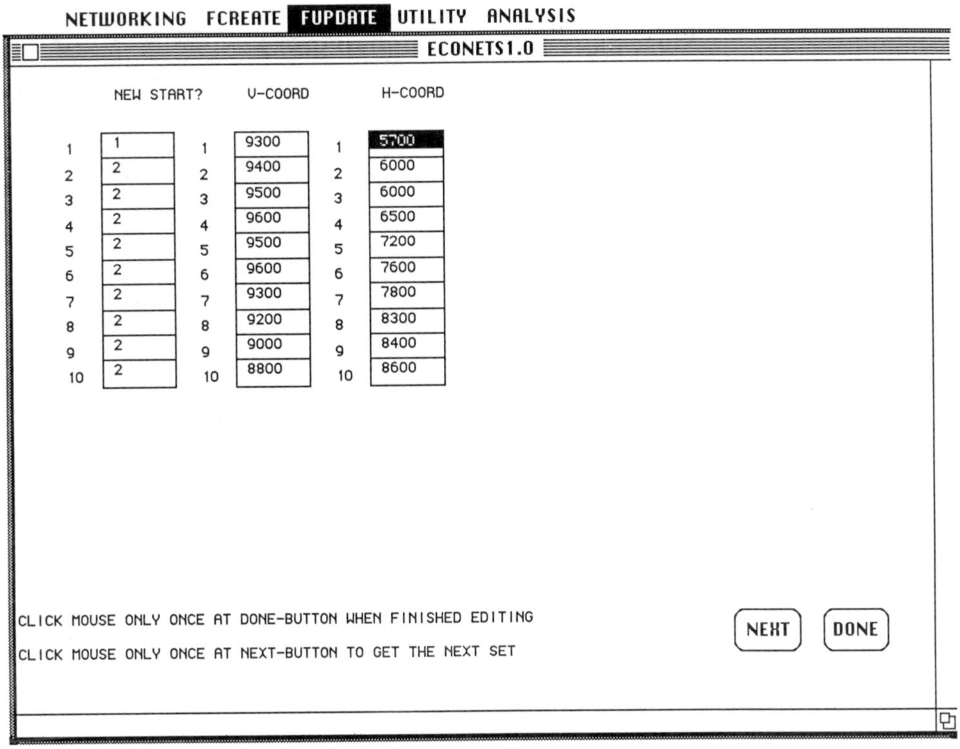

Figure 6.6. MAP file data in a structured form, as obtained from using the FUPDATE menu.

1. Rename File
2. Duplicate File
3. VHD Files Merge
4. VHD Traffic Multiplier
5. MAP Files Merge
6. FTD Files (or FTFs) Merge
7. View Output Files
8. View Help Files
9. View Topology

6.5.3.1 Rename File

This allows the designer to rename any input file, such as VHD, LINK, MAP, NLT, TARIFF, SDF, FTF, and LATA, for achieving consistency. This capability also enables the designer to save some output files under different names before being overwritten by the networking tasks.

6.5.3.2 Duplicate File

This allows the designer to create a backup file for any input file, such as VHD41 and NAMES. The designer is encouraged to create such backup file, since some rare com-

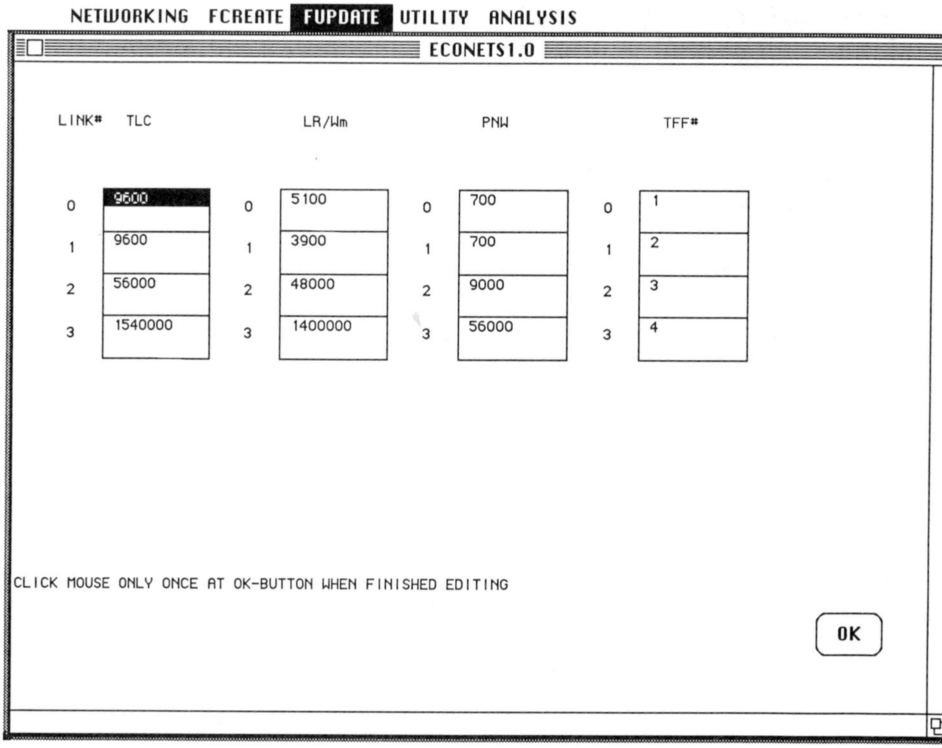

Figure 6.7. NLT file data in a structured form, as obtained from using the FUPDATE menu.

puter/disk malfunctions and manual mishaps can destroy the original files. The backup files may be uniformly named "Original B."

6.5.3.3 VHD Files Merge

The task of creating a large VHD file may be divided among several workers. Each worker creates a small VHD file (e.g., for the Eastern U.S.A. region). When all of the subfiles have been created, the VHD subfiles can be merged two at a time to create the final unified VHD file with proper nodal numbering.

6.5.3.4 VHD Traffic Multiplier

There are occasions when the designer will want to multiply the traffic loads of each location by a fixed multiplier, in order to create a new VHD file corresponding to a Traffic Growth Factor (TGF) = 1. This will create a new reference VHD file. Since each VHD file is a double-integer data file, it can only store integers. Therefore, in order to eliminate zeroes for traffic intensities, the new stored values are adjusted to [INT(Original Value * Multiplier) + 1]. VHD files with real numbers would have increased the storage requirements and increased the program execution times significantly.

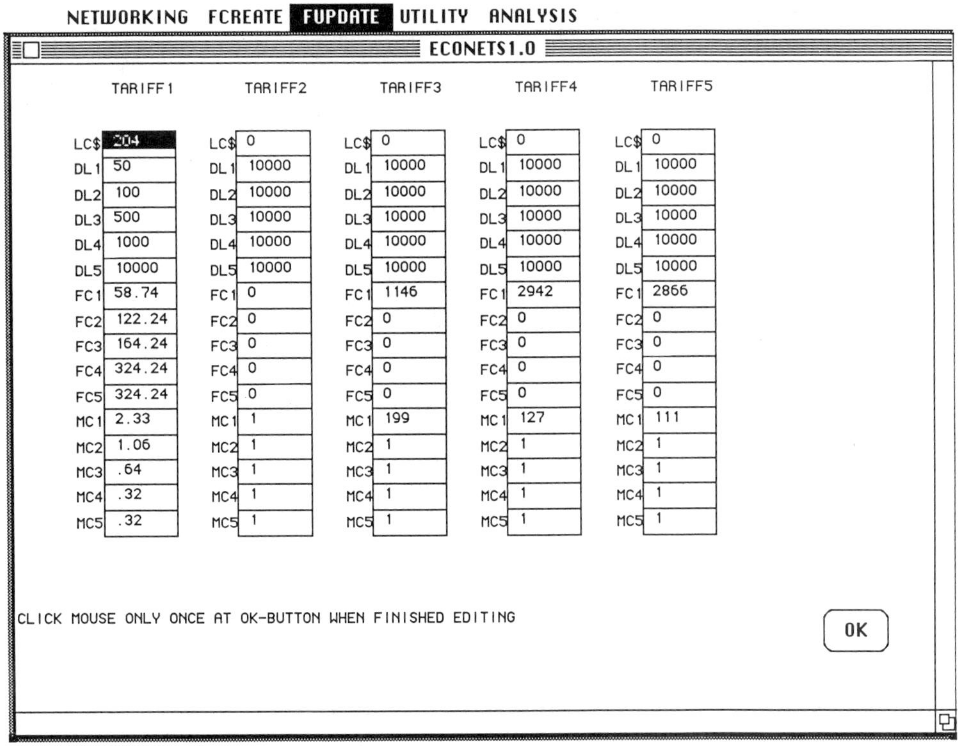

Figure 6.8. TARIFF file data in a structured form, as obtained from using the FUPDATE menu.

6.5.3.5 MAP Files Merge

This task is similar to that for the VHD Files Merge program. Several map files can be created to handle the outer boundary of the country, the state boundaries, and other meaningful partitions (e.g., RBHCs, sales regions, political subdivisions). Some of these separate map files can be created by different workers in parallel. When all of these separate map files are ready, one can create a composite MAP file through the process of merging two map files at a time.

6.5.3.6 FTD Files Merge

Since an FTD file is also a data file, and it can also be a large file with 3-element vectors arranged sequentially, the above Files Merge program can be used for merging several FTD files into a single file. This scheme allows several workers to prepare many FTD files separately and in parallel.

6.5.3.7 VIEW Output Files

This capability allows the designer to view the output files that describe the output results of any networking task. The designer can scroll up and down to view any part of the

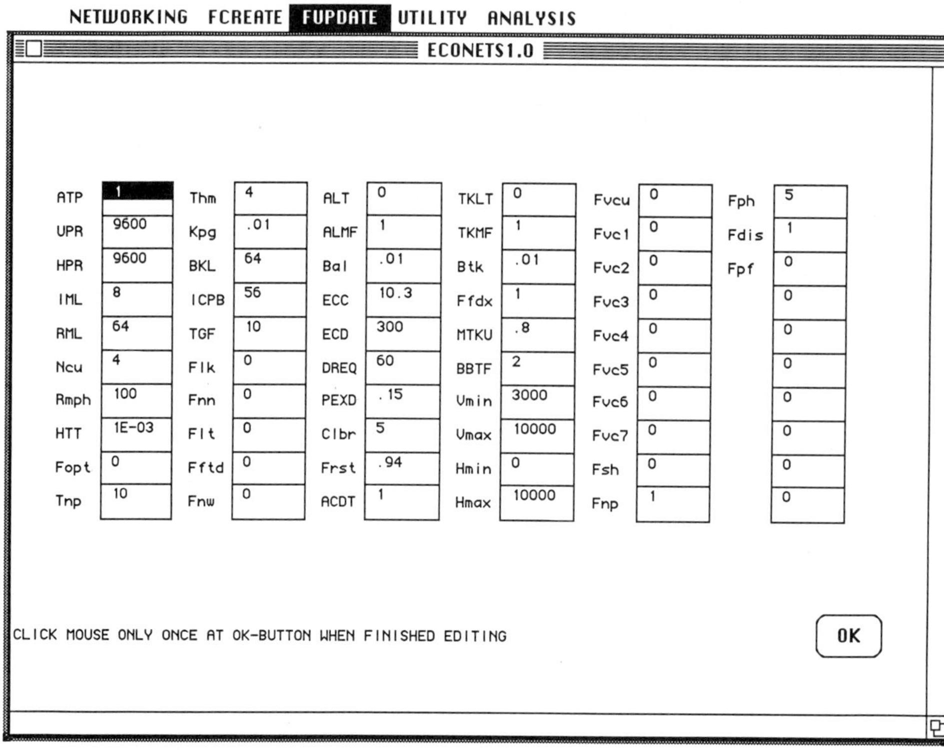

Figure 6.9. SDF file data in a structured form, as obtained from using the FUPDATE menu.

output file. When the task is chosen, a choice is provided to select any output file present in the working folder. When the designer is through studying the file, the mouse should then be pointed at the "close window" button of the window and clicked. The current window sent to the background and a new choice for viewing another output file is provided. Up to six output files can be viewed concurrently in this manner. By clicking on the Cancel Button, one can quit the Utility menu altogether.

6.5.3.8 VIEW Help Files

This capability allows the designer to view the Help files defining the various networking tasks. The designer can scroll up and down to view any part of the Help file. Otherwise, the operation is similar to that for the previous task. A separate item is provided for viewing Help files, since these files employ a different font and font size when compared to those used in the standard output files.

6.5.3.9 VIEW Topology

This task will enable the designer to view any network topology already saved in memory. Several topologies can be viewed concurrently, using the approach discussed ear-

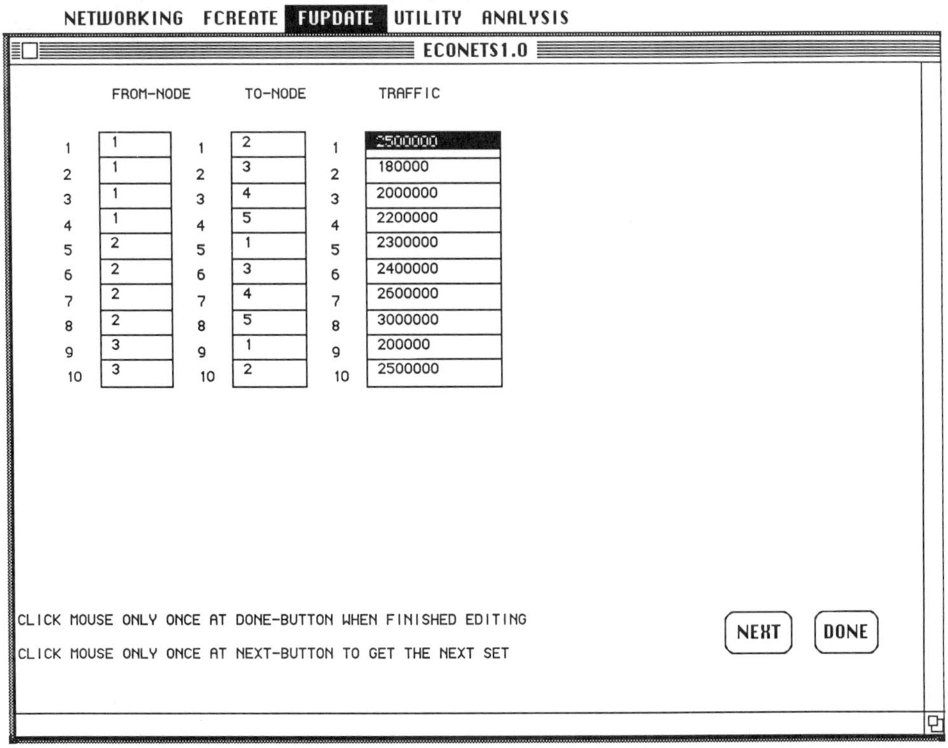

Figure 6.10. FTD file data in a structured form, as obtained from using the FUPDATE menu.

lier. By clicking the mouse on the Cancel Button, one can quit the Utility menu altogether.

6.5.4 ANALYSIS MENU

The ECONETS package provides several proven tools that can enable the designer to analyze several networking situations in terms of sizing, performance and reliability, delay distributions, and returns-on-investment, without leaving the application. All of these tools are based on the material covered in Chapters 2, 3, and 4. These tools are particularly handy to (1) check the results as contained in the output file, (2) create a model of a tariff for a privately owned facility, (3) compare the performance of LANs, and (4) compute the equivalent monthly costs of owning hardware and/or the pay-off period. The various items under the ANALYSIS Menu are described in the following paragraphs.

6.5.4.1 Find B-ERLB

This analysis tool will enable the designer to compute the blocking factor, B, for a loss system characterized by A erlangs of traffic intensity and N servers, using the Erlang-B formula. The maximum allowable value of N is 4990 servers.

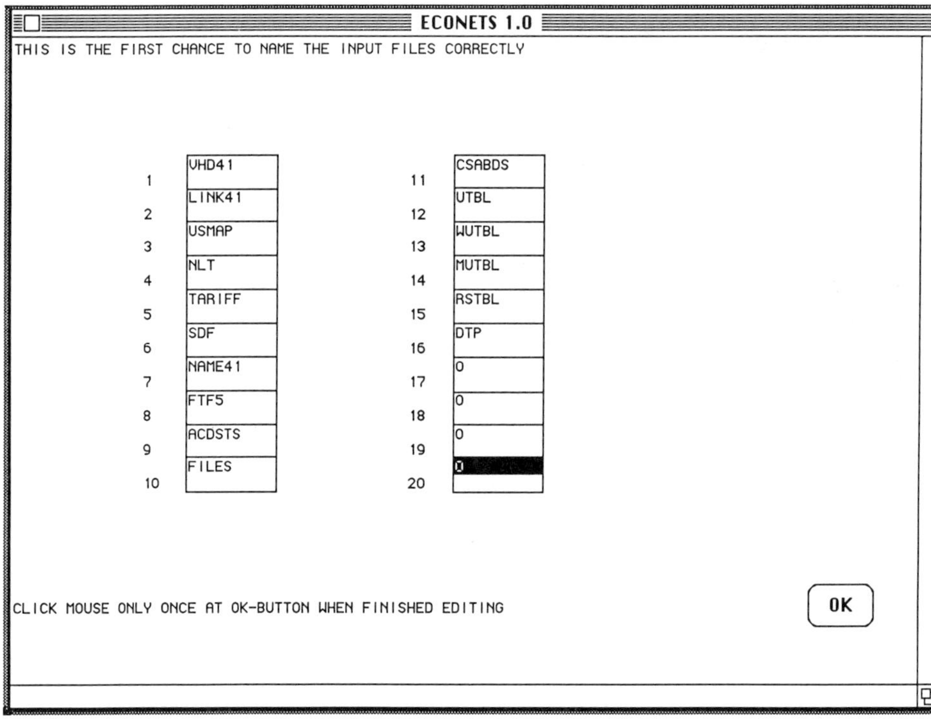

Figure 6.11. FILES text data in a structured form, as obtained from using the FUPDATE menu.

6.5.4.2 Find N-ERLB

This analysis tool will enable the designer to compute the number of servers for a loss system characterized by A erlangs of traffic intensity, and a desired blocking factor of B, using the Erlang-B formula. The maximum allowed traffic intensity is about 4990 erlangs.

6.5.4.3 Find N-Poisson

This analysis tool will enable the designer to compute the number of servers for a loss system characterized by A erlangs of traffic intensity and a desired blocking factor of B, using the Poisson's formula. There is no limit as to the maximum allowed traffic intensity value.

6.5.4.4 Find N-ERLC

This analysis tool will enable the designer to compute the number of servers for a delay system characterized by the number of calls handled during a peak hour, expected call duration (ECD), the required delay (DREQ), and desired probability of observed delay exceeding DREQ, using the Erlang-C formula.

6.5.4.5 ERLC Delay Analysis

This analysis tool will enable the designer to compute the delay distribution for a delay system characterized by A erlangs handled and the expected call duration (ECD) for integer values of servers exceeding A. This analysis can be allowed to continue until the value of the probability of observed delay exceeding a certain value becomes extremely small. The delay distribution is described in terms of the probability of observed delay exceeding integral multiples of (ECD/8).

6.5.4.6 Composite ACD Analysis

This analysis tool provides the probability of observed delay exceeding a required delay (DREQ) and the probability of blocking on the associated trunk bundle for a composite ACD system (which includes the ACD delay system and the trunk-related loss system), when the values of ECD and the number of busy hour calls are known. This is a very useful tool for computing the grade-of-service (GOS) of a composite ACD system.

6.5.4.7 Multipriority Queue Analysis

This analysis tool enables the designer to compute the average queuing delay in a multipriority system when the number of priorities and the respective loads are known. The outputs are meaningful only when the sum of the queuing loads is less than 1. This tool is valid for one or more priorities and an M/M/1 type queue characterized by random arrivals, negative exponentially distributed execution times, and a single server.

6.5.4.8 Average Message Expansion Factor (AMEF)

This analysis tool enables the designer to compute the factor by which the raw message transmission time must be multiplied in order to compute the actual message receipt time at a receiving channel that is also being contended for by additional sources of data. Assuming that each message consists of N packets, S sources of traffic all vying for the receiving channel with a loading ρ and first-in-first-out service discipline pertaining to a M/M/1 queue, one can compute the value of AMEF. This approach is essential when the bid-to-end-of-message delay is required.

6.5.4.9 LAN Performance Analysis

This analysis tool provides a performance comparison among the three most popular LAN architectures—token ring, token bus, and CSMA/CD. The designer must provide the message length in bits, propagation delay on the LAN media, processing delay on the LAN interface unit, number of stations served by the LAN, and the transmission capacity of the LAN media in bps in order to obtain the maximum transmission rate on the LAN media. The answers are provided for two operational situations: (1) only one station is busy, and (2) all stations are busy. These answers can then be used to estimate the average access time on the LAN.

6.5.4.10 Packet Switching System (PSS) Delay Analysis

This analysis tool provides an end-to-end delay distribution for a PSS characterized by several M/M/N type queues in series. Several types of transactions with unique execution times are taken into account. Whereas the M/M/1 type queue is valid for a PS node, the M/M/N type queue is valid for an output transmission line system characterized by N internodal trunks. The output distribution is expressed as 90, 99, and 99.9 percentile values of the end-to-end delay. Up to 5 subsystems in series can be handled.

6.5.4.11 Reliability Analysis

This analysis tool computes the reliability of each subsystem and the entire system for the known values of mean-time-between-failures (MTBF) and mean-time-to-repair (MTTR) of each component, and their configuration topology (Serial, Parallel, or Required K-out-of-N). The reliability analysis is in the form of subsystem/system availability, MTBF, and MTTR. Up to two subsystems arranged in series and up to five elements per subsystem can be handled. Using this approach, larger systems can also be modelled for reliability.

6.5.4.12 Return-On-Investment (ROI) Analysis

This analysis tool provides the equivalent monthly costs for hardware purchased at a fixed price, and the cost of money when the purchase price, the assumed lifecycle of the system, and annual interest rate are known. This figure must be added to the monthly transmission costs computed using the ECONETS package. The ROI analysis generally deals with the payoff period, which can be computed as the ratio of [the cost of money computed using this tool] to [the monthly savings achieved by using ECONETS outputs]. This tool can also be used to model the equivalent monthly tariff for a privately owned transmission facility (e.g., fiber optical or a digital microwave facility or a VSAT hub).

6.5.5 NETWORKING MENU

The NETWORKING Menu allows the designer to model, analyze, and design data, voice, and integrated voice/data networks. In order to model, analyze, and design a network, one must first define all the applicable input files in FILES. The format of each input file has been defined in Section 6.3. Up to 20 input files can be defined by FILES. The order in which the input files are named is very important. Any error or alert message related to an input file always mentions the file number. For example, if the VHD file is not available according to the name specified in FILES, an alert message will inform the designer that the file number 1 is not available. The following capabilities are provided for modeling, analysis, and design:

1. FIND COGs for locations as candidates for switches and concentrators.
2. STARNET for the modeling, analysis, and design of data networks with a star topology.
3. DLNET for the modeling, analysis, and design of data networks with a directed link (DL) topology.

4. MDNET for the modeling, analysis, and design of data networks with multidrop topology.
5. VOICENET for the modeling, analysis, and design of two-level voice networks with star topology.
6. IVDNET for the modeling, analysis, and design of integrated voice and data networks.
7. MSTNET for the modeling, analysis, and design of an unconstrained minimal spanning tree network.
8. ACD NETWORK for the modeling, analysis, and design of a voice network for incoming call management (also known as the Automatic Call Distribution).
9. BACKBONE NET for the modeling, analysis, and design of a backbone network using any arbitrary sets of links, including the T-1/T-3 type.
10. CREATE DB for creating a data base of inputs, tariffs, and design parameters for any run.
11. QUIT for exiting the ECONETS application and going to the Desktop.

6.5.5.1 FIND COGs

This requires only three input files related to VHD, MAP, and SDF, as named in FILES. The program first requests that the designer input the number of decompositions, d, for finding COGs. The correct inputs are 0, 1, 3, 7, . . . (or $d = 2^x - 1$), where the variable x is any integer 0, 1, 2, 3, 4, . . . so on. The program then requests the type of the first decomposition—horizontal or vertical. For a country shaped as a flattened rectangle (e.g., U.S.A. or Canada), it is recommended to start with the horizontal type. Of course, the designer can experiment with the other type and analyze the penalty, if any. The program provides 2 * d + 1 COGs as possible candidates for locating switches and/or concentrators. The program then provides the capability to save the network map and the COG-related data.

6.5.5.2 STARNET

This models, analyzes, and designs multilevel data networks based on the connected star topology. It requires at least five input files, namely VHD, MAP, NLT, TARIFF, and SDF, as defined in FILES. Additional files related to LINK, NAME, FTF, and LATA may be needed for some applications. This program employs bits per second as the measuring unit for representing average peak traffic intensity. The program requests that the designer first input the number of switches, and then define the number related to each switch location. To minimize the RAM requirements, the maximum number of switch/concentrator locations are limited to 16/32 for the smallest version. When entering the number of SW locations, the designer may be confused as to the number entered previously. In that case, the designer should enter a "0" at the next query. The cycle for entering the SW numbers will be repeated. Each point-to-point netlink serves exactly two nodes, one of these being the SW or concentrator node. The traffic at each node is noted and [INT(Traffic/Wm) + 1] links are used to serve that node. The value Wm is defined in the NLT file for each LINK type as the allowable rate in bps. The node is connected to the closest SWITCH from among all of the switches. If no LINK file is used, the STARNET has only two levels—the netlink level and the backbone trunk level. By using the LINK file, one can employ several concentrator levels. To illustrate, some nodes may have LINK numbers as zeroes and some have ones. The nodes with LINK numbers as ones will be considered as concentrators. At the first pass, the nodes with LINK numbers as zeroes will be connected to the closest concentrator according to a star topology. During the next pass, the concentrator nodes will be connected to the closest SWITCH according to the star topology, and the backbone trunk network will then be designed. The NLT file defines the actual tariffs that apply to each node as determined by its LINK number. If the FTF is not specified, the backbone trunk is designed assuming a symmetrical traffic flow proportional to the SWITCH node traffic loads. By using the FTD file, the program will ignore the traffic loads as specified in the VHD file and recompute the new nodal traffic loads and the correct inter-Switch traffic flows before designing the backbone network. The program produces the final network topology with a summary list of AL and trunk costs, and the average response time. The program also prints the cost of transporting a Megabit of data on the network based on the assumptions of F_{ph} peak hours per day and 20 business days per month. The program finally provides the capability of saving the network map, the AL output data and trunk output data files for later analysis and printout. By varying the design parameters, such as Traffic Growth Factor (TGF) and BackBone Trunking Factor (BBTF), the designer can study their effects on system costs and topology.

6.5.5.3 DLNET

This models, analyzes, and designs a data network with netlinks based on the directed link topology. The modeling, analysis, and design process is identical to that described for a STARNET. The outputs also provide the average, maximum per subnet, and maximum per entire network, values of end-to-end queuing delay, turnaround delay on a netlink, and bid-to start delay (also called the access delay) in milliseconds. The design parameter ATP determines the type of response time analysis: (1) for an ASYNC point-of-point, (2) for a virtual multipoint (VMPT), and (3) for a multipoint BSC/SDLC protocols.

6.5.5.4 MDNET

This models, analyzes, and designs a data network with netlinks based on the multidrop topology. The modeling, analysis, and design process is identical to those described for STARNETs and DLNETs. See Chapter 8 for a comparison of several data network topologies.

6.5.5.5 VOICENET

This models, analyzes, and designs a voice network with ALs connected to the nearest switches, according to a star topology. Millierlangs are employed to represent traffic intensity. To model, analyze, and design a voicenet, one also needs at least five input files, namely VHD, MAP, NLT, TARIFF, and SDF. Additional input files, such as NAME, FTF, and LATA may be required for some applications, in a manner described for the STARNET. The LINK file is not required if one needs to model, analyze, and design a voicenet. The designer needs only to input the number of switches and their respective designations. Important design parameters are the Economical Call Cost (ECC), Expected Call Duration (ECD), Traffic Growth Factor (TGF), Peak-Hour Factor (F_{ph}), and BackBone Trunking Factor (BBTF). Since the ECC represents the cost of a 5-minute call on the virtual private network (also on a software defined network), one can control the number of private lines leased for ALs and trunks. If the ECC is artificially made too high, the entire voicenet will employ dedicated lines. If the ECC is too low, the entire network will be synthesized using virtual facilities. The cost of a backbone trunk network will also depend on the value of BBTF. The maximum trunking cost will correspond to BBTF = 1. The larger values of BBTF will imply higher intranodal voice traffic and lower internodal traffic. The value of TGF can be varied to see its influence on the network cost. The values of AL blocking (B_{al}) and trunk blocking (B_{tk}) can also be varied to study the relationship between network cost and grade-of-service. The output consists of a network topology map, summary AL and trunk costs, and average cost of a call minute based on the assumptions of F_{ph} peak hours per day and 20 business days per month. The program also requests of the designer whether a need exists to save the related map and output files. See Chapter 7 for a study of meaningful voicenets using this module.

6.5.5.6 IVDNET

This models, analyzes, and designs an integrated voice and data network based on superimposed topologies. The program always needs the LINK file, in addition to the five essential input file described earlier. The program first designs a two-level voice network involving locations with a "0" LINK attribute and the switch locations specified by the designer. The design parameters described in VOICENET will influence the network costs. The program then designs a data network using the multidrop topology and several levels of concentration, as determined by the LINK assignments. First the locations with the "1" LINK attributes are chosen to design a low-level data network using the nodes, with "2" LINK attributes as concentrators. The process is repeated with nodes with "2" LINK attributes, until no more nodes are available. In the last pass, the program asks for the number of switches and their numerical designations. It is recommended that the switch locations correspond with those for the voicenet. The colocation of the voice and data switches will generally lower the cost of an IVDNET. The colocation of the two types of voice and data switches will also allow the use of T1 multiplexers. The trunk output for an IVDNET show the incremental cost values of the data trunks, but with the composite trunk matrix. By comparing the values of the various trunk matrices related to the voice network and IVD network, one can easily understand the outputs.

6.5.5.7 MSTNET

This program connects all of the locations according a minimal spanning tree topology, assuming no capacity constraints. The STARNET, DLNET, and MDNET were characterized by the link capacity constraint of W_m. The program requires the five input files, as mentioned earlier. The MSTNET module provides the AL costs based on a specified tariff (according to the NLT file), in addition to the topology map. The outputted response time summary corresponds to the case when all nodes form a single netlink. Although the analysis outputs may be meaningless to some, the module does provide valuable insights to the designer as to the manner in which various nodes get interconnected.

6.5.5.8 ACD NETWORK

This program provides a cost analysis of a corporate network designed to manage incoming voice calls using virtual facilities such as the 800- and Megacom-Services, and to handle outgoing calls using the Out-WATS facilities. The program requires several additional input files in addition to the five input files mentioned above. The ACD NETWORK program should never employ the FTD file, due to the use of virtual circuits. The program employs an LATA file that assigns to each location the corresponding state/region number. It employs a large table to compute the customer service area, based on the originating and destination state/region numbers of the two locations involved. It is a 62 × 62 matrix and is called the CSATBL. It also requires a usage table for 800 or WATS or Megacom, a rate-step table, and a daily traffic profile table to apportion the incoming calls to the 24 hours of any day of the month. Besides computing the cost of the virtual facilities, the program also computes the number of agents (or voice response units, if applicable) required to handle the incoming calls during each hour. The designer can select any number of switches (or ACD centers), design and grade-of-service parameters such as expected call duration (ECD), delay required (DREQ), probability of delay greater than DREQ, and trunk blocking factor. The designer can also define the cost of a labor hour to compute the total cost of agent labor. The designer can combine the ACDNET with a VOICENET/IVDNET package to compute the cost of a corporate network using both dedicated and virtual facilities.

6.5.5.9 BACKBONE NET

This package can be used to model, analyze, and design a digital backbone T-1 network using an FTF file in conjunction with the five input files described earlier. It employs bits per second as the measuring unit to define the traffic flow intensities between any pair of nodes. It should employ an FTD file, thus creating new traffic loads different from those given in the applicable VHD file. The designer can use the CREATE DB program to study the new applicable nodal traffic rates. By changing the FTF data, the designer can vary the connectivity of the backbone network. The NLT file must be updated to define the applicable T-1 tariffs and maximum useful data rates on the links employed. Most T-1 multiplexer vendors specify upper limits on the use of T-1 or T-3 type links. The program provides the standard network topology map and the summary network costs. The designer has the capability to save the map and any related outputs for later editing and printing. See Chapter 9 for some useful results obtained by using this module.

6.5.5.10 CREATE DB

This task will allow the designer to create a data base of VHD, LINK, NLT, TARIFF, and SDF files for each modeling, analysis, and design run. The designer can save the data base for later viewing, editing, and printing. It is a very useful exercise, especially when one is varying a large number of design parameters to see their combined effect on the network cost and performance. Since the output file is saved under a relatively fixed title (e.g., VHD59DB), the designer should rename this output file quickly before it is overwritten by another DB file.

6.5.5.11 QUIT

This allows the designer to quit the ECONETS application altogether to go to the Finder.

6.6 A DESCRIPTION OF THE OUTPUT FILES

ECONETS creates output files containing five categories of design data, defined as follows:

1. Access Line (AL) design data for networks with star topology. This covers both voice and data networks falling in the categories of STARNET and VOICENET.
2. Access Line (AL) design data for data networks falling in the categories of DLNET and MDNET.
3. Backbone trunking design data for both voice and data networks.
4. A data base for each network modeled, analyzed, or designed.
5. COG data.
6. Topology-related, object-oriented, PICT file

Each of these output files can be viewed by using the VIEW Outputs or VIEW Topology items under the UTILITY menu. The EDIT or the MacWrite applications can also be used to view, update, and print the text output files. We will now describe the structure of each output table.

6.6.1 OUTPUT FILE FOR STAR TOPOLOGY NETWORK ACCESS LINES (ALs)

As soon as the network topology is completely shown on the screen, the program asks the designer to save an output file(s) for AL data, trunking data, and topology.

Each output file starts with a delineation of some important design parameters such as AL blocking factor (B_{al}), economical call cost (ECC), AL link type (ALT), expected call duration (ECD), and AL multiplexing factor (ALMF) for voicenet, maximum traffic weight or load allowed per data AL (W_m), and traffic growth factor (TGF).

Each output table then presents its data with 6 columns, including the node #, the switch node # that serves the node, the number of links (voice or data) in the AL bundle, the mileage of the AL, the cost of each AL, and the total cost of each AL bundle. There are as many entries as there are AL bundles. The cost of each AL bundle is determined by the LINK type of the AL bundle, which determines, in turn, the tariff type through the NLT file. For voicenets, the number of ALs is determined by the AL blocking factor and the AL multiplexing factor. In some cases, the number of ALs may be zero, while the AL bundle cost may be non-zero. This suggests a virtual private network (or software defined network) option chosen by the program to save money.

This data is saved for each level of the network design. For example, a STARNET will allow the designer to design an AL network for each Link type, starting with the

"0" type. The designer should design a LINK input file in such a way that the nodes with a "0" Link type will be connected to concentrators located at a node with Link type "1," and so on. The final switches can be located anywhere and must be selected by the designer.

The output file also provides a summary of the results for each switch/concentrator at each level. The summary presents useful information in four columns entitled the Switch #, the Switch Size (in terms of terminations), total traffic (bps or millierlangs) handled by the switch, and the total cost of ALs terminated on that switch). Additionally, the total AL cost (this cost is identical to that appearing on the network topology map) is listed in the summary, which can also print out the grand average of all response times on all netlinks of a data STARNET.

6.6.2 OUTPUT FILE FOR MULTIDROP (DLNET OR MDNET) DATA NETWORK ACCESS LINES (ALs)

The program asks the designer for a command to save an output file for AL data in a manner identical to that described earlier. The designer must also provide a name under which the output file will be saved.

A typical output file first presents a summary data about each network link (or netlink) in terms of the Switch number, the Link (or netlink) number, actual traffic load handled in bps by the netlink, average queuing time (AQT) in MS on a turnaround basis on the netlink, the average total turn-around time (ATAT) in MS on the netlink, expected value of response time $E(T_r)$ on the netlink, and the list of nodes that are served on the particular netlink. The summary also includes the maximum and the average values of T_r observed on each subsystem (served by a switch) and a system average value for T_r.

The output file then presents detailed data for each connection in the form of six columns entitled the From Node number, To Node number, the circuit length in miles, the circuit cost, the running total cost, and the running total for circuit mileage for each netlink. There are as many lines of data as there are multidrop connections.

The output file then lists the data about any special links created to connect certain nodes with the switch to handle traffic loads in excess of W_m (the maximum allowable data rate for the netlink). Each special link is defined in a manner similar to that for a connection of a netlink, as described above. It must be emphasized that the special links also influence the response time computations.

The output file presents the summary costs for all of the AL connections and special links. It also presents some of the important design parameters that were used to compute system performance and costs.

It should be pointed out that the above data is presented for each network level separately. First the nodes with Link Type of "0" are considered, then the nodes with a Link Type of "1" are considered, and so on.

6.6.3 OUTPUT FILE FOR BACKBONE TRUNKING DESIGN DATA

The program also asks the designer for a command to save an output file for the backbone trunking data. This follows soon after the network topology is fully designed and plotted on the active window.

This output file first lists some of the most important design parameters, such as the Trunk-Link Type (TKLT), Trunk Multiplexing Factor (TKMF), BackBone Traffic Factor (BBTF), Traffic Growth Factor (TGF), Economical Call Cost (ECC) for a virtual private network, allowed trunk blocking (B_{tk}), grand total (GT) of internodal and intranodal traffic (erlangs for voice and bps for data), and the number of switching centers (N_c) and their location site numbers.

The output file then shows the original ($N_c \times N_c$) trunk traffic matrix. Each T_{ij} element defines the traffic intensity flowing from the ith switch center to the jth switch center. The units for matrix elements are always erlangs for a voicenet and bps for a datanet. If an FTD file was not used, the internodal traffic values are derived using a simplified traffic model with symmetric but proportional data flows. According to this model, the traffic intensity on the i-j trunk can be expressed as $(TT_i * TT_j)/GTT$, where TT_i and TT_j are the total traffic handled by the ith and jth nodes, respectively, and GTT is the grand total of all traffic (inter- and intranodal) handled by the backbone network. The computed values of internodal traffic must then be divided by BBTF and the difference added to the intranodal traffic. The larger the value of BBTF, the lower the value of the internodal traffic. It should be clear that BBTF is a simple device to influence the amount of internodal trunk traffic.

If an FTD file was employed for network design, only the exact values of traffic handled by each node and trunk are computed. In that case, the value T_{ij} may not equal T_{ji}. The value of BBTF (SDF parameter number 35) is also ignored when computing the actual values of traffic flows. The next ($N_c \times N_c$) matrix presents the modified values of traffic. For a voicenet, the traffic loads T_{ij} and T_{ji} are simply added. For data networks, the values are added only if the trunk connecting the nodes i and j is not the full duplex (FDX) variety. One can define the trunks to be HDX or FDX through the 33rd SDF parameter. The intranodal traffic values, along with one side of the internodal value, are zeroed, since these do not contribute to trunk requirements.

The next ($N_c \times N_c$) matrix presents the modified values of the trunks required between each pair of nodes. Only the top right part of the matrix is populated with values. The other side does not add to the information. Since the intranodal traffic does not require trunks, the diagonal values are also zeroed. For voice, the Erlang-B formula is employed to compute the number of trunks required for a given value of trunk blocking, as defined by the 32nd SDF element. For data networks, the number of trunks are computed using the allowed maximum trunk utilization factor (MTKU per the 34th SDF element). The last matrix defines each trunk bundle by specifying the two associated switch nodes, the circuit miles, the cost per circuit, and the total cost of each trunk bundle.

The last lines of the output file for trunking data summarizes the total access line cost, the total trunking costs, and the total system costs. The cost of a single five-minute voice call or moving 1 megabit of data on the network is also listed. The contents of the output file for trunking data can be viewed by the VIEW Outputs item of the Utility menu. The EDIT or MacWrite application can also be used to modify the file contents before printing.

6.6.4 OUTPUT FILE FOR A DATA BASE

The output file representing an integrated data base for any set of input files listed in FILES can be created by using the Networking menu item CREATE DB. One may also make the 15th, 16th, and 17th parameters of the SDF file equal to "1" to read the listed

LINK, NAME, and LATA files, so that their data can also be incorporated into the data base. A complete data base presents the entire network design data in a very structured form that is easy to read and refer to at a later date when analyzing all of the output files dealing with access lines and trunks.

The Output file for a data base first lists the entire VHD file in a structured fashion, with nodal number, the associated V and H coordinates, nodal traffic (actual or computed from an FTF file as determined by SDF parameter number 18), LATA number, LINK type, and NAME. At the end of this nodal data, the DB also summarizes the total nodal traffic handled by the system during a peak period.

The DB then lists the entire NLT file in a manner that is easy to read. Each line lists the link capacity (e.g., bps), maximum traffic weight or allowable data rate on the link (W_m), the potential data rate per each link/drop, and the associated tariff number. There are as many lines as there are node/link types (≤ 5).

The DB then lists each of the five tariffs contained in the TARIFF file used for the network design. For each tariff, the monthly local-loop costs for both ends of the leased circuit, five mileage band limits, five fixed monthly costs for each mileage band, and five costs per mile for each mileage band are listed in a manner that is easy to read. Finally, the DB lists the 60 elements of the SDF input file.

The contents of a DB can be viewed by the Utility Menu Item View Outputs. Of course, one can also use the EDIT and MacWrite applications to view/update and print the DB-related output files.

6.6.5 OUTPUT FILE FOR COG DATA

An output file can also be created while working with the FIND COGs Networking Menu item. It lists the V and H coordinated of each mathematical COG, the closest network node/site, and its associated V and H coordinate. Such an output file should be created and consulted for selecting switch/concentrator locations. This output file can also be viewed by using the View Output item of the UTILITY menu. Again, only the EDIT or a MacWrite application can enable one to modify and/or print this or any other output file.

6.7 COMPUTING THE V AND H COORDINATES FOR COUNTRIES OTHER THAN THE U.S.A.

It was in 1960 when AT&T began using the V and H coordinate system to determine the cost of interstate DDD calls and leased private lines for North America, which includes Canada and the U.S.A. V and H coordinates introduced a great deal of simplicity in computing the line-of-sight mileage. Previously, one used the latitudes and the longitudes and a complicated formula to compute the surface mileage. Data bases are now available that relate the NPA-NXX portion of each telephone number to the V and H coordinates of an associated wire (or rate) center. Therefore, it is an easy matter to compute the V and H coordinates of U.S. and Canadian locations for the VHD input file and map the boundaries of the two countries and their associated states or subdivisions for the MAP input files. Although the need for designing private networks for other countries and continents is increasing, no unified data bases are available for associating telephone numbers to a V and H-like coordinate system for other countries. An approximate method can be employed for creating simple V and H coordinate maps for other countries in order to avoid the use of latitudes and longitudes. It is described in the following paragraph.

Obtain a good atlas with maps of countries drawn in any one of the appropriate projection systems (e.g., Bonne, Conic, etc.) with linear scales of miles and kilometers. Overlay a transparent grid with each inch divided into ten intervals. Number the major scales (i.e., inches) in such a way that the vertical and horizontal values range between 0 and 10,000, similar to the V and H maps for North America. The boundaries of the country and its provinces or states can thus be read to create MAP files. One can also read the V and H coordinates for all of the locations to create any VHD file. The above method was employed to create MAP and VHD files for Australia and India. It was found that one required a multiplication factor of 1.557 to compute the distances in kilometers for Australia, and a multiplying factor of 0.86 to compute the distances in miles for India, if one used the maps appearing on pages 40 and 34 respectively, of the 1980 Edition of the Hammond World Atlas. It was a coincidence that the two projections employed for Australia and India provided distances that are closer to miles than kilometers using the approximate method. The multiplier for distance computation and the ranges for drawing map boundaries can be specified by the SDF parameter number 51 (F_{dis}) and design parameters number 36, 37, 38, and 39 (V_{min}, V_{max}, H_{min}, and H_{max}), respectively. See Figures 6.12 and 6.13 for maps obtained for Australia and India, respectively, using the above technique.

6.8 CREATING MODELS OF TARIFFS

It should now be clear that ECONETS is primarily an interactive tool for optimizing network topologies very rapidly. Such a goal was partly achieved by avoiding the use of (1) a large tariff data base during the topological optimization process, and (2) using meaningful but simple models for computing circuit costs on a monthly basis.

In order to avoid the use of a large tariff data base, we employ the computed V and H coordinates for each site (or node) directly in the VHD file. This minimizes the computation of site's V and H coordinates each time a circuit is priced by analyzing the NPA-NXX part of the site's telephone number. Since the network topology is primarily affected by the architectural considerations (e.g., use of T-1 circuits or SNA or Doelz or Datakit or ISDN), models of a few meaningful tariffs will be employed. To illustrate, not all applicable tariffs (as offered by AT&T, U.S. Sprint, or MCI) for leasing an analog VG circuit will be considered during the early phase of topology optimization. Instead, one can employ AT&T's Tariffs 9, 10, and 11 for a VG line or 56 Kbps DDS circuit, or a ACCUNET T1.5 circuit can be employed for studying various topologies. Only when a final network topology is achieved, can one use a private line pricing (PLP) package to compare all of the available tariffs, in order to select a particular set of carriers for leasing lines.

Situations also arise when an enterprise must consider owning its own transmission facilities. In particular, state and federal agencies may already own rights on laying the digital fiber optical (DFO) cables or digital microwave (DMW) routes on certain paths. By creating simplified models of one-time costs and using the Return-on-Investment

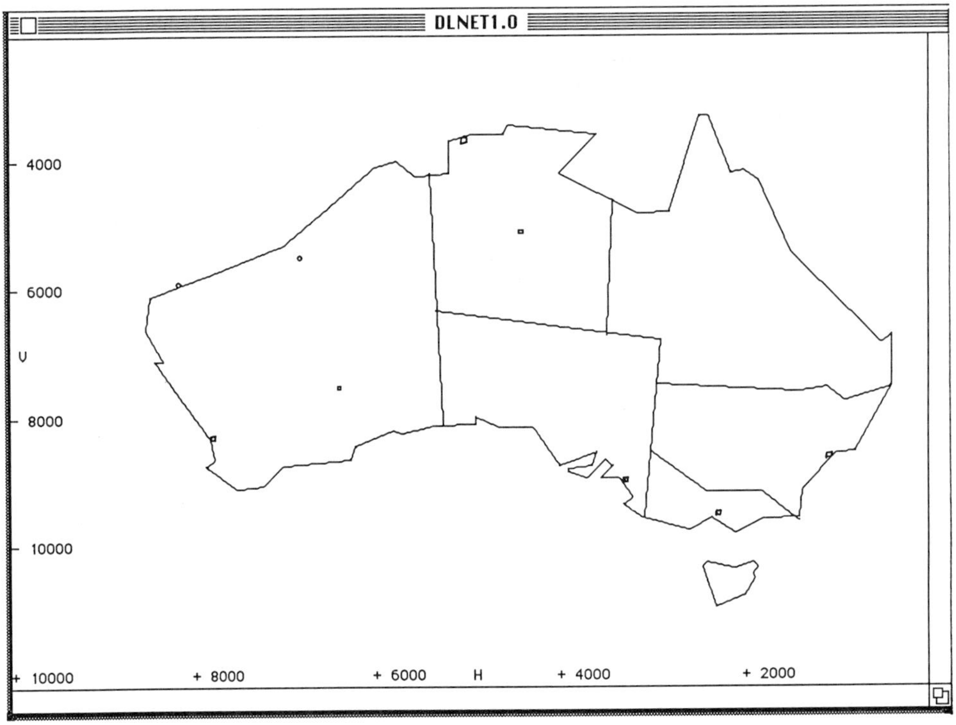

Figure 6.12. A map of Australia obtained using arbitrary V and H coordinates.

model already described, one can compute the equivalent monthly costs. Each model generally consists of a fixed value, two multiplying variables for the cable/path length and the number of equivalent VG circuits (or 64 Kbps channels) carried. This is illustrated by the following expression:

$$\text{Monthly cost}(\$) = \text{constant} + a * (\text{circuit/path length in miles}) + b * (\#\text{VGCs Carried}) \qquad (6.1)$$

Figure 6.14 compares AT&T's Tariff 9 for a VG, 56 Kbps DDS channel, and ACCUNET T1.5 circuit with owned DFO and DMW facilities. Similar models can be created for other possible tariffs. The existing structure (i.e., five distance intervals, five fixed costs, and five mileage charges for each interval) of FCC's approved tariffs and SDF file's 52nd parameter can be employed to handle a large variety of possible tariffs. SDF file's 52nd parameter is non-zero only for owned facilities. Each tariff file can consist of five useful tariffs for topology optimization. Additional tariff files can prepared for other sets of tariffs.

6.9 PERFORMANCE METRICS FOR ECONETS

A designer is generally interested in the execution times and the random access memory (RAM) requirements for any given software package. Excessive execution times will

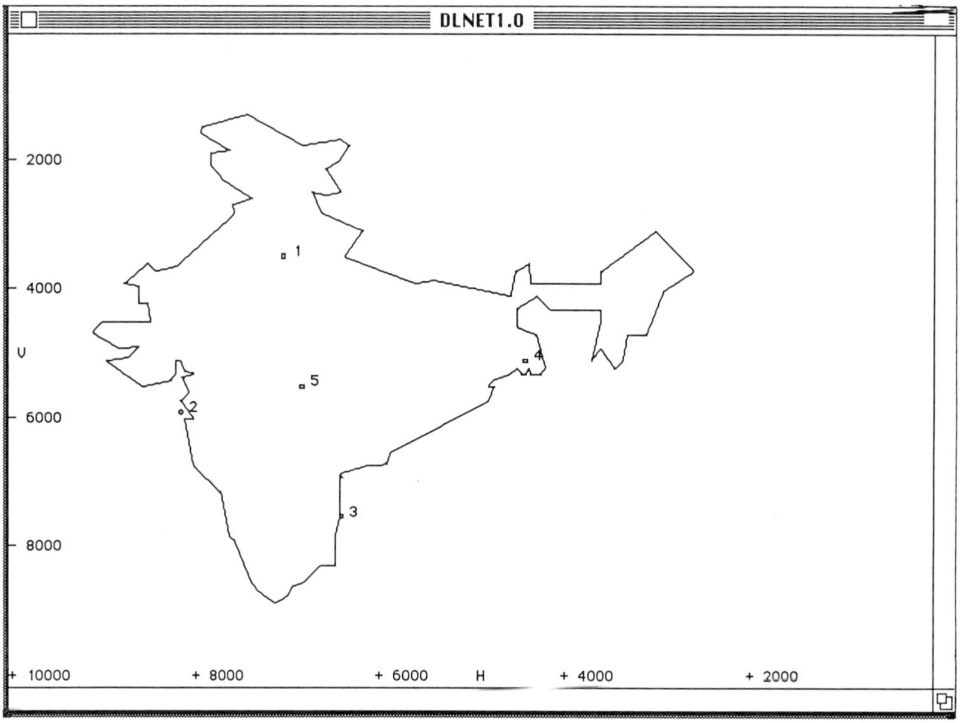

Figure 6.13. A map of India obtained using arbitrary V and H coordinates.

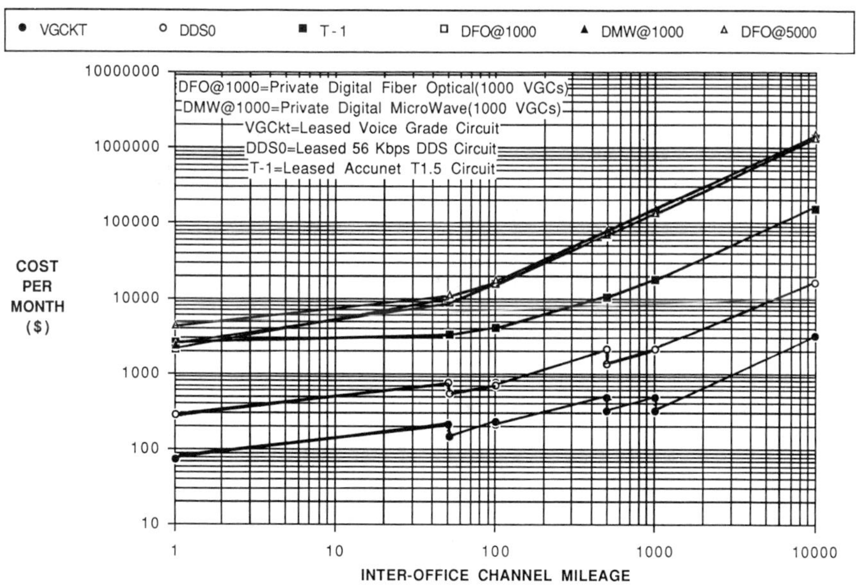

Figure 6.14. Cost models of some leased and owned transmission facilities.

prevent the software package from being called an interactive one. Excessive memory requirements may force the package to be either impossible to implement on a desktop computer, or too expensive to afford despite the recent drops in RAM prices. The following paragraphs define two useful performance metrics for ECONETS that are valid for a Macintosh II desktop computer.

Several typical network configurations were run to compute the execution times for various network topologies and architectures related to the NETWORKING Menu. The results are plotted on Figure 6.15 as a function of the number of locations (N) and for either one switch ($S = 1$) or an appropriate number of switches exceeding 2 ($S > 2$). Various network topologies, such as directed link (DL), multi-drop (MD), StarNet (voice/data), ACDNet, and Minimal Spanning Tree (MST) with no capacity constraint are considered. Execution times for the optimized DLNet and MDNet topologies achieved by using Sharma's exchange algorithm (by making $F_{opt} = 1$) are also plotted (with the suffix OPT).

The curves show that the star topologies require the least time for execution. Even the largest networks generally require less than a minute. ACDNets require about two to five minutes for a typical large system. Non-optimized DL, MD with one or more switches, and optimized DL topologies with multiple switches can be synthesized within less than 20 minutes even for large networks ($N \approx 400$). An optimized DLNet with a single switch may require up to an hour for a large network ($N \approx 400$). The optimized versions of MDNets will generally require a good deal of computer time. It is interesting

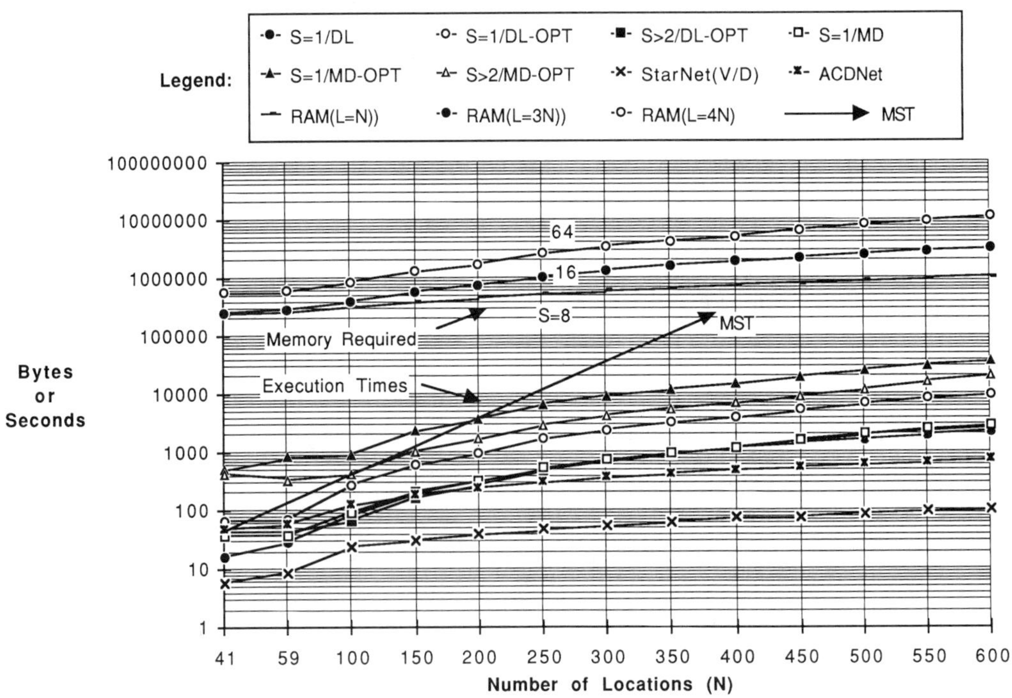

Figure 6.15. Execution times and memory requirements for the ECONETS package.

to note that the execution times for most networks with less than 200 locations are less than 20 minutes. Some very large networks can be decomposed into subnetworks and synthesized separately and quickly without increasing the network cost appreciably. Furthermore, the designer should not use the optimization capability until a suitable network topology has been synthesized by considering alternative network architectures and feasible design parameters.

Random access memory (RAM) requirements have also been plotted in Figure 6.15 for some typical network configurations characterized by the number of locations (N), number of switches (S), number of netlinks (L), and a fixed number of concentrators (C = 64). End user's networking requirements must be considered very carefully before designing the Dimension Statements in the source code. If this is not done, the system may simply hang up or, in many cases, provide a meaningless result. The source code is minimally designed to handle up to 200 nodes, 8 switches, 64 concentrators, and about 600 netlinks (both regular and special types) to prevent the RAM requirements from exceeding the 1 megabytes level. An additional 1–3 megabytes are generally recommended to allow (1) the design of larger networks and/or (2) multitasking.

6.10 CONCLUDING REMARKS

An attempt has been made to define (1) inputs required by the ECONETS package, (2) the operational capabilities of the ECONETS package in terms of the means to create input files and view/update input/output files in order to model, design, and analyze telecommunications networks in quick succession, and (3) outputs that describe the computational data. Two useful performance metrics (i.e., execution times and computer memory required) were also discussed for ECONETS. Now we are ready to apply the analytical tools of Chapters 2, 3, and 4, the algorithms of Chapter 5, and the ECONETS software package to obtain some insights related to corporate voice, data, and integrated voice/data networks. This task will be accomplished in the next three Chapters.

Chapter 7 will study only voice networks, and will present results obtained by varying several key design parameters in the NLT and SDF input files. Chapter 8 will concentrate on data networks with an emphasis on the multidrop topologies. The dependence of network topologies on the algorithms employed and the number of switches/concentrators will be studied in detail. Chapter 9 will focus on all of the design and implementation considerations required for achieving a significant saving by integrating voice, data, and video applications on shared transmission facilities. Each chapter will clearly show the way in which design parameters are selected, and outputs interpreted, for each network type.

7

Synthesis of Voice Network Topologies

7.1 INTRODUCTORY REMARKS

The network design algorithms of Chapter 5, and a software tool described in Chapter 6, will be used to design private voice networks for on-net traffic only, and private ACD networks for off-net traffic only. It is true that most private voice networks handle circuit-switched voice calls between the on-net and off-net locations, and that the off-net traffic generally exceeds the on-net traffic. However, the design methodologies for voice and ACD can be combined to handle an arbitrary mixture of the two traffic types for any private voice network. The methodology presented here should be used for optimizing the network topology of both local area and wide area networks. Most designers will be surprised by the savings resulting from the strict application of this methodology for all LANs and WANs.

Wide area networks (WANs) can be synthesized by enhancing the software of certain PBXs to act as tandem switches, and connecting them via a desired number of leased trunks facilities. The remaining PBXs are then connected to the nearest enhanced PBXs via ALs, resulting in the multistar, multicenter network topology of Figure 1.8. In some cases, it may be preferable to employ a certain number of stand-alone tandem switches. Still another way to synthesize a WAN is to employ a proper number of intelligent T1 multiplexers that are interconnected via T1 highways. Chapter 9 presents a methodology for selecting the attributes of T1 multiplexers (and also voice and data switches) for designing an integrated voice/data network. In the future, when the public telephone networks are upgraded to offer ISDN-based services (see Appendix B), it will be possible to synthesize a WAN by using the Primary Rate Access. Which approach will dominate the evolving marketplace, only time will tell. No matter what happens, one should be able to optimize voice network topologies by using the design methodologies of this chapter and the correct tariffs.

7.2 THE BASIC DESIGN PROCESS FOR A VOICE NETWORK

In order to design a voice network topology, the network design package, as described in the previous chapter, allows the execution of all of the steps defined in Chapter 5, and in particular Subsections 5.4.1 and 5.4.2. The network design software package also allows the designer to vary a large number of design parameters for synthesizing an economical voice network.

The ALT and TKLT parameters define the types of access links (sometimes used to lower the transmission cost by means of multiplexing several VG circuits) and trunk links. The NLT file is then used to choose the particular tariff employed for each link type to compute the circuit cost. We will use a tariff file that defines at least three tariffs: the Tariffs 9, 10, and 11 for the voice grade (VG), 56-kbps DDS, and T-1 lines. Such tariffs were defined in Table 2.4. We will also assume that the average monthly cost of the two local loops and termination charges equals an amount of $102.00 per VG, $62.76 per DDS, and $0.00 per T1 circuit, respectively. This implies no bypass for the VG circuits and some form of bypass for the DDS and full bypass for T1 circuits. Although most of the analyses assume some bypass for the DDS and full bypass for T1 circuits, an attempt is made near the end of this chapter to show the comparison between the network cost with, and without, bypass of the local exchange carrier.

The parameter ALMF defines the number of VG circuits that can be multiplexed on to an AL, and TKMF defines the number of VG circuits that can be multiplexed on to a backbone trunk. The choice of ALT, TKLT, ALMF, and TKMF allows one to compute the cost of a voice network for any mix of AL and trunk circuit types.

Another important design parameter called the economical voice call cost (ECC) defines the threshold above which either a virtual private network (VPN) or a software defined network (SDN) option becomes cheaper. If the cost of a call with an expected call duration (ECD) in seconds becomes greater than ECC by using an AL bundle, then it is advisable to employ either a VPN or an SDN offering from an OCC. Such an option will make a location, such as an off-net location, to be served by the OCC. The traffic presented to the OCC is switched by their public network to any continental U.S.A. location. As a result, this traffic will not be handled by the transmission facilities of the private network. The analysis presented here uses a value for the ECC of $1.31. Any other value can be chosen by fixing the SDF parameter number 23. On the other hand, by arbitrarily increasing the value of ECC ($13.31 or more), one can force every location to be connected to the private network via an AL bundle. In other words, each location becomes an on-net location. It is therefore quite easy to compute the cost of a hybrid network that employs both private leased lines and SDN offerings. It is recommended that the cost of two extreme options always be compared, for instance, a totally private network with leased lines only, and a hybrid network with an economical AL design. This provides a cost basis for selecting the hybrid network.

One can compute the cost of a voice call by assuming F_{ph} busy hours per business day, 20 business days per month, and (3600/ECD) calls handled per busy hour (BHR) per line. The total cost per call minute (TCPM) on a given network handling "A" busy hour originating erlangs can be computed as follows:

$$TCPM = TMC/(A * K)/2 \tag{7.1}$$

where

A = Total busy hour erlangs handled by all customer nodes (e.g., see Table 7.1 for such a value at the end of the table)

TMC = Total monthly cost = AL$ + TK$

$$
\begin{aligned}
K &= (20 \text{ business days/mo.}) * (\mathrm{F_{ph}}) * (3600/\mathrm{ECD}) \\
&= \text{Expected number of calls per month per unit erlang} \\
&= 1200 \text{ for ECD} = 300 \text{ seconds and } \mathrm{F_{ph}} = 5 \\
ECD &= \text{Expected call duration in seconds}
\end{aligned}
$$

The performance of a voice network is greatly determined by the amount of traffic handled. The software design package allows one to vary the traffic volume by changing the traffic growth factor (TGF). The analyses presented here generally use three values: 1, 5, and 10. The parameter TGF multiplies the traffic intensities as defined by the VHD input file.

A design parameter called BBTF, which determines the amount of traffic flowing on the backbone network trunks, also influences the cost of the network. The analysis presented in Subsection 2.4.7 assumes a uniform community-of-interest among all of the locations. We will consider two values for the backbone traffic factor (BBTF) for a symmetric situation. The value of 1 represents a uniform community-of-interest. The value of 2 for the community-of-interest creates more intraswitch traffic at the expense of interswitch traffic. According to our model, when BBTF = 2, it almost halves the cost of interswitch trunks. The amount of traffic that does not flow on a trunk becomes part of the intraswitch traffic. By using the values of BBTF of 1 and 2, one can study the two extreme situations that generally characterize most private networks.

In the analysis presented later, each voice network is synthesized to handle only the on-net to on-net traffic. In some cases, only when the cost of an AL bundle exceeds a certain threshold do we employ the VPN or the SDN options to handle that location's traffic. In order to handle the off-net traffic presented on either outgoing WATS or incoming 800 traffic, one is advised to study the design process for ACD networks, as presented in a latter section.

7.2.1 SYNTHESIS OF A 41-NODE VOICE NETWORK

We will now synthesize a voice network serving 41 locations as defined by the VHD41 data file. It is the first file required by the program. See Table 7.1 for a printout of a related data base. It represents an actual user and defines the V and H coordinates and the time-consistent, peak period traffic loads in erlangs for each of the locations. (Some new versions of ECONETS use millierlangs as a unit.) One can assume that each location is served by a PABX. Such a PABX handles not only the local traffic but also acts as a gateway for the interlocation traffic. The objective is to synthesize an economical network topology for the available tariffs. The program then requires the names of the applicable Map, NLT, Tariff, and SDF files. The program also requires the V and H boundaries for the map.

The data in Table 7.1 and some other tables shows additional editing when compared to the raw data of Table 7.6, for example. To achieve the desired amount of editing for a proposal, for example, one must first drag the output file icon into a word processing folder. By first opening the word processing program, one should then open the text-based output file and perform any amount of desired editing. The Edit program employed by ECONETS allows only one font type and font size for the entire document. The word processing programs (e.g., Claris' MacWrite) generally allow many options for editing.

Table 7.1 Network Design Data Base for VHD41 Input File.

#	-V-	-H-	LOAD (Erlangs)	LATA	LINK	NAME
1	7027	4203	7	524	0	KANM0
2	9213	7878	7	730	0	LAXCA
3	5015	1430	1	224	0	NWKNJ
4	4687	1373	1	920	0	HFDCT
5	6263	2679	1	922	0	CINOH
6	5972	2555	1	324	0	COLOH
7	4997	1406	9	132	0	NYCNY
8	8492	8719	5	722	0	SFOCA
9	7501	5899	1	656	0	DENCO
10	7489	4520	2	532	0	WICKA
11	7947	4373	1	536	0	OKLOK
12	8436	4034	1	552	0	DALTX
13	8266	5076	1	546	0	AMLTX
14	8938	3536	1	560	0	HOUTX
15	8483	2638	1	490	0	NOLLA
16	8549	5887	3	664	0	ALBNM
17	6807	3482	6	520	0	STLMO
18	7010	2710	1	470	0	NASTN
19	8173	1147	1	952	0	TMPFL
20	8351	527	2	460	0	MIAFL
21	7260	2083	2	438	0	ATLGA
22	9345	6485	1	668	0	TCSAR
23	9135	6748	3	666	0	PHXAR
24	8486	8695	1	722	0	OAKCA
25	8665	7411	1	721	0	LASNE
26	5986	3426	5	358	0	CHIIL
27	7707	4173	1	538	0	TULOK
28	5536	2828	1	340	0	DETMI
29	4422	1249	3	128	0	BOSMA
30	5574	2543	1	320	0	CLEOH
31	6261	4021	2	635	0	CDRIA
32	6272	2992	1	336	0	INDIN
33	6529	2772	1	462	0	LOUKY
34	6657	1698	1	422	0	CHTNC
35	6113	2705	1	328	0	DATOH
36	5251	1458	5	228	0	PHLPA
37	5622	1583	3	236	0	WASDC
38	5510	1575	1	238	0	BALMA
39	5621	2185	6	234	0	PITPA
40	5363	1733	1	226	0	HARPA
41	5166	1585	1	228	0	ALNPA

Total Nodal Traffic = 95 erlangs

NLT#= 0 TARIFF#= 1
NLT#= 1 TARIFF#= 1
NLT#= 2 TARIFF#= 3
NLT#= 3 TARIFF#= 4

TRAFFIC GROWTH FACTOR (TGF) = 1

Using the VHD file of a 41-node network, one should first derive the nodal distribution of 41 locations over the Continental U.S.A. (CONUS), as shown in Figure 7.1. It is up to the designer to choose whether the nodal numbers are to be printed or not. It shows

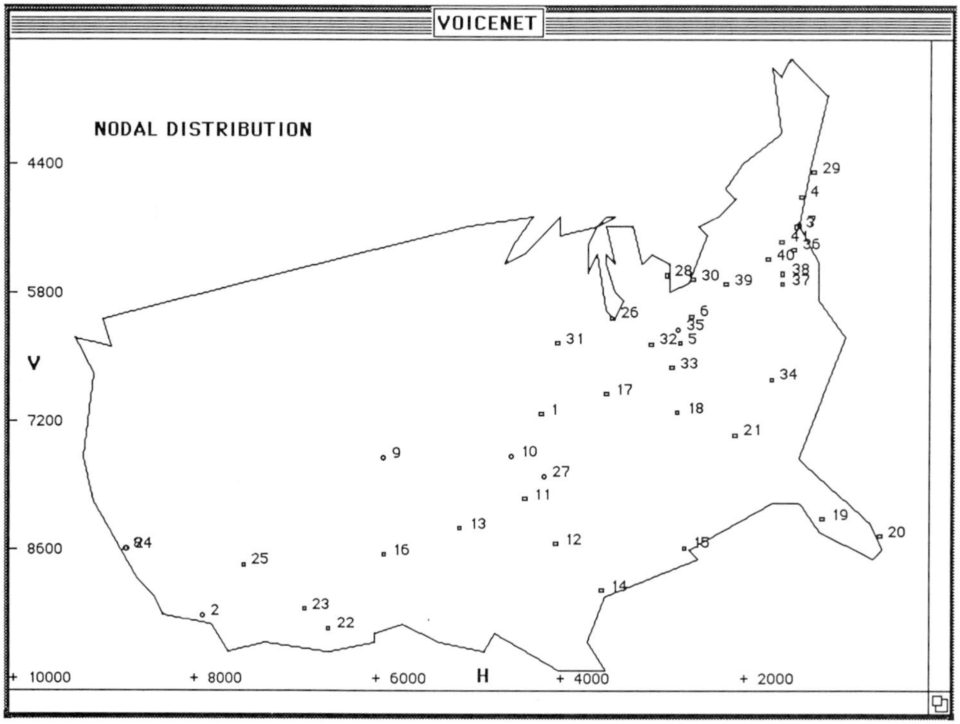

Figure 7.1. Nodal distribution for the VHD41 input file.

a concentration of nodes in the northeast corner of the country. That region can be expanded by merely providing the desired V- and H-coordinate limits for the map boundary. For this study we can ignore map expansions and the use of LATA or LINK files. First, we shall study the seven switch locations, as based on the COGs (see Chapter 5). The output recommends that the locations (17), (16, 39), (25, 1, 18, and 41) as possible switch sites. See Figure 7.2 for illustrations of such COGs. The various groupings imply the particular nodal decomposition employed. The node 17 is recommended for a single switch. The nodes 16 and 39 are recommended for two switch locations. One should use nodes 16, 17, and 39 for locating three switches. Nodes 25, 1, 18, and 41 should be selected as sites for four switches. The recommended sites for five switches are at nodes 25, 1, 17, 18, and 41.

The data base of Table 7.1 also shows that a total of 95 erlangs are handled by all the sites. Using the nomenclature of Chapter 3, A = 95 can be defined for the entire network. Half of these erlangs are of the originating type. This value also determines the number of completed calls in the system during a busy hour. If the expected call duration (ECD) of a voice call is expressed in seconds, then the number of busy hour call completions can be defined as follows:

$$N_{bhc} = (A/2) * (3600/ECD) \tag{7.2}$$

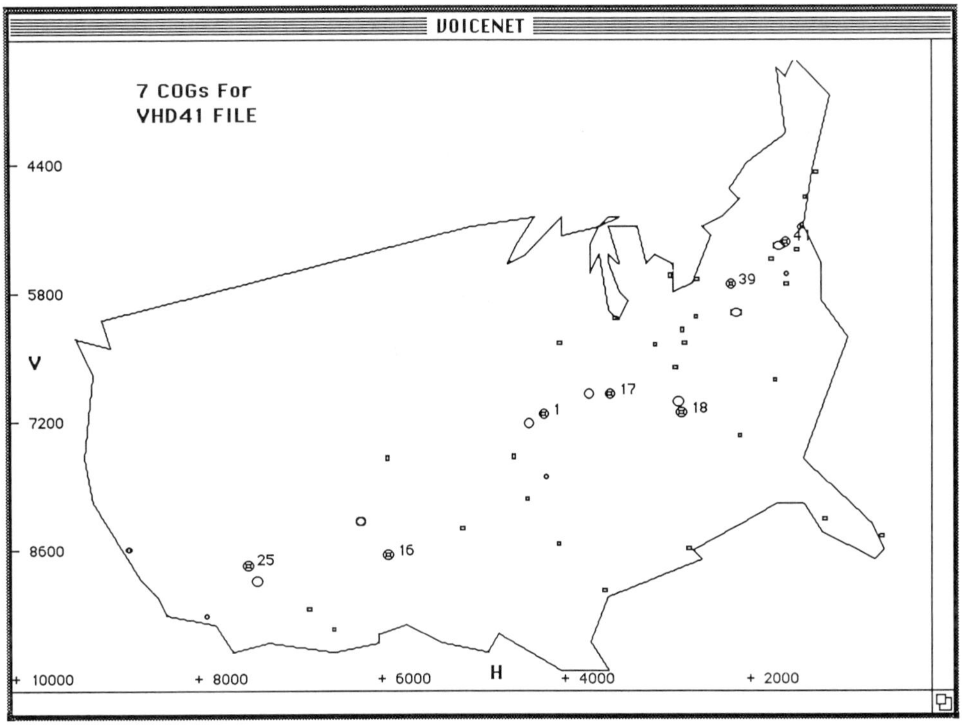

Figure 7.2. Graph showing seven COGs based on three decompositions for VHD41.

Since each call is handled by exactly two switches for the case of a fully connected backbone mesh network, the total erlangs in Table 7.1 does represent the total call intensity handled by the network switches.

The cost of transmission is summarized for both access lines (AL$) and trunks (TK$) on the network plot. The output graph also shows the average cost of a voice call-minute. Tables 7.2, 7.3, and 7.4 represent the printouts of AL costs for one, two, and three switches respectively. Figures 7.3, 7.4, and 7.5 represent the synthesized network topologies for one, two, and three switches respectively. Although the computations were also performed for four and five switches, the results are not included here, for the sake of brevity. The peak hour erlangs handled by each site, as defined by the fourth column in Table 7.1, were used to compute the number (N) of ALs. A grade-of-service, as defined by the blocking probability (B) equal to 0.01 for both access lines and trunks according to the Erlang B formula, was employed. The values of N can be compared with those obtained from Figures 3.1 through 3.6 that plot *N* required circuits versus A erlangs and several useful values of blocking probabilities, B. SDF parameters numbers 22 and 32, define the blocking probabilities for ALs and trunks, respectively. These parameters can be modified according to the simple technique defined in Chapter 6.

Table 7.5 represents the trunking analysis for trunk bundle dimensioning and transmission costs for two switches. Each one of the first three matrices are of Ns × Ns size,

Table 7.2 Cog and Access Line (AL) Connections Printout

A. COG PRINTOUT

NO. OF DECOMPOSITIONS= 3
INITIAL DECOMPOSITION FACTOR= 1 (O FOR VERT, 1 FOR HORIZ)
COGs WITH THEIR V-H COORDs & CLOSEST NODES WITH THEIR V-H COORDs

6814	3711	17	6807	3482
8190	6197	16	8549	5887
5936	2125	39	5621	2185
8834	7319	25	8665	7411
7131	4353	1	7027	4203
6896	2746	18	7010	2710
5210	1655	41	5166	1585

B. ACCESS LINE CONNECTION ANALYSIS

NO. OF SWITCHES= 1 AT 17

AL BLOCKING FACTOR= .01 ECONOMICAL CALL COST= 13.31

AL TRUNK LINK TYPE= 0 AL MULTIPXG. FACTOR= 1

TRAFFIC GROWTH FACTOR= 1 EXP.CALL DURATION-SEC= 300

#	SW	ALs	MIs	$/AL	$/ALB
1	17	14	238	520	7290
2	17	14	1584	1035	14494
3	17	5	861	803	4019
4	17	5	945	830	4154
5	17	5	306	564	2822
6	17	5	394	620	3103
7	17	17	870	806	13717
8	17	11	1739	1084	11934
9	17	5	795	782	3913
10	17	7	392	619	4336
11	17	5	457	661	3305
12	17	5	543	702	3511
13	17	5	683	746	3734
14	17	5	674	743	3719
15	17	5	593	718	3590
16	17	8	939	828	6629
18	17	5	252	529	2648
19	17	5	855	801	4009
20	17	7	1054	865	6059
21	17	7	465	665	4660
22	17	5	1243	926	4630
23	17	8	1268	934	7472
24	17	5	1731	1082	5412
25	17	5	1374	968	4840
26	17	11	260	534	5882
27	17	5	358	597	2989
28	17	5	452	657	3287
29	17	8	1033	858	6870
30	17	5	490	681	3409
31	17	7	242	523	3664
32	17	5	229	515	2575
33	17	5	241	522	2612
34	17	5	566	709	3546
35	17	5	329	579	2895
36	17	11	807	786	8652
37	17	8	707	754	6037
38	17	5	729	761	3808
39	17	13	555	706	9178
40	17	5	717	757	3788
41	17	5	793	782	3910

SW. LOC	#VTERMs	SW. COST
1	276	207102

TOTAL ACCESS LINE COST 207102

Table 7.3 Access Line (AL) Connections Printout

NO. OF SWITCHES= 2 AT 16 39

AL BLOCKING FACTOR= .01 ECONOMICAL CALL COST= 13.31

AL TRUNK LINK TYPE= 0 AL MULTIPXG. FACTOR= 1

TRAFFIC GROWTH FACTOR= 1 EXP.CALL DURATION-SEC= 300

#	SW	ALs	MIs	$/AL	$/ALB
1	16	14	717	757	10610
2	16	14	663	740	10368
3	39	5	306	564	2820
4	39	5	391	618	3093
5	39	5	256	532	2660
6	39	5	161	471	2357
7	39	17	315	570	9693
8	16	11	895	814	8963
9	16	5	331	580	2901
10	16	7	547	703	4922
11	16	5	515	693	3465
12	16	5	587	716	3580
13	16	5	271	542	2710
14	16	5	753	769	3846
15	39	5	916	821	4107
17	39	13	555	706	9178
18	39	5	469	668	3343
19	39	5	871	807	4035
20	39	7	1010	851	5960
21	39	7	519	694	4860
22	16	5	314	569	2848
23	16	8	329	578	4631
24	16	5	888	812	4062
25	16	5	483	677	3387
26	39	11	409	630	6930
27	16	5	603	721	3607
28	39	5	205	499	2497
29	39	8	481	676	5408
30	39	5	114	441	2206
31	39	7	614	724	5074
32	39	5	327	578	2890
33	39	5	341	587	2935
34	39	5	362	599	2999
35	39	5	226	513	2565
36	39	11	257	533	5866
37	39	8	190	490	3920
38	39	5	196	493	2468
40	39	5	164	473	2367
41	39	5	238	520	2603

SW LOC	#VTERMs	SW COST
1	99	69900
2	169	102834

TOTAL ACCESS LINE COST = $172734

where Ns is the number of switches. The first such matrix represents the erlangs flowing between any two switches. For example, the first number at the top left of the matrix implies the erlangs flowing from one AL bundle to another AL bundle, which terminate on the first Switch at 16 in Table 7.5. The number of its right implies the erlangs flowing from Switch 16 to Switch 39. The number below the top left number represents the erlangs flowing from Switch 39 to Switch 16. The remaining number represents the intranodal erlangs at Switch 39. The second matrix represents a modified erlang table

Table 7.4. Access Line Connection Analysis

NO. OF SWITCHES= 3 AT 16 17 39
AL BLOCKING FACTOR= .01 ECONOMICAL CALL COST= 13.31
AL TRUNK LINK TYPE= 0 AL MULTIPXG. FACTOR= 1
TRAFFIC GROWTH FACTOR= 1 EXP.CALL DURATION-SEC= 300

#	SW	ALs	MIs	$/AL	$/ALB
1	17	14	238	520	7290
2	16	14	663	740	10368
3	39	5	306	564	2820
4	39	5	391	618	3093
5	39	5	256	532	2660
6	39	5	161	471	2357
7	39	17	315	570	9693
8	16	11	895	814	8963
9	16	5	331	580	2901
10	17	7	392	619	4336
11	17	5	457	661	3305
12	17	5	543	702	3511
13	16	5	271	542	2710
14	17	5	674	743	3719
15	17	5	593	718	3590
18	17	5	252	529	2648
19	17	5	855	801	4009
20	39	7	1010	851	5960
21	17	7	465	665	4660
22	16	5	314	569	2848
23	16	8	329	578	4631
24	16	5	888	812	4062
25	16	5	483	677	3387
26	17	11	260	534	5882
27	17	5	358	597	2989
28	39	5	205	499	2497
29	39	8	481	676	5408
30	39	5	114	441	2206
31	17	7	242	523	3664
32	17	5	229	515	2575
33	17	5	241	522	2612
34	39	5	362	599	2999
35	39	5	226	513	2565
36	39	11	257	533	5866
37	39	8	190	490	3920
38	39	5	196	493	2468
40	39	5	164	473	2367
41	39	5	238	520	2603

SW LOC	#VTERMs	SW COST
1	58	39870
2	91	54790
3	106	59482

TOTAL ACCESS LINE COST 154142

that is used to compute the number of trunks between the switches. The third matrix represents the actual number of computed trunks between the switches.

When computing the number of trunks for a given bundle, we assume that the two-way type, four-wire trunks were used to handle the sum of the erlangs flowing in two directions. If only one-way type trunks are used to avoid the phenomenon of *glare*, the trunks must be computed for each direction separately. This will result in a higher cost

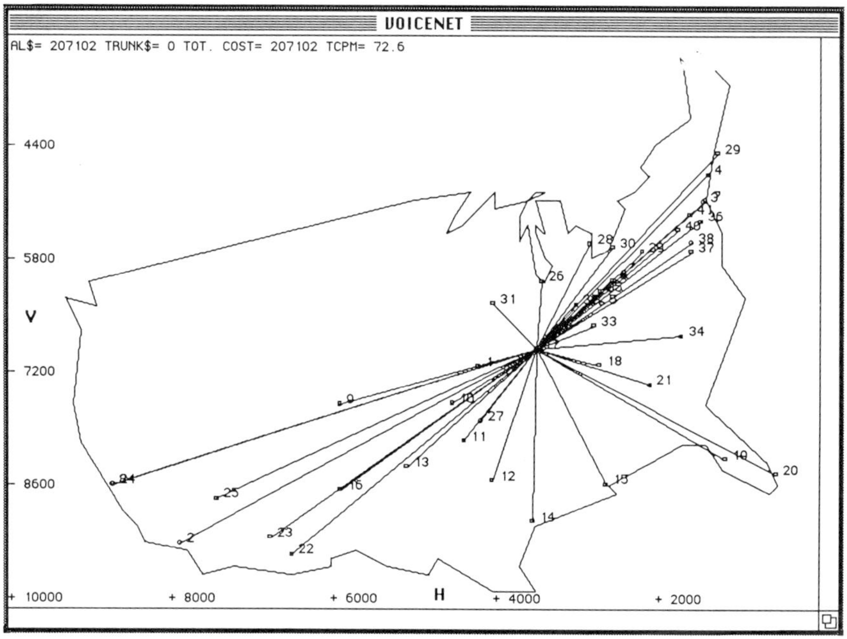

Figure 7.3. A voice network topology with one switch and all on-net locations: VHD41, TGF = 1, and BBTF = 2.

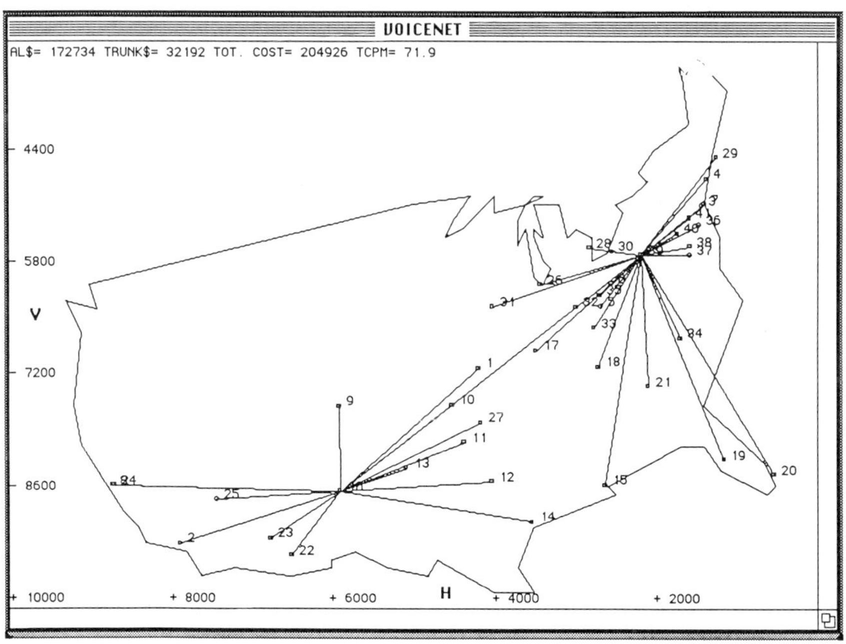

Figure 7.4. A voice network topology with two switches and all on-net locations: VHD41, TGF = 1, and BBTF = 2.

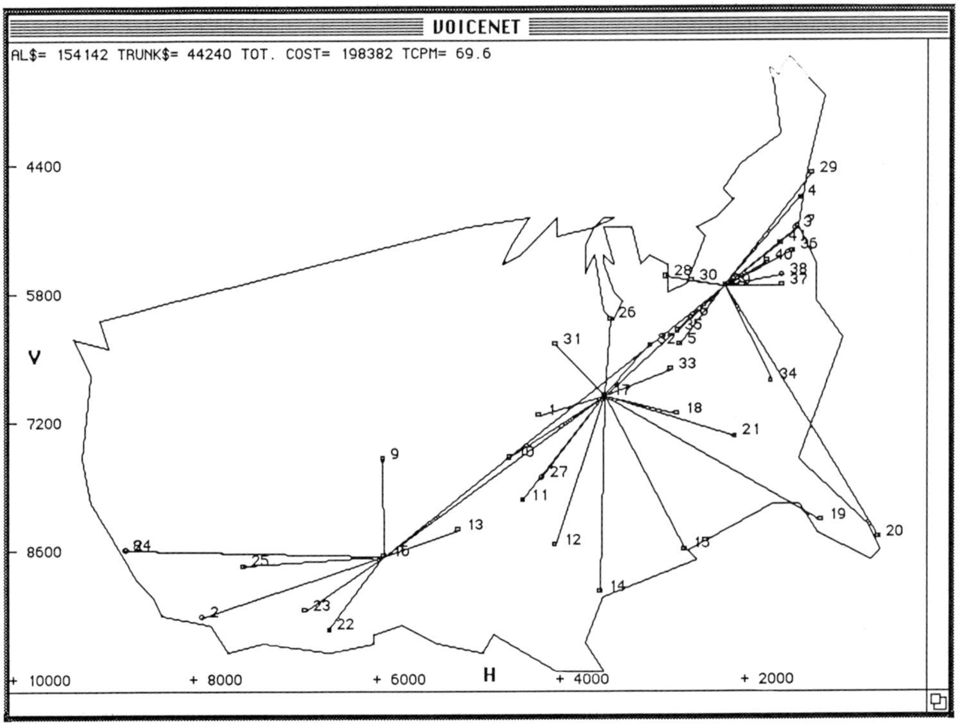

Figure 7.5. A voice network topology with three switches and all on-net locations: VHD41, TGF = 1, and BBTF = 2.

of trunks. The rationale behind this conclusion is described by the curves of Figure 3.7 in Chapter 3. The fourth matrix actually lists each trunk bundle, the associated nodes, the trunk mileage, per trunk cost, and the total cost of each trunk bundle. These tables should familiarize the way computations are tabulated for any number of switches.

Figures 7.6 and 7.7 plot the cost per call-minute and the number of off-net locations, respectively, as a function of traffic growth factor (TGF), and for two design constraints: only VG lines chosen on the basis of either all on-net ALs (the worst case) or an economical AL design with 8 VG circuits multiplexed per 56 KBPS DDS AL/trunk (the best case), respectively. The results show that the need for a hybrid network is eliminated only for large values of TGF (>10) while using the design option of 8 VG circuits multiplexed per 56 Kbps AL/TK. More than half of the locations will require a VPN/SDN type of service, even for TGF = 20, when only VG circuits are employed on an economical AL basis. Figures 7.8 through 7.10 represent the total transmission costs (AL$ + TK$) of the best networks plotted as a function of two values of BBTF, and for three values of TGF, respectively. The results show that, when BBTF = 1, it generally makes a single switch network to be the most optimum case. See Figure 7.11 for a plot of the network cost distribution when we add the monthly costs of NMC and hardware. For brevity, we plot the results for only one set of values: TGF = 10 and BBTF = 2.

Table 7.5 Trunking Analysis

TRUNK LINK/NODE TYPE 0

TRUNK MULTIPLEXING FACTOR 1

BACKBONE TRAFFIC FACTOR= 2

TRAFFIC GROWTH FACTOR= 1

ECONOMICAL CALL COST= 13.31

TRUNK BLOCKING = .01

TOTAL INTER/INTRA NODAL TRAFFIC= 95

NO. OF SWITCHES= 2 AT 16 39

ORIGINAL TRUNK TRAFFIC MATRIX AS FOLLOW:

24	11
11	47

MODIFIED TRAFFIC MATRIX IS AS FOLLOWS:

0	22
0	0

MODIFIED TRUNK MATRIX IS AS FOLLOWS:

0	32
0	0

TRUNK COST MATRIX IS AS FOLLOWS:

FROM	TO	MILES	COST	TOTALS
16	39	1493	1006	32192

*******SUMMARY OF SYSTEM COSTS*******

TOTAL ACCESS LINE COST($)= 172734

TOTAL TRUNK COST($)= 32192

TOTAL SYSTEM COST($)= 204926

TOTAL COST PER 5-MIN. CALL($) = 3.60

The model employed for computing the cost of hardware requires a knowledge of AL and trunk terminations, as shown in Tables 7.2 through 7.5. The total hardware cost can then be defined as follows:

$$C_{th} = (N_{alt} + N_{tt}) * 800 \ldots \$ \tag{7.3}$$

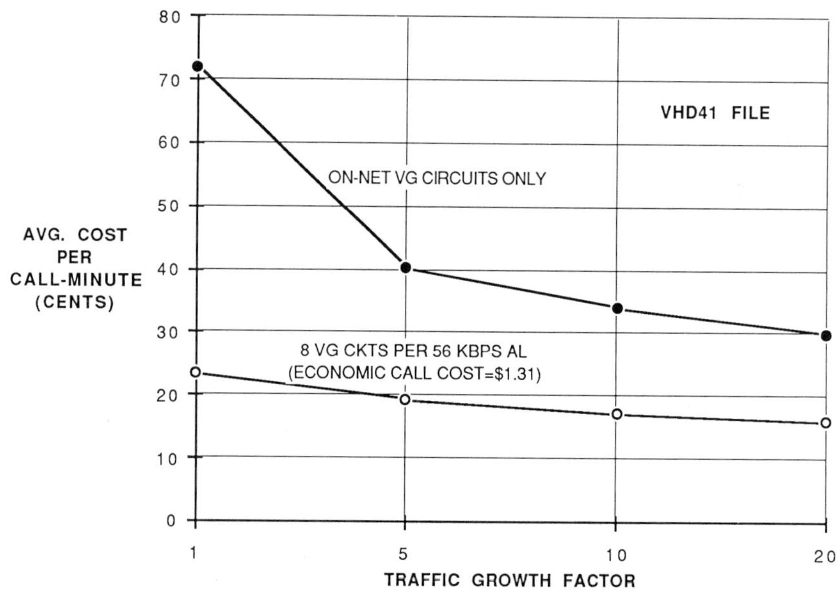

Figure 7.6. Average cost per call minute (ACPM) versus traffic growth factor: VHD41 and one switch.

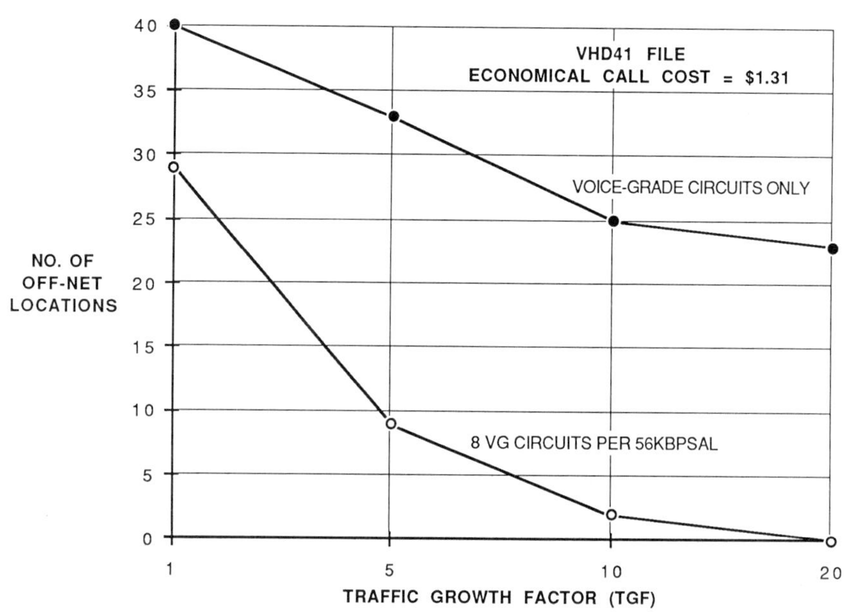

Figure 7.7. Number of off-net locations versus: TGF: VHD41, ECC = $1.31, and one switch.

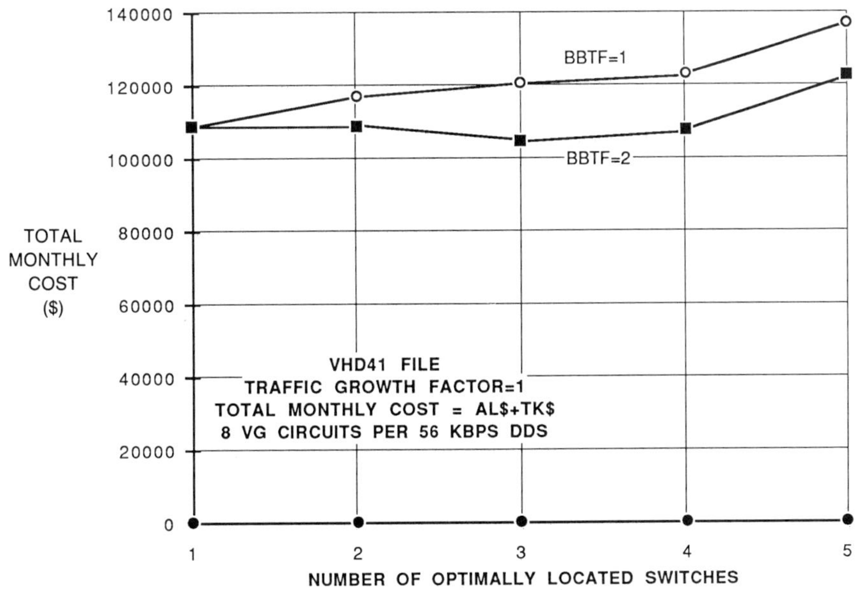

Figure 7.8. Total monthly cost for network transmission versus the number of optimally located switches: VHD41, TGF =1, BBTF = 1 or 2, and eight VG circuits per economically designed AL/TK.

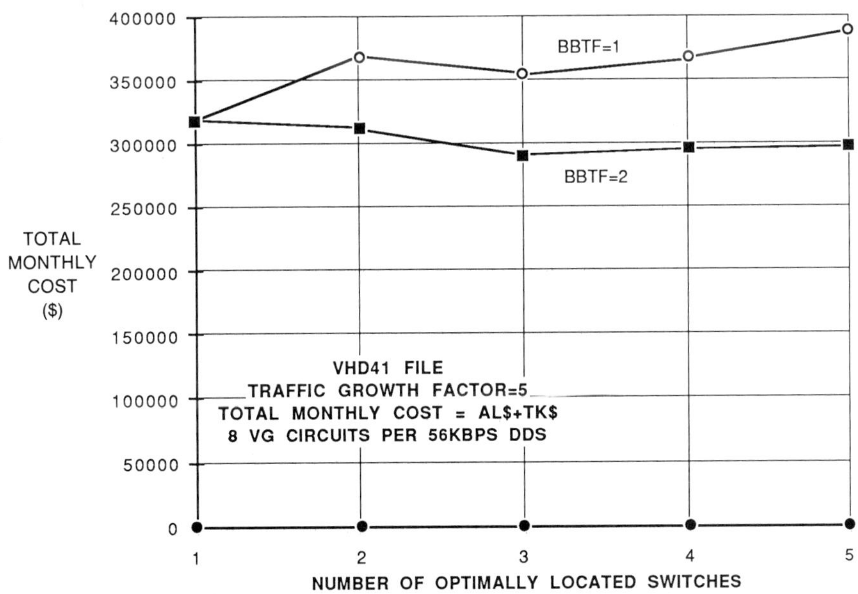

Figure 7.9. Total monthly cost for network transmission versus the number of optimally located switches: VHD41, TGF = 5, and BBTF = 1 or 2, and eight VG circuits per economically designed AL.

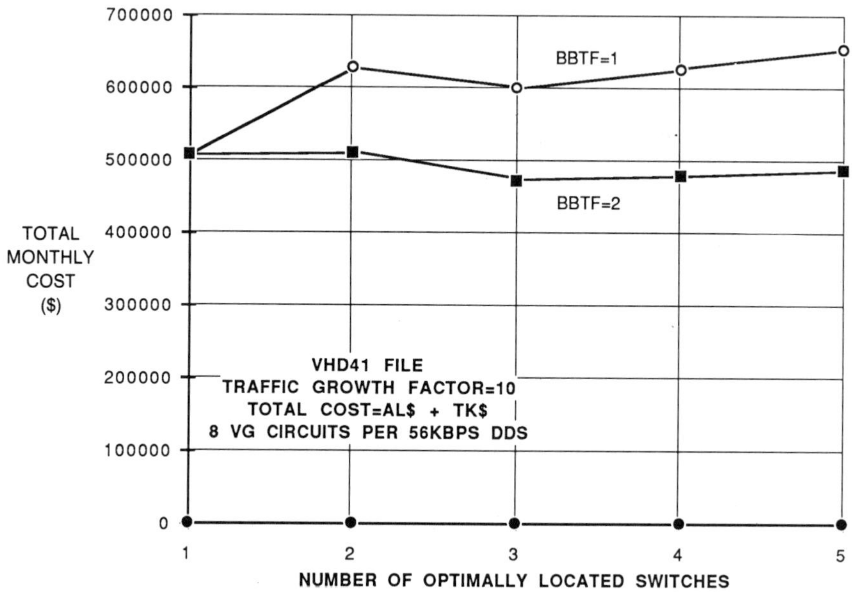

Figure 7.10. Total monthly cost for network transmission versus the number of optimally located switches: VHD41, TGF = 10, BBTF = 1 or 2, and eight VG circuits per economically designed AL.

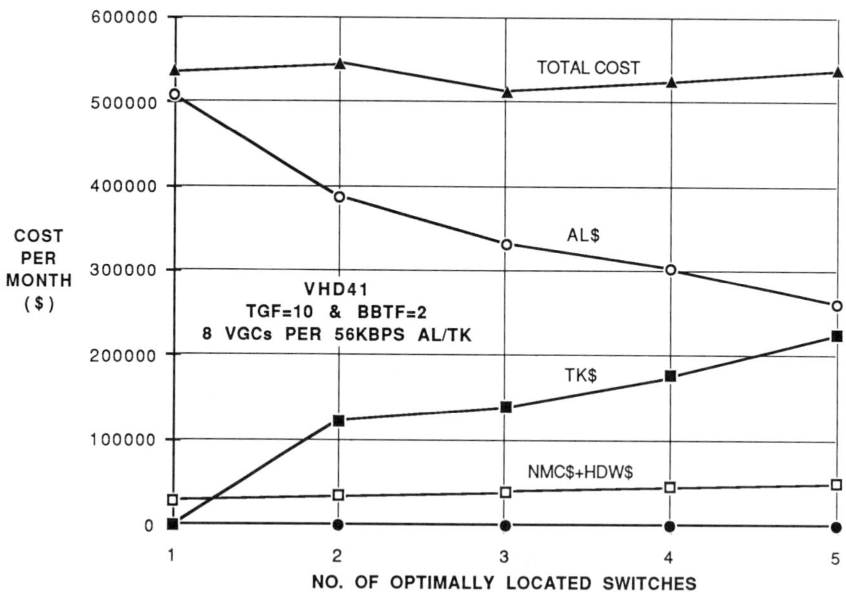

Figure 7.11. Distribution of total monthly costs versus the number of optimally located switches: VHD41, TGF = 10, BBTF = 2, and eight VG circuits per economically designed AL or TK.

The cost of hardware on a monthly basis can be computed by using the methodology of Subsection 2.4.4, and assuming a seven-year lifecycle and 10 percent interest rate. The resulting value of hardware cost per month as employed in our analysis is as follows:

$$C_{hm} = C_{th}/60.237 + 3000 * N_s \ldots \$/MO \tag{7.4}$$

The monthly cost of network management and control can be determined by considering the cost of network maintenance, assuming that the switches have a dual redundancy. We will use a sample formula based on practice, and it includes the labor and overhead associated with personnel needed to manage a private network.

$$C_{nmc} = 7500 + 4500 * N_s \ldots \$/MO \tag{7.5}$$

7.2.2 SYNTHESIS OF A 59-NODE VOICE NETWORK

See Table 7.6 for a definition of the VHD59 data base. This data base also represents an actual user. Such a network handles 140 erlangs (In and Out) during the peak period, when compared to the 95 erlangs handled by the 41-node network. Therefore, the 59-node network should cost more than the 41-node network. Figure 7.12 illustrates the nodal distribution for the VHD59 file. A startling fact is apparent: the nodes are clustered around Atlanta (node number 1), which is also the firm's headquarters. Figure 7.13 illustrates the seven recommended COGs. Figures 7.14 through 7.16 represent some useful network topologies. The solution for a single switch at node 1 is cheaper by $5000 when compared to a single switch at location 37, as recommended by the COG analysis. This can be explained by the fact that node number 1 handles a large number of erlangs. If node number 1 is used as a switch, the associated AL bundle is eliminated. Consequently, the designer must always temper the COG analysis with such an observation. See Tables 7.7 and 7.8 for samples of AL cost analysis for 8 VG circuits per 56 Kbps AL/TK. A "0" for number of ALs signifies the use of a VPN/SDN type service. See Figure 7.17 for the plots of call-minute cost versus TGF for two useful set of design parameters. Due to the clustering of nodes around the major switch location number 1, the call-minute costs of a 59-node network are generally lower than that for a 41-node network. See Figure 7.18 for the plot of off-net locations versus TGF and two useful designs.

See Figures 7.19 through 7.21 for the plots of transmission costs as a function of the number of switches and two values of BBTF. See Figure 7.22 for the distribution of all network-related costs versus the number of switches and two values of BBTF. Due to the extreme clustering of nodes around the major switch location of 1, the network rarely demands more than two switches. A single switch is generally more effective even for BBTF = 2.

7.2.3 SYNTHESIS OF A 100-NODE VOICE NETWORK

See Table 7.9 for a definition of a randomly produced data base for 100 nodes. Such a network handles 270 erlangs (In and Out) during a busy hour for TGF = 1. This value is approximately twice that for the 59-node network for a similar TGF. Figure 7.23

Table 7.6 Network Design Data Base for VHD59 Input File

#	-V-	-H-	LOAD (Erlangs)	LATA	LINK	NAME
1	7260	2083	12	438	0	ATLGA
2	7089	1674	2	442	0	AGSGA
3	6749	2001	1	420	0	AVLNC
4	5510	1575	1	238	0	BALMD
5	7518	2446	2	476	0	BHMAL
6	8806	3298	1	562	0	BPTTX
7	8476	2874	1	492	0	BTRLA
8	6901	1589	2	434	0	CAESC
9	7098	2366	1	472	0	CHATN
10	5986	3426	6	358	0	CHIIL
11	7021	1281	1	436	0	CHSWV
12	6261	4021	4	635	0	CIDIA
13	6657	1698	2	422	0	CLTNC
14	5972	2555	1	324	0	CMHOH
15	7556	2045	2	438	0	CSGGA
16	6263	2679	4	922	0	CVGOH
17	8436	4034	6	552	0	DALTX
18	6113	2705	1	328	0	DAYOH
19	5622	1583	3	236	0	WASDC
20	7501	5899	3	656	0	DENCO
21	5536	2828	2	340	0	DETMI
22	8409	3168	1	486	0	ESFLA
23	6729	3019	1	330	0	EVVIN
24	5942	2982	1	334	0	FWAIN
25	8479	4122	6	552	0	GSWTX
26	7827	3554	1	528	0	HOTAR
27	8938	3536	3	560	0	HOUTX
28	6272	2992	2	336	0	INDIN
29	8035	2880	2	482	0	JANMS
30	7649	1276	1	452	0	JAXFL
31	8665	7411	1	721	0	LASNV
32	9213	7878	2	730	0	LAXCA
33	6459	2562	1	466	0	LEXKY
34	7721	3451	1	528	0	LITAR
35	7364	1865	1	446	0	MCNGA
36	7899	2639	1	482	0	MEIMS
37	7471	3125	5	468	0	MEMTN
38	7692	2247	1	478	0	MGMAL
39	8351	527	11	460	0	MIAFL
40	7027	4203	2	524	0	MKCMO
41	8148	3218	1	486	0	MLOLA
42	5781	4525	2	628	0	MSPMN
43	8483	2638	5	490	0	MSYLA
44	4997	1406	8	132	0	NYCNY
45	7954	1031	1	458	0	ORLFL
46	8165	606	1	460	0	PBIFL
47	5251	1458	1	228	0	PHLPA
48	6982	3088	1	468	0	PUKTN
49	7266	1379	2	440	0	SAVGA
50	6529	2772	1	462	0	SDFKY
51	9468	7629	1	732	0	SDOCA
52	8492	8719	2	722	0	SFOCA
53	7310	3836	1	522	0	SGFMO
54	8272	3495	2	486	0	SHVLA
55	7496	1340	1	440	0	SSIGA
56	6807	3482	3	520	0	STLMO
57	5704	2820	1	326	0	TOLOH
58	8173	1147	2	952	0	TPAFL
59	6801	2251	2	474	0	TYSTN
TOTAL NODAL TRAFFIC LOAD=			140 ERLANGS			

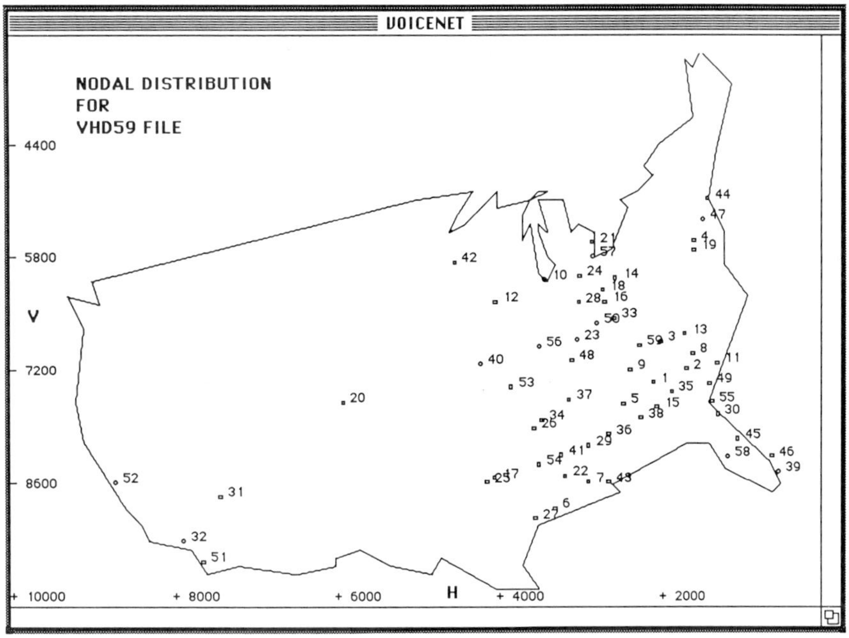

Figure 7.12. Nodal distribution for the VHD59 input file.

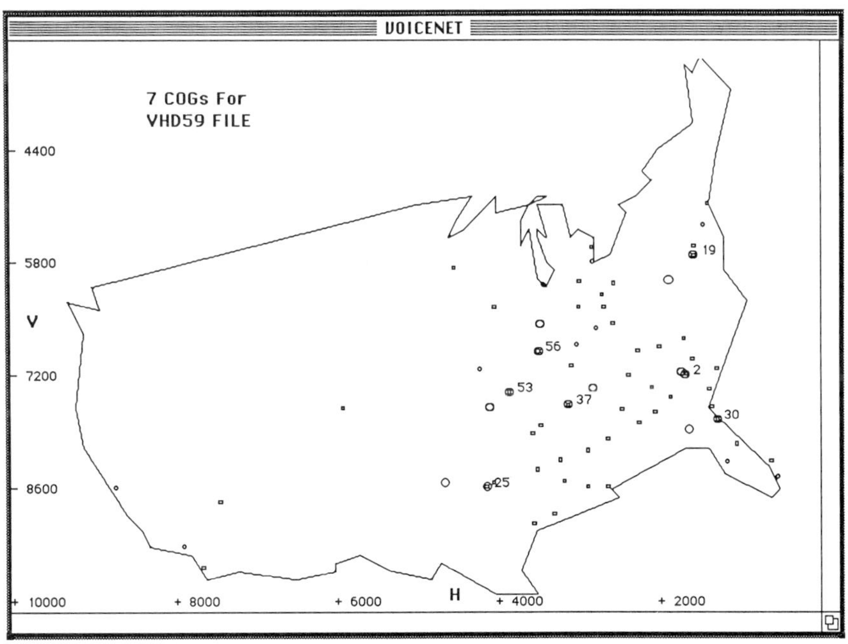

Figure 7.13. Graph showing seven COGs based on three decompositions for VHD59.

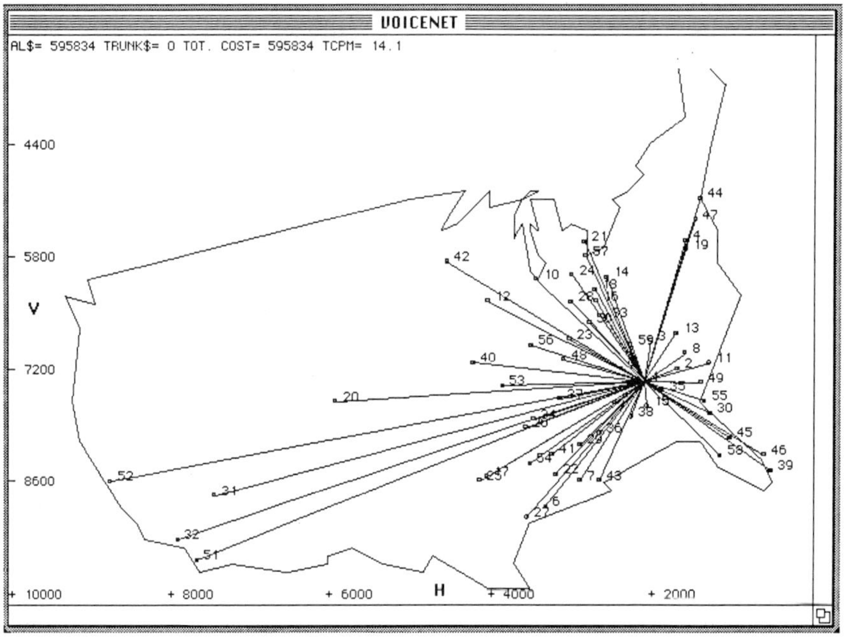

Figure 7.14. A voice network topology with one switch: VHD59, TGF = 10, and eight VG circuits per economically designed 56kbps AL.

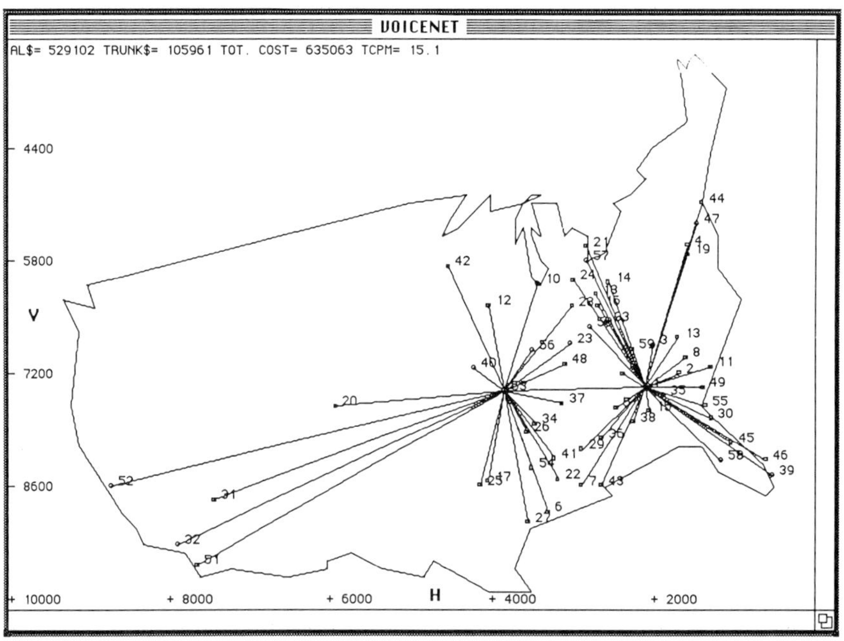

Figure 7.15. A voice network topology with two switches: VHD59, TGF = 10, BBTF = 2, and eight VG circuits per economically designed 56kbps AL or TK.

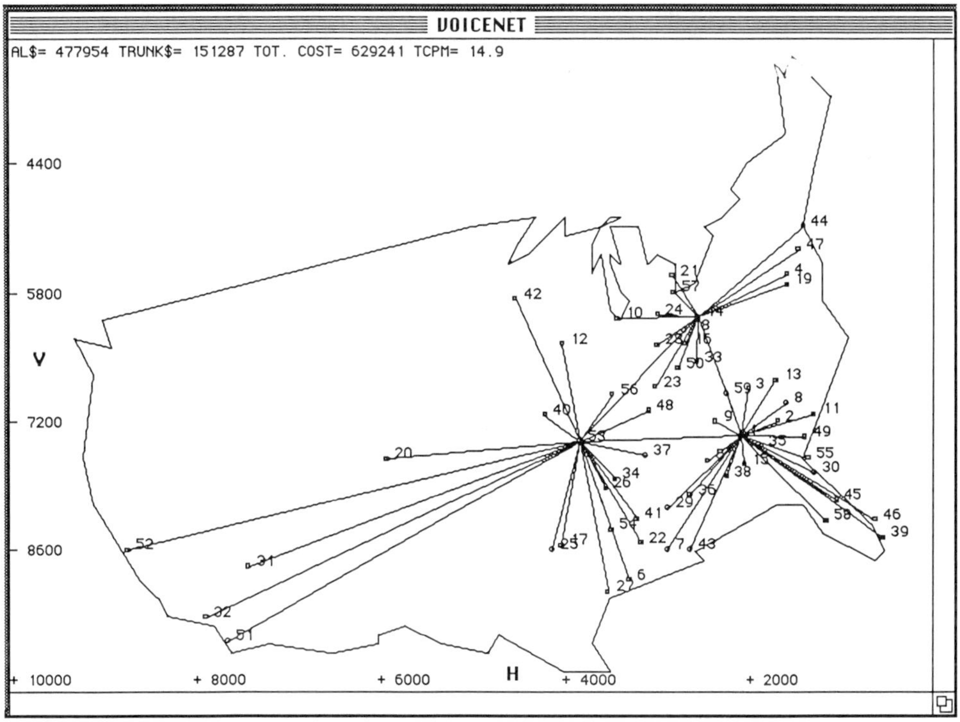

Figure 7.16. A voice network topology with three switches: VHD59, TGF = 10, BBTF = 2, and eight VG circuits per economically designed 56kbps AL or TK.

shows the nodal distribution for such a network. Figure 7.24 illustrates the seven recommended COGs for such a network. Figures 7.25 through 7.27 represent some useful network topologies. Figure 7.28 plots the call-minute cost as a function of the TGF for two sets of design parameters. Figure 7.29 plots the number of off-net locations as a function of the TGF for two designs. Figures 7.30–7.32 plot the network transmission costs as a function of the TGF and two values of BBTF. See Figure 7.33 for the distribution of all network-related costs versus the number of switches for BBTF = 2.

Although a study of the above three networks should be adequate to form some conclusions, a network serving 200 nodes was also synthesized for few judicious design parameters, and the results will be discussed to make a point. Table 7.10 defines a randomly produced DB for a 200-node network. Such a network generates 483 total erlangs (In and Out). Figures 7.34 and 7.35 represent some useful network topologies for a practical design. Figures 7.36 and 7.37 represent total network costs as a function of the number of switches for two values of TGF and BBTF.

7.2.4 SUMMARY OF VOICE NETWORK SYNTHESIS AND CONCLUSIONS

If all of the design variables, such as ALMF, TKMF, BBTF, TGF, the number of switches, AL, and trunk blocking, were simultaneously addressed, the designer would

Table 7.7 Access Line Connection Analysis

NO. OF SWITCHES= 1 AT 1
AL BLOCKING FACTOR= .01 ECONOMICAL CALL COST= 1.31
AL TRUNK LINK TYPE= 1 AL MULTIPXG. FACTOR= 8
TRAFFIC GROWTH FACTOR= 10 EXP.CALL DURATION-SEC= 300

#	SW	ALs	Mls	$/AL	$/ALB
2	1	4	140	1228	4913
3	1	3	163	1299	3897
4	1	3	576	2444	7332
5	1	4	140	1230	4921
6	1	3	621	2520	7562
7	1	3	458	2190	6571
8	1	4	193	1388	5553
9	1	3	103	1116	3349
10	1	10	585	2459	24594
11	1	3	264	1604	4813
12	1	7	689	2635	18447
13	1	4	226	1488	5953
14	1	3	433	2115	6345
15	1	4	94	1080	4320
16	1	7	367	1914	13401
17	1	10	720	2687	26875
18	1	3	412	2051	6153
19	1	6	541	2385	14312
20	1	6	1209	3513	21081
21	1	4	593	2473	9895
22	1	3	499	2314	6943
23	1	3	340	1832	5498
24	1	3	504	2322	6968
25	1	10	751	2739	27397
26	1	3	498	2310	6932
27	1	6	701	2656	15938
28	1	4	424	2087	8349
29	1	4	351	1866	7467
30	1	3	283	1660	4981
31	1	0	1742	0	7860
32	1	0	1933	0	15720
33	1	3	295	1696	5089
34	1	3	456	2183	6551
35	1	3	76	993	2981
36	1	3	267	1614	4842
37	1	9	336	1820	16382
38	1	3	146	1246	3739
39	1	17	600	2485	42258
40	1	4	674	2609	10439
41	1	3	455	2181	6544
42	1	4	902	2995	11983
43	1	9	424	2087	18790
44	1	13	746	2732	35522
45	1	3	398	2008	6026
46	1	3	547	2395	7187
47	1	3	665	2594	7783
48	1	3	329	1800	5402
49	1	4	222	1477	5910

Table 7.7 *(Continued)*

#	SW	ALs	MIs	$/AL	$/ALB
50	1	3	317	1764	5293
51	1	0	1887	0	7860
52	1	0	2134	0	15720
53	1	3	554	2407	7222
54	1	4	549	2398	9594
55	1	3	246	1549	4649
56	1	6	465	2209	13257
57	1	3	544	2390	7170
58	1	4	413	2053	8214
59	1	4	154	1271	5087

SW LOC	#VTERMs	SW COST
1	1853	595834

TOTAL ACCESS LINE COST 595834

Table 7.8 Access Line Connection Analysis

NO. OF SWITCHES= 2 AT 53 1
AL BLOCKING FACTOR= .01 ECONOMICAL CALL COST= 1.31
AL TRUNK LINK TYPE= 1 AL MULTIPXG. FACTOR= 8
TRAFFIC GROWTH FACTOR= 10 EXP.CALL DURATION-SEC= 300

#	SW	ALs	MIs	$/AL	$/ALB
2	1	4	140	1228	4913
3	1	3	163	1299	3897
4	1	3	576	2444	7332
5	1	4	140	1230	4921
6	53	3	502	2319	6959
7	1	3	458	2190	6571
8	1	4	193	1388	5553
9	1	3	103	1116	3349
10	53	10	438	2128	21289
11	1	3	264	1604	4813
12	53	7	336	1822	12756
13	1	4	226	1488	5953
14	1	3	433	2115	6345
15	1	4	94	1080	4320
16	1	7	367	1914	13401
17	53	10	361	1896	18969
18	1	3	412	2051	6153
19	1	6	541	2385	14312
20	53	6	655	2577	15464
21	1	4	593	2473	9895
22	53	3	406	2033	6099
23	53	3	317	1762	5287
24	1	3	504	2322	6968
25	53	10	380	1954	19543
26	53	3	186	1367	4102
27	53	6	523	2354	14128
28	53	4	423	2082	8330

Table 7.8 (*Continued*)

#	SW	ALs	MIs	$/AL	$/ALB
29	1	4	351	1866	7467
30	1	3	283	1660	4981
31	53	0	1208	0	7860
32	53	4	1412	3857	15430
33	1	3	295	1696	5089
34	53	3	178	1342	4028
35	1	3	76	993	2981
36	1	3	267	1614	4842
37	53	9	230	1501	13512
38	1	3	146	1246	3739
39	1	17	600	2485	42258
40	53	4	146	1247	4990
41	53	3	329	1799	5398
42	53	4	530	2366	9465
43	1	9	424	2087	18790
44	1	13	746	2732	35522
45	1	3	398	2008	6026
46	1	3	547	2395	7187
47	1	3	665	2594	7783
48	53	3	258	1585	4755
49	1	4	222	1477	5910
50	1	3	317	1764	5293
51	53	0	1379	0	7860
52	53	0	1588	0	15720
54	53	4	322	1779	7119
55	1	3	246	1549	4649
56	53	6	194	1392	8355
57	1	3	544	2390	7170
58	1	4	413	2053	8214
59	1	4	154	1271	5087

SW LOC	#VTERMs	SW COST
1	810	237418
2	1025	291684

TOTAL ACCESS LINE COST= 529102

end up with a huge collection of plots. Only a small fraction of them should provide a useful insight into voice networking. The plots already obtained for the four VHD files and for some useful values of TKMF, BBTF, and number of tandem switches, should be adequate to provide a useful understanding of voice networking.

The first observation, derived from Figures 7.7, 7.18, and 7.29, implies that a private network must employ VPN- or SDN-like services to be cost effective at low traffic volumes. These figures show that, if economical AL design is employed using an ECC = \$1.31, most locations will require some off-net services to remain cost effective. Similar conclusions can be drawn from Figures 7.6, 7.17, and 7.28, which plot the average cost of a call-minute as a function of TGF for two designs: one based on all on-net voice circuits, and the other based on an optimum mix of on-net and off-net facilities, with 8

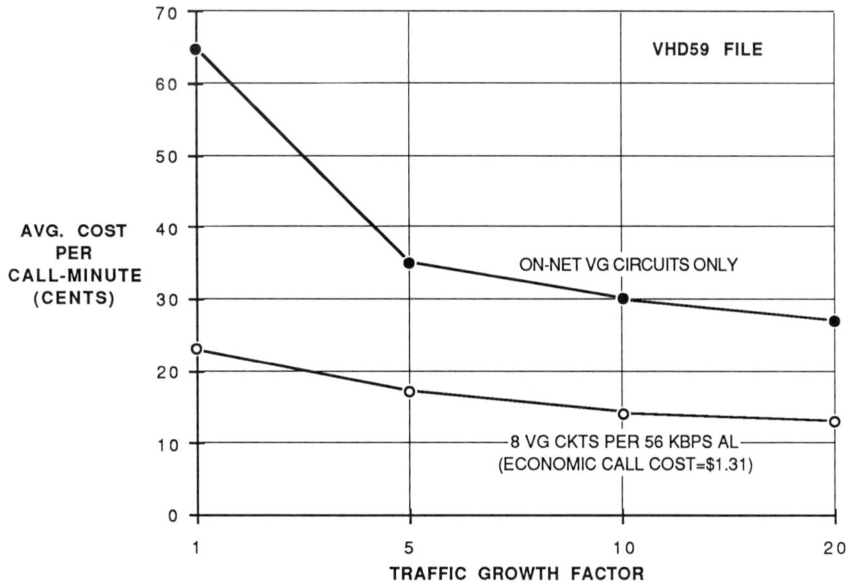

Figure 7.17. Average cost per call minute versus the TGF for an optimized network: VHD59, BBTF = 2, and economical call costs of \$13.31 and \$1.31.

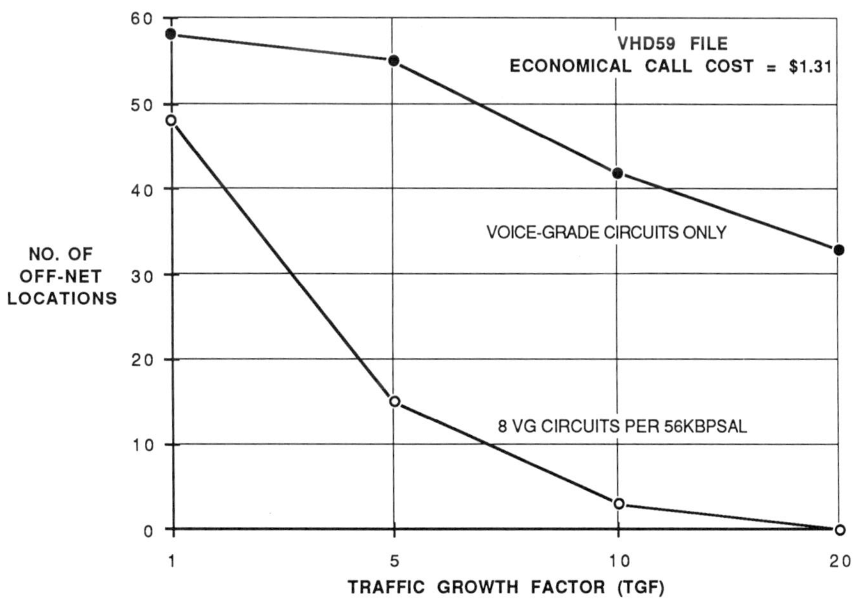

Figure 7.18. Number of off-net locations versus the TGF for an optimized network: VHD59, BBTF = 2, and economical call cost of \$1.31.

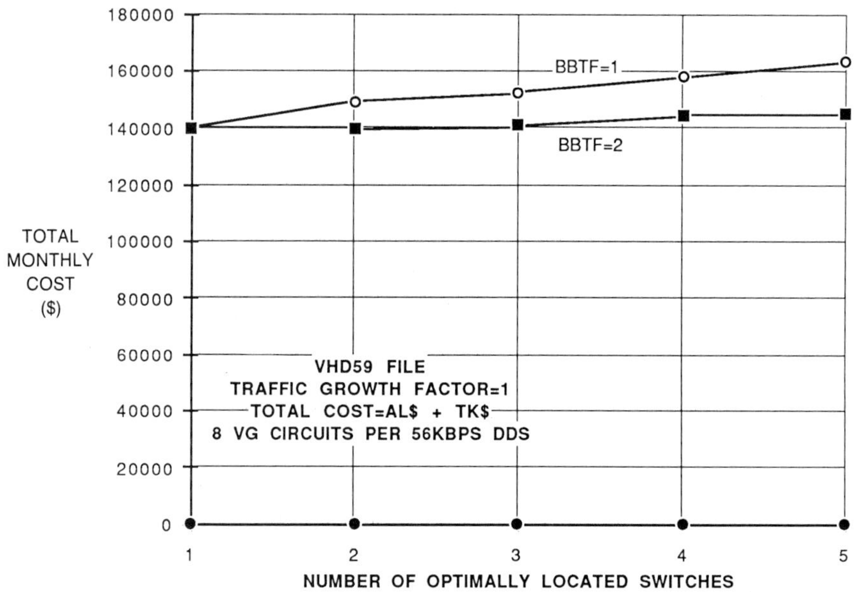

Figure 7.19. Total monthly costs for network transmission versus the number of optimally located switches: VHD59, TGF = 1, BBTF = 1 or 2, and eight VG circuits per economically designed AL/TK.

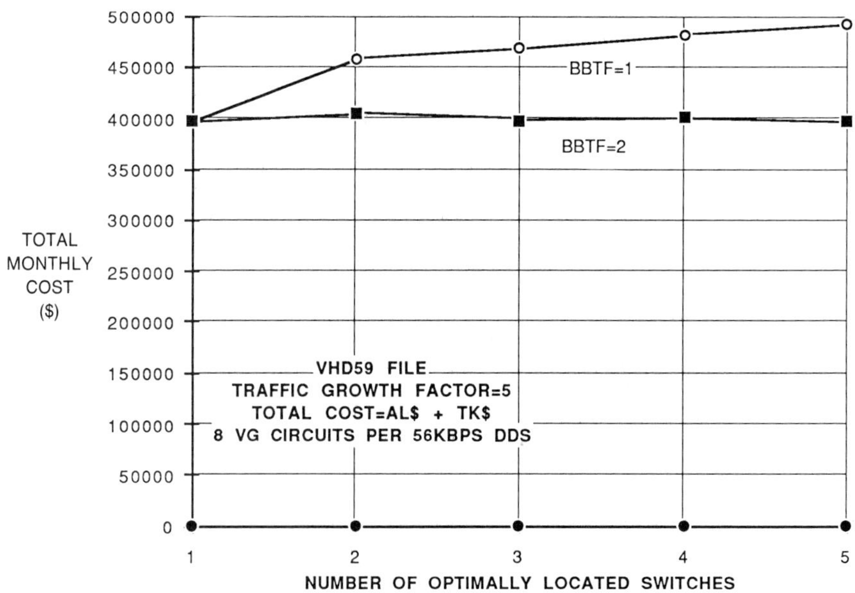

Figure 7.20. Total monthly costs for network transmission versus the number of optimally located switches: VHD59, TGF = 5, BBTF = 1 or 2, and eight VG circuits per economically designed AL/TK.

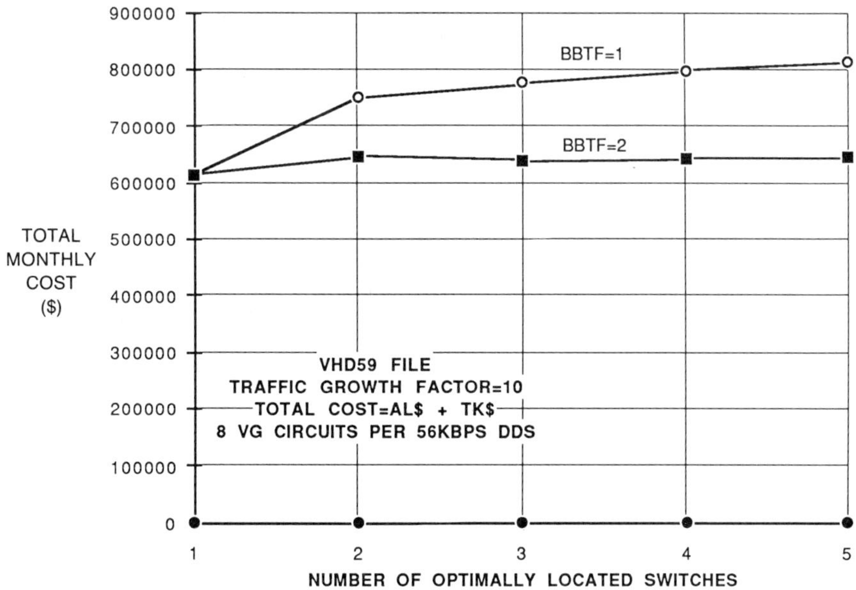

Figure 7.21. Total monthly cost for network transmission versus the number of optimally located switches: VHD59, TGF = 10, BBTF = 1 or 2, and eight VG circuits per economically designed AL or TK.

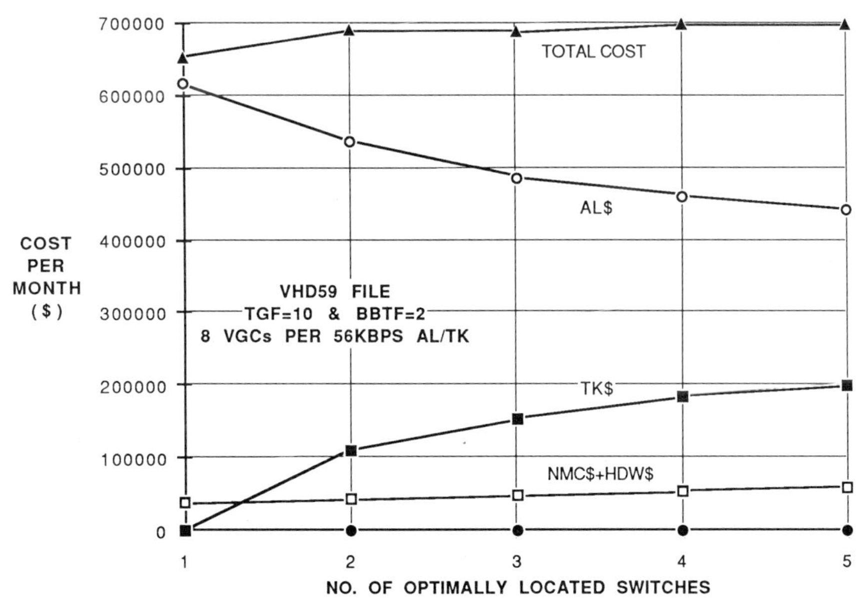

Figure 7.22. Distribution of total monthly costs versus the number of optimally located switches: VHD59, TGF = 10, BBTF = 2, and eight VG circuits per economically designed AL or TK.

Table 7.9 Network Design Data Base for VHD100 Input File

#	-V-	-H-	LOAD (Erlangs)	LATA	LINK
1	6053	1503	1	0	0
2	6807	1551	4	0	0
3	7262	2762	2	0	0
4	6230	7031	1	0	0
5	7731	8494	4	0	0
6	6189	6192	1	0	0
7	7273	3024	2	0	0
8	5026	4496	3	0	0
9	6660	4249	4	0	0
10	7488	2566	2	0	0
11	8509	4905	1	0	0
12	5587	5771	4	0	0
13	7966	6606	4	0	0
14	6525	1653	3	0	0
15	5414	1499	2	0	0
16	7469	4943	4	0	0
17	6498	8705	4	0	0
18	9159	5733	4	0	0
19	5816	2549	2	0	0
20	8940	3945	2	0	0
21	8617	8046	1	0	0
22	7741	1659	2	0	0
23	5554	5528	3	0	0
24	7705	8603	2	0	0
25	8320	2532	4	0	0
26	6418	1915	1	0	0
27	6153	1601	2	0	0
28	6343	6014	3	0	0
29	8173	1147	4	0	0
30	7236	2104	3	0	0
31	6017	6721	3	0	0
32	7345	8087	3	0	0
33	5143	3203	1	0	0
34	5013	2230	4	0	0
35	8618	3012	3	0	0
36	7186	4742	1	0	0
37	7526	6805	2	0	0
38	6801	5433	4	0	0
39	7894	2331	3	0	0
40	8018	6063	2	0	0
41	6167	8590	3	0	0
42	5493	5736	2	0	0
43	7187	6475	1	0	0
44	6161	4713	3	0	0
45	6900	4811	4	0	0
46	5991	2238	3	0	0
47	7987	7595	4	0	0
48	7317	1829	4	0	0
49	5454	2975	3	0	0
50	8337	538	1	0	0
51	5306	3144	4	0	0
52	9420	6266	2	0	0
53	5419	3706	3	0	0
54	7281	8340	1	0	0
55	7421	3060	2	0	0
56	5260	2880	1	0	0
57	7096	2534	4	0	0
58	5462	5410	3	0	0
59	8075	2360	3	0	0
60	7154	8706	2	0	0
61	6085	5037	4	0	0

Table 7.9 (*Continued*)

#	-V-	-H-	LOAD	LATA	LINK
62	8521	4680	1	0	0
63	5710	3050	1	0	0
64	5900	7580	3	0	0
65	7406	5950	3	0	0
66	6385	3240	4	0	0
67	7340	4173	2	0	0
68	7002	8292	3	0	0
69	7396	3910	3	0	0
70	8574	4784	2	0	0
71	9011	4392	3	0	0
72	9297	4734	1	0	0
73	6095	7726	4	0	0
74	8782	4000	1	0	0
75	9393	3742	1	0	0
76	6226	9256	1	0	0
77	5597	2282	3	0	0
78	5837	3537	3	0	0
79	7597	4733	3	0	0
80	5533	5945	4	0	0
81	8278	7304	3	0	0
82	7695	4471	2	0	0
83	5705	1659	4	0	0
84	7450	4789	1	0	0
85	8866	4387	4	0	0
86	7962	4721	1	0	0
87	8869	6056	3	0	0
88	7396	6516	4	0	0
89	8210	5927	3	0	0
90	5131	2540	4	0	0
91	7267	5263	2	0	0
92	9240	7833	4	0	0
93	6383	3739	4	0	0
94	7237	8655	2	0	0
95	5453	3478	2	0	0
96	6257	6593	4	0	0
97	7701	1956	4	0	0
98	5323	2628	3	0	0
99	7718	6720	4	0	0
100	8692	8268	4	0	0
TOTAL NODAL TRAFFIC LOAD=			270 Erlangs		

VG circuits multiplexed per 56 Kbps AL, or a trunk. The case of multiplexing 48 VG channels per T-1 AL or a trunk was also considered, but such a network was generally more costly than the best case. These two designs yield the same cost per call-minute only for very large values of TKMF. For very large values of TGF (>10), all of the networks studied tend to provide full connectivity to all locations via privately leased lines.

Figures 7.8, 7.9, 7.10, 7.19, 7.20, 7.21, 7.30, 7.31, and 7.32 plot total network transmission costs as a function of the number of optimally placed switches (N_{SW}), and two values of BBTF. The results show that a network with uniform community-of-interest (BBTF $= 1$) requires only a single tandem switch for a minimum cost criterion. However, most networks show some cyclical variation of network transmission costs as a function of N_{SW}. The only exception is the 59-node network, which was characterized by nodal clustering around node 1. For that case, a single switch generally yields the

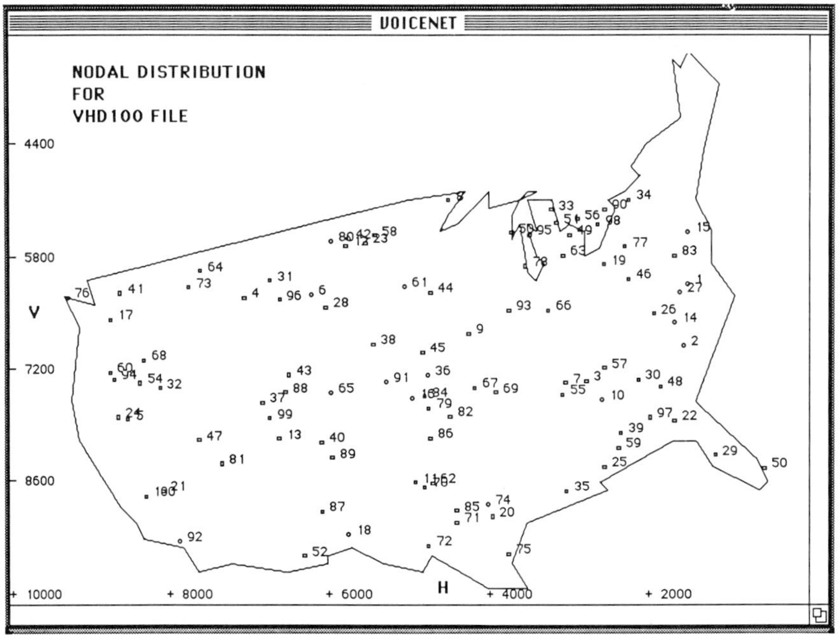

Figure 7.23. Nodal distribution for the VHD100 input file.

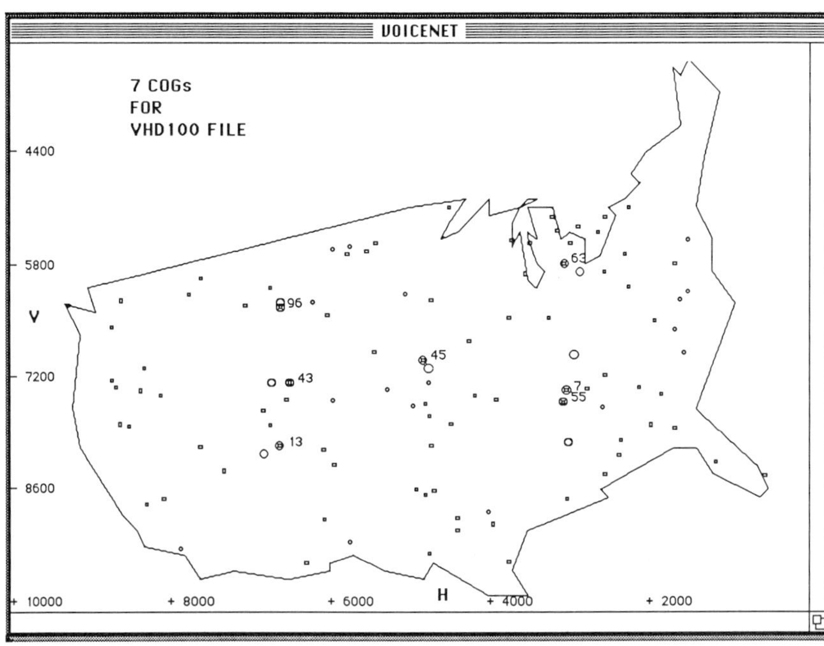

Figure 7.24. Graph showing seven COGs based on three decompositions for VHD100.

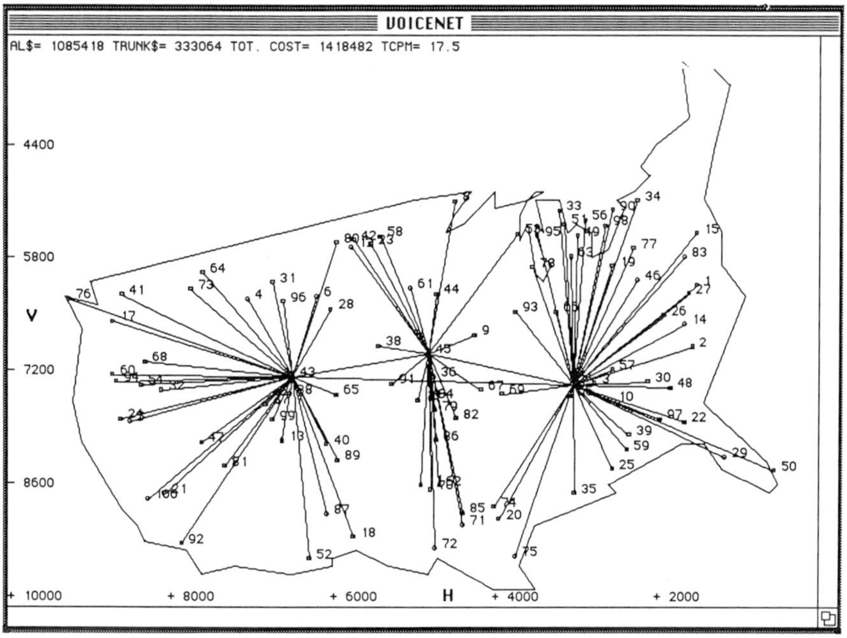

Figure 7.25. A voice network topology with three switches: VHD100, TGF = 10, BBTF = 2, and eight VG circuits per 56kbps AL or TK.

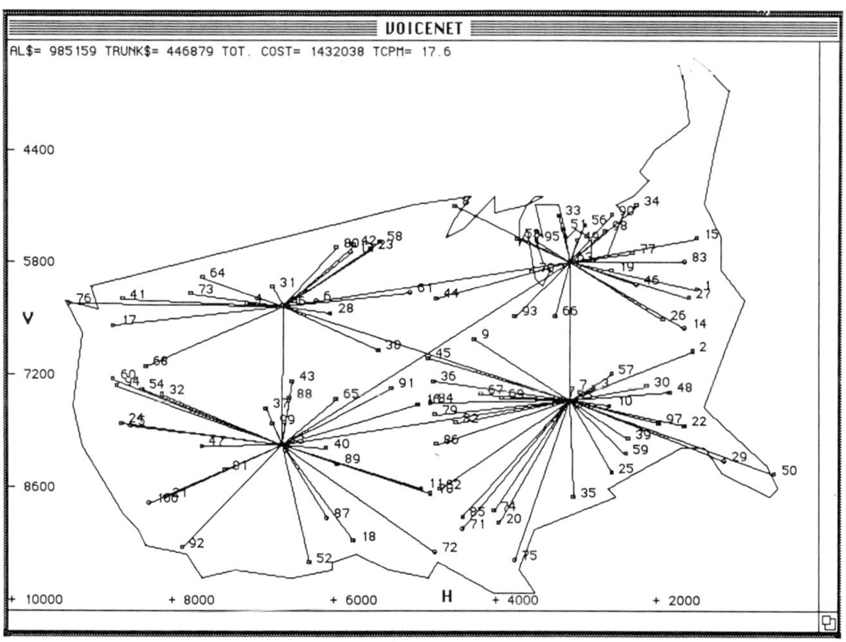

Figure 7.26. A voice network topology with four switches: VHD100, TGF = 10, BBTF = 2, and eight VG circuits per 56kbps AL or TK.

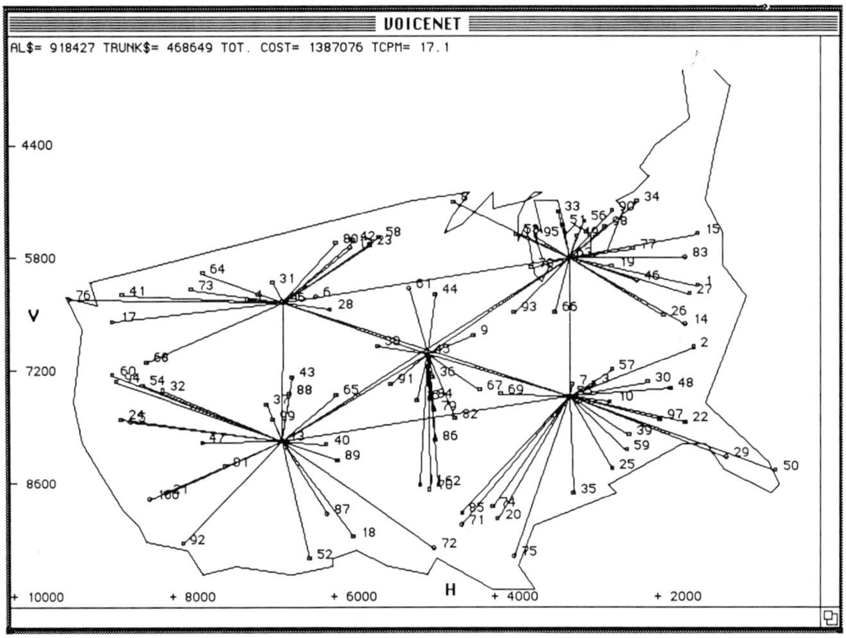

Figure 7.27. A voice network topology with five switches: VHD100, TGF = 10, BBTF = 2, and eight VG circuits per 56kbps AL or TK.

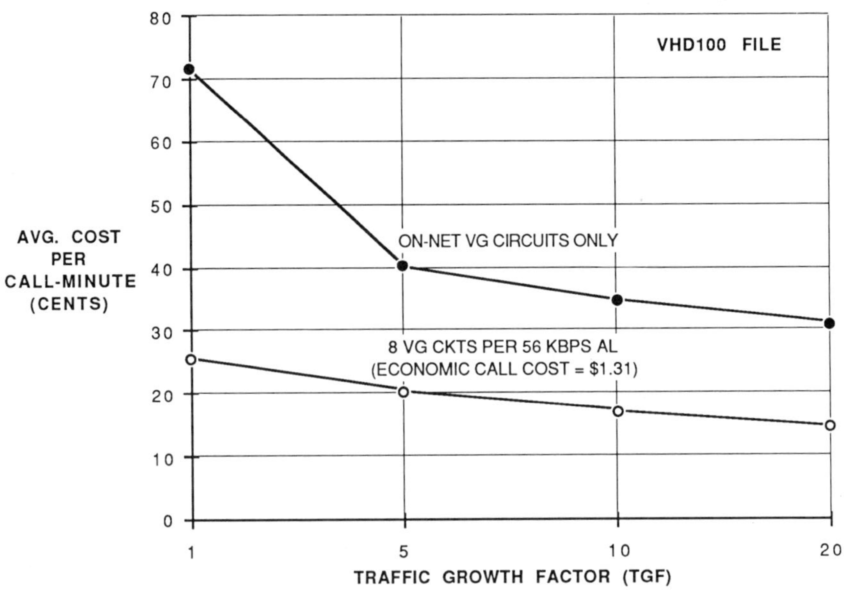

Figure 7.28. Average cost per call minute versus the TGF for an optimized network: VHD100, and economical call costs of $13.31 and $1.31.

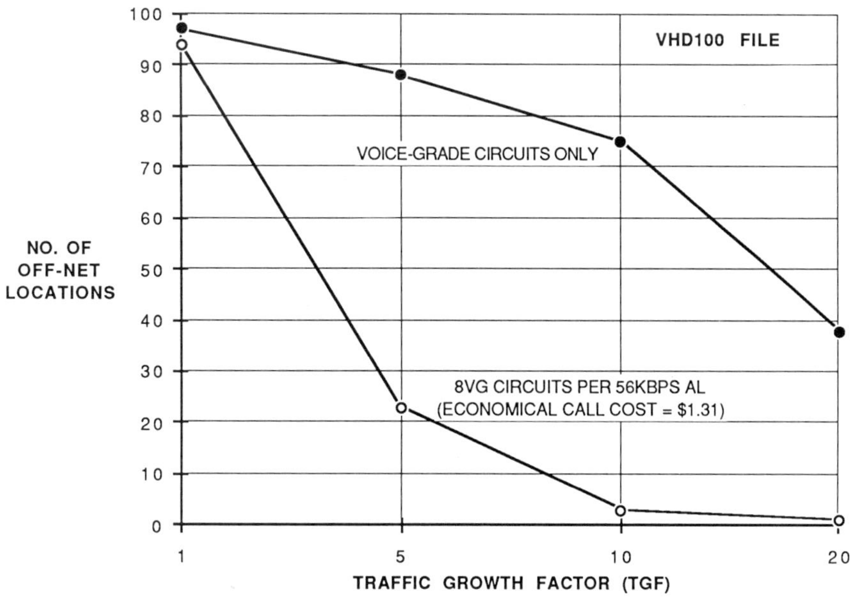

Figure 7.29. Number of off-net locations versus the TGF: VHD100, ECC = \$1.31 and one switch.

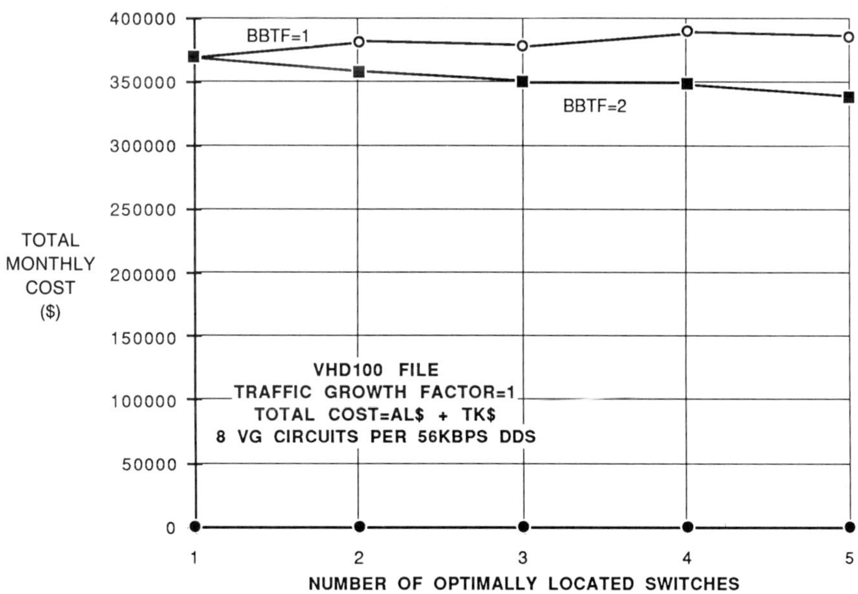

Figure 7.30. Total monthly cost for network transmission versus the number of optimally located switches: VHD100, TGF = 1, BBTF = 1 or 2, and eight VG circuits per economically designed AL/TK.

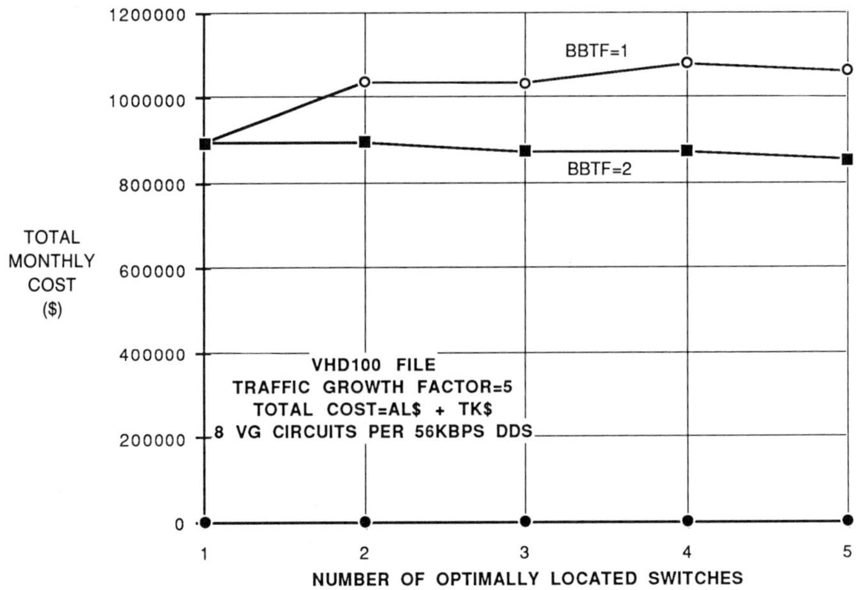

Figure 7.31. Total monthly cost for network transmission versus the number of optimally located switches: VHD100, TGF = 5, BBTF = 1 or 2, and eight VG circuits per economically designed AL/TK.

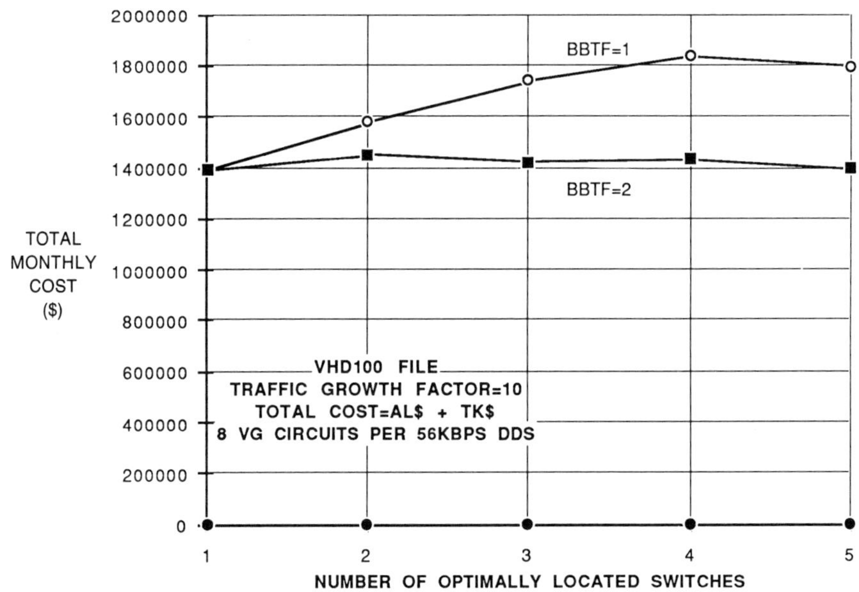

Figure 7.32. Total monthly cost for network transmission versus the number of optimally located switches: VHD100, TGF = 10, BBTF = 1 or 2, and eight VG circuits per economically designed AL or TK.

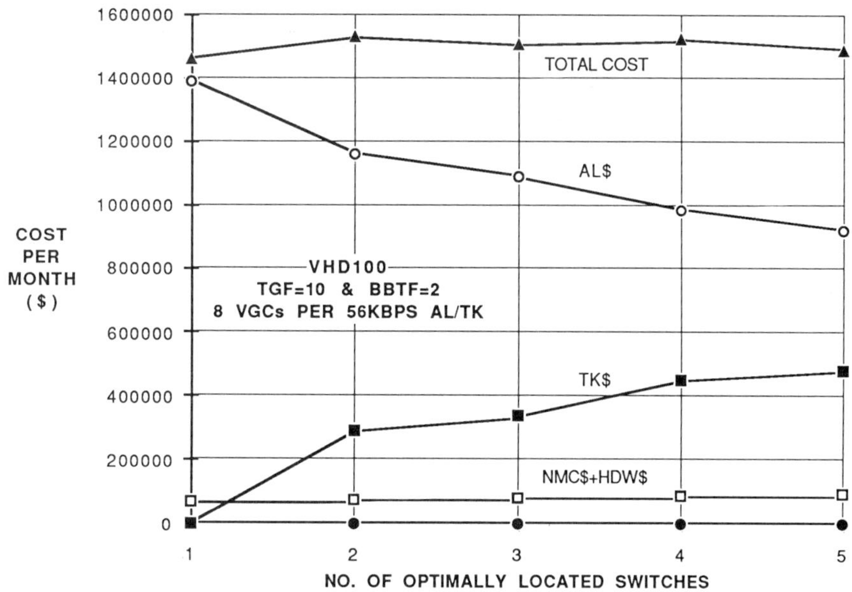

Figure 7.33. Distribution of total monthly costs versus the number of optimally located switches: VHD100, TGF = 10, BBTF = 2, and eight VG circuits per economically designed AL or TK.

best solution. Only for a large value of TGF were two switches shown to be optimum. For networks with 100 or 200 nodes, five switches seem to be appropriate, although further analysis needs to be performed.

In order to select the final design, plots of Figures 7.11, 7.22, and 7.33 are suggested. These plots exhibit costs for not only the ALs and trunks but also for the switching and NMC-related hardware on a monthly basis. An interesting fact emerges: the hardware costs increase the network costs by only one or two cents per call-minute. Transmission costs are still the major factors.

A study of network costs suggests that the costs per call minute are almost identical for very large values of TGF for most networks studied. Only the 59-node network is slightly cheaper. It appears that costs per call-minute reach their asymptotic value as TGF reaches about 20, for most properly designed networks for given tariffs. Another approach for reaching major conclusions is to display some of the output data in the form as shown in Table 7.11. The printout shows TGF, ALMF (AL multiplexing factor), TKMF (trunk multiplexing factor), and cents/CM (cost of a call-minute).

It is interesting to note that a 100-node private voice network that employs all on-net VG circuits is very costly. The transmission costs range between 35 and 29 cents per call minute, even when TGF is varied between the high values of 10 and 40. Therefore, the network designer must employ some kind of digital multiplexing on AL and trunk circuits. Another interesting aspect of the data is that ALMF = 8 and TKMF = 8 yield the cheapest networks. The perceived quality of end-to-end transmission in the network must be studied carefully before employing 56 Kbps DDS circuits for both ALs and trunks. The data of Table 7.11 also shows that the cost of a call minute is not reduced

Table 7.10 Network Design Data Base for VHD200 Input File

#	-V-	-H-	LOAD (Erlangs)	LATA	LINK
1	6375	1919	4	0	0
2	9063	7322	1	0	0
3	5550	1786	3	0	0
4	5878	1408	1	0	0
5	6774	8875	1	0	0
6	6859	1711	3	0	0
7	5204	1370	4	0	0
8	6309	6818	4	0	0
9	6410	1731	3	0	0
10	6726	5527	2	0	0
11	6411	2431	4	0	0
12	9301	6816	1	0	0
13	5237	5454	4	0	0
14	7331	6802	3	0	0
15	7753	2748	3	0	0
16	8267	2327	4	0	0
17	5731	5090	4	0	0
18	7285	5559	1	0	0
19	5367	1996	2	0	0
20	7557	2836	1	0	0
21	6780	3572	4	0	0
22	5982	6680	1	0	0
23	8853	6918	1	0	0
24	6095	7733	4	0	0
25	7597	2240	3	0	0
26	9045	7737	3	0	0
27	8187	2924	4	0	0
28	5736	2809	3	0	0
29	6652	4493	4	0	0
30	6330	2377	4	0	0
31	7348	4859	3	0	0
32	7995	3965	1	0	0
33	7371	5936	1	0	0
34	8100	3190	3	0	0
35	8257	6049	3	0	0
36	6798	3127	3	0	0
37	5938	7411	1	0	0
38	9492	6114	4	0	0
39	8501	8787	4	0	0
40	8636	3780	2	0	0
41	8553	8152	2	0	0
42	8229	5933	1	0	0
43	8976	5173	3	0	0
44	8451	4011	3	0	0
45	7730	7076	1	0	0
46	8981	7684	4	0	0
47	5487	2264	3	0	0
48	8556	4222	1	0	0
49	9449	7188	1	0	0
50	7338	8522	3	0	0
51	7389	6223	2	0	0
52	5547	3849	4	0	0
53	6031	3529	2	0	0
54	7533	5603	2	0	0
55	8113	3389	1	0	0
56	6957	1536	2	0	0
57	5100	4099	1	0	0
58	8141	3528	2	0	0
59	6568	7444	1	0	0
60	7766	7118	2	0	0
61	5809	7385	3	0	0
62	6168	5105	2	0	0
63	7065	3947	1	0	0
64	6300	2475	2	0	0
65	8250	580	4	0	0
66	7549	2006	2	0	0
67	6251	6962	2	0	0
68	8005	1493	2	0	0
69	6354	9224	1	0	0
70	5790	2633	4	0	0
71	7440	3971	3	0	0
72	8751	5813	1	0	0
73	7565	2565	4	0	0
74	9391	5914	3	0	0
75	6394	7889	1	0	0
76	6192	7056	1	0	0
77	7593	3348	4	0	0
78	5182	3265	4	0	0
79	8250	540	4	0	0
80	8083	6676	2	0	0
81	9311	4048	3	0	0
82	6370	4455	1	0	0

Table 7.10 (*Continued*)

#	-V-	-H-	LOAD	LATA	LINK
83	9094	7703	3	0	0
84	5646	4061	3	0	0
85	7721	8195	4	0	0
86	5626	3606	1	0	0
87	8052	8885	3	0	0
88	7325	1604	2	0	0
89	7208	2733	4	0	0
90	7597	1813	2	0	0
91	7553	5983	1	0	0
92	7627	8627	4	0	0
93	8710	5670	1	0	0
94	5303	1812	3	0	0
95	8006	8934	1	0	0
96	7615	4964	1	0	0
97	9001	8394	1	0	0
98	8064	5872	1	0	0
99	5575	6063	1	0	0
100	8655	3881	1	0	0
101	7943	3687	1	0	0
102	9262	5267	4	0	0
103	6480	8521	3	0	0
104	7154	7953	1	0	0
105	5687	6496	1	0	0
106	6147	3748	1	0	0
107	5280	4182	2	0	0
108	5913	3851	3	0	0
109	6153	6886	2	0	0
110	6331	8623	3	0	0
111	8978	6181	2	0	0
112	8992	7485	1	0	0
113	8535	7874	3	0	0
114	6366	2377	1	0	0
115	5071	4279	2	0	0
116	7262	2338	3	0	0
117	5934	5384	1	0	0
118	6579	7701	4	0	0
119	7760	4949	2	0	0
120	7302	1711	1	0	0
121	6047	7159	4	0	0
122	6114	5346	3	0	0
123	7441	8279	2	0	0
124	6427	929	2	0	0
125	6200	2615	3	0	0
126	6556	1313	1	0	0
127	6245	2807	4	0	0
128	6560	4456	1	0	0
129	6016	6584	1	0	0
130	7267	6105	3	0	0
131	7768	9076	4	0	0
132	9310	6425	4	0	0
133	6529	6230	1	0	0
134	8146	725	2	0	0
135	7975	7018	2	0	0
136	5726	3313	2	0	0
137	8363	3830	1	0	0
138	7811	1415	3	0	0
139	8721	4771	2	0	0
140	8185	2225	1	0	0
141	6027	5569	3	0	0
142	8239	3032	3	0	0
143	7968	3485	1	0	0
144	8441	2707	2	0	0
145	6328	7292	4	0	0
146	7012	5775	1	0	0
147	6689	4816	4	0	0
148	7705	4045	4	0	0
149	5168	3177	2	0	0
150	6439	2651	3	0	0
151	7694	4000	3	0	0
152	7087	5562	2	0	0
153	6607	3805	1	0	0
154	7434	1924	3	0	0
155	7347	8414	4	0	0
156	9318	6815	2	0	0
157	8526	4247	2	0	0
158	6556	2175	3	0	0
159	7731	1053	2	0	0
160	7070	1747	4	0	0
161	7953	5624	2	0	0
162	9064	7599	3	0	0
163	6786	3129	1	0	0
164	8067	3836	1	0	0

Table 7.10 *(Continued)*

#	-V-	-H-	LOAD	LATA	LINK
165	9416	6310	3	0	0
166	6651	8820	1	0	0
167	7321	2156	2	0	0
168	8914	8094	1	0	0
169	8084	7033	3	0	0
170	6772	8464	2	0	0
171	5265	5524	1	0	0
172	6394	3182	2	0	0
173	6725	2120	3	0	0
174	5802	4191	2	0	0
175	7450	2233	4	0	0
176	6059	3909	1	0	0
177	8565	8610	3	0	0
178	8055	1079	2	0	0
179	7008	7709	3	0	0
180	7026	7878	1	0	0
181	5700	4235	2	0	0
182	6893	1343	3	0	0
183	6667	1681	2	0	0
184	7102	1191	4	0	0
185	6423	8454	3	0	0
186	7064	8251	3	0	0
187	8834	3133	4	0	0
188	7998	1398	4	0	0
189	6179	5937	3	0	0
190	7481	7681	1	0	0
191	5936	6994	4	0	0
192	7862	1268	4	0	0
193	8581	8385	2	0	0
194	6093	5675	4	0	0
195	8315	2999	3	0	0
196	6267	2993	3	0	0
197	6777	7594	2	0	0
198	6541	6837	4	0	0
199	8870	6407	4	0	0
200	8145	8951	2	0	0
TOTAL NODAL TRAFFIC LOAD=			483 Erlangs		

by employing 56 Kbps DDS-ALS with ALMF = 8 and T-1 trunks with TKMF = 48 (valid for ADPCM), when compared to the cheapest network for ALMF and TKMF = 8. In fact, the cost of the network goes up somewhat when T-1 circuits are used for both ALs and trunks.

Since a 100-node network handles about 8.1 million call-minutes in a month for TGF = 10, even a ten cent saving per call-minute from the VPN/SDN cost of about 27.5 cents per call-minute implies a monthly savings of about $810,000. Such a savings can pay for a large amount of hardware.

Additional savings can be achieved through the use of facilities from other OCC's and bulk-transmission carriers. In some cases, a fractional T1 can now be leased with less than twenty four 64 Kbps PCM channels. By using the low-bit rate voice compression technique, the desired number of VGCs can be carried on a fractional T1. The methodology presented here should be adequate to study any arbitrary tariff, as opposed to always be dependent upon a fixed large tariff data base.

One can also achieve a significant savings by lowering the grade-of-service on ALs and trunks. Special results were computed for a 100-node network for TGF = 10, and these are shown in Table 7.12 for only the best designs, that is, for the proper number of optimally located switches.

Since the analyses presented earlier assume bypass of the local BOCs for the 56 Kbps and T1 circuits, it should be useful to study the effects of no bypass on 56 Kbps and T1 circuits. This can be done by adding $737 and $2000 respectively to the "0th" fixed value in Tariff file for these two types of circuits. These values were dictated by the

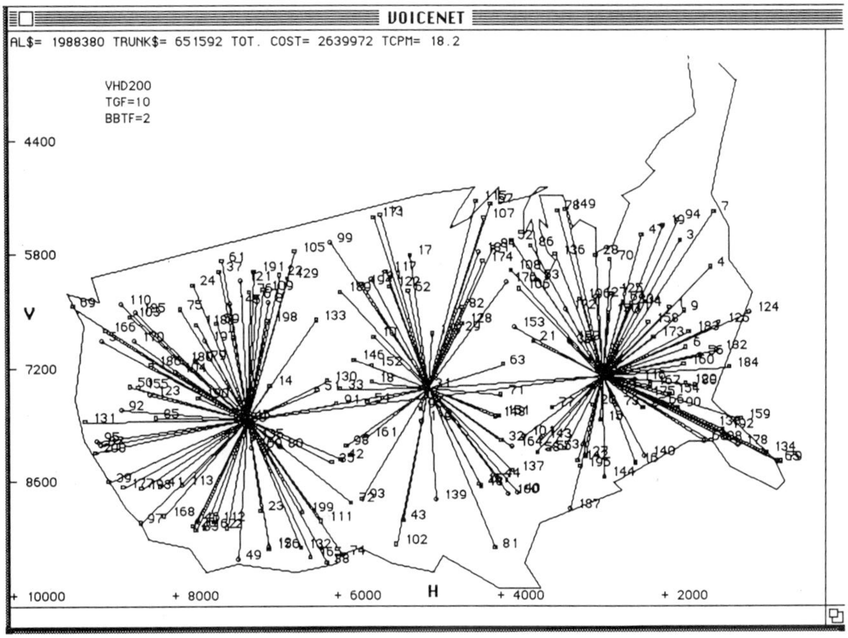

Figure 7.34. A voice network topology with three switches: VHD200, TGF = 10, BBTF = 2, and eight VG circuits per 56kbps AL or TK.

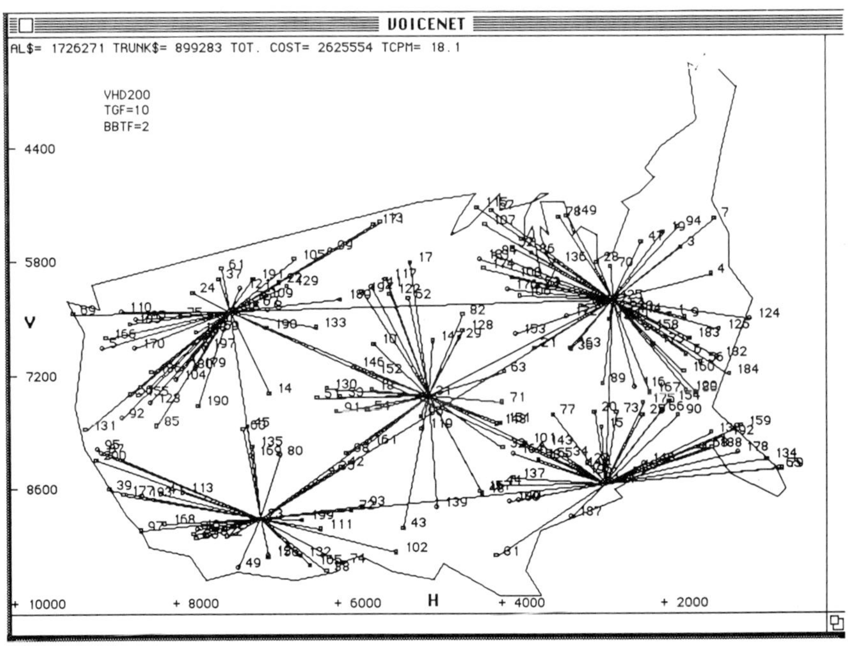

Figure 7.35. A voice network topology with five switches: VHD200, TGF = 10, BBTF = 2, and eight VG circuits per 56kbps AL or TK.

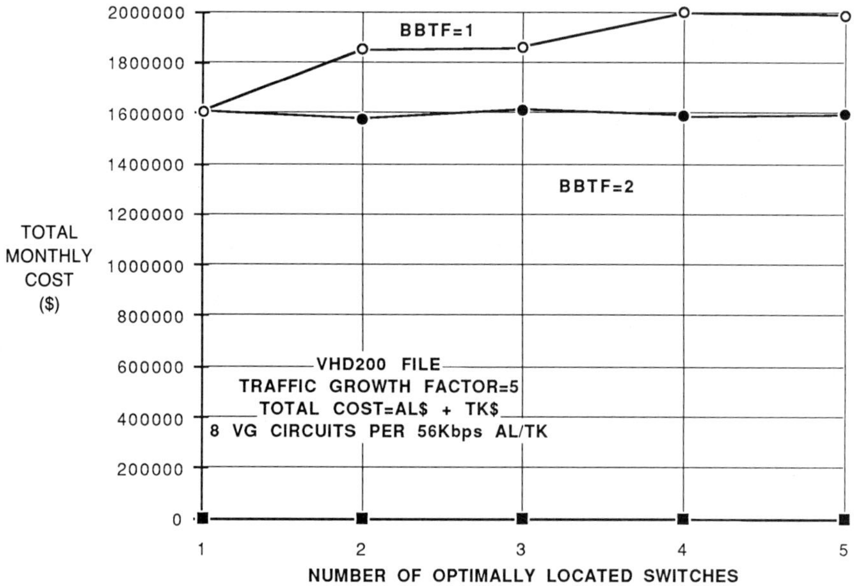

Figure 7.36. Total monthly cost for network transmission versus the number of optimally located switches: VHD200, TGF = 5, BBTF = 1 or 2, and eight VG circuits per economically designed AL or TK.

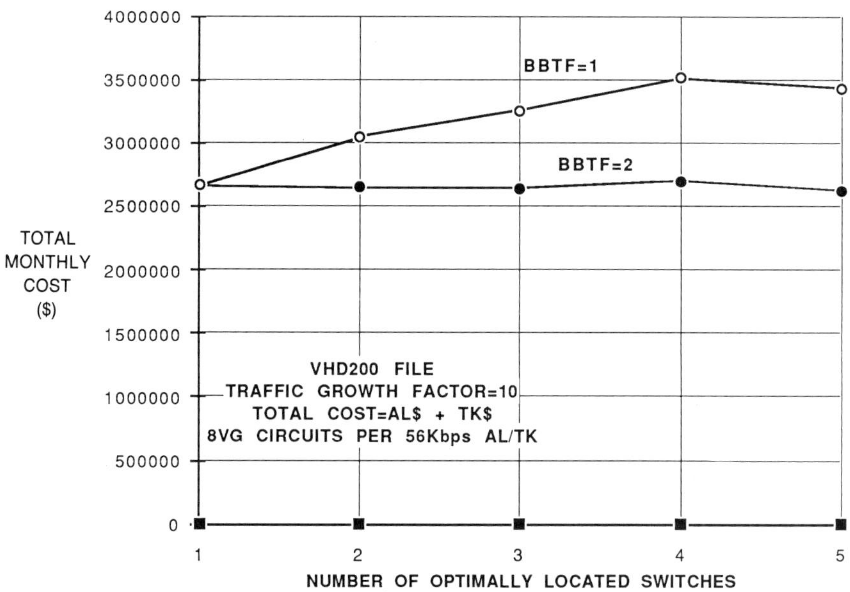

Figure 7.37. Total monthly cost for network transmission versus the number of optimally located switches: VHD200, TGF = 10, BBTF = 1 or 2, and eight VG circuits per economically designed AL or TK.

Table 7.11 Costs per Call-minute vs. TGF and ALMF and TKMF for a 100-Node: AL and Trunk blocking factors of .01 and AT&T's Tariffs 9, 10 and 11

TGF	ALMF	TKMF	Cents/CM
10	1(All On-Net VG)	1	34.8
20	1(All On-Net VG)	1	31.0
40	1(All On-Net VG)	1	28.6
10	8(56Kbps AL)	8(56Kbps Trunk)	17.1
20	8(56Kbps AL)	8(56Kbps Trunk)	15.1
10	8(56Kbps AL)	48(T-1 Trunk)	17.6
20	8(56Kbps AL)	48(T-1 Trunk)	15.1
10	48(T-1 AL)	48(T-1 Trunk)	18.1
20	48(T-1 AL)	48(T-1 Trunk)	15.3

facts that the average length of the DDS and T1 LECs were found to be 20 and 10 miles, respectively, for VHD41 and VHD59 files. See Figure 7.38 for the plots of costs per call-minute versus the number of network nodes and two multiplexing techniques, and for TGF = 10. The cost difference is about two to five cents per call-minute. Networks with 41,100, and 200 nodes are almost identical in costs. The network for VHD59 files is slightly cheaper, due to the fact that nodes are clustered around the switch node number 1. Since networks with only voice grade circuits are always designed with no bypass, curves were not plotted. The bypassing of a few VG circuits is generally not worth the trouble.

The results presented above should enable the designer to make some predictions as to the expected costs for their future networks. The results should also help an enterprise to assess the performance of their existing networks. The results should never be taken as a substitute for an in-depth analysis of their existing or future needs, since each user has a unique situation.

7.3 THE BASIC DESIGN PROCESS FOR AN ACD NETWORK

An Automatic Call Distribution (ACD) system is employed to manage incoming and/or outgoing voice calls. Some businesses, such as airlines, car rentals, and hotel chains,

Table 7.12 Cost per Call-minute versus AL and Trunk Blocking Factor (VHD100, TGF=10, ALMF=8, TKMF=8 and AT&T's Tariffs 9, 10 & 11)

AL Blocking	Trunk Blocking	Cost per Call-Minute
.01	.01	17.1 cents
.01	.1	15.1 cents
.1	.01	16.1 cents
.1	.1	14.0 cents

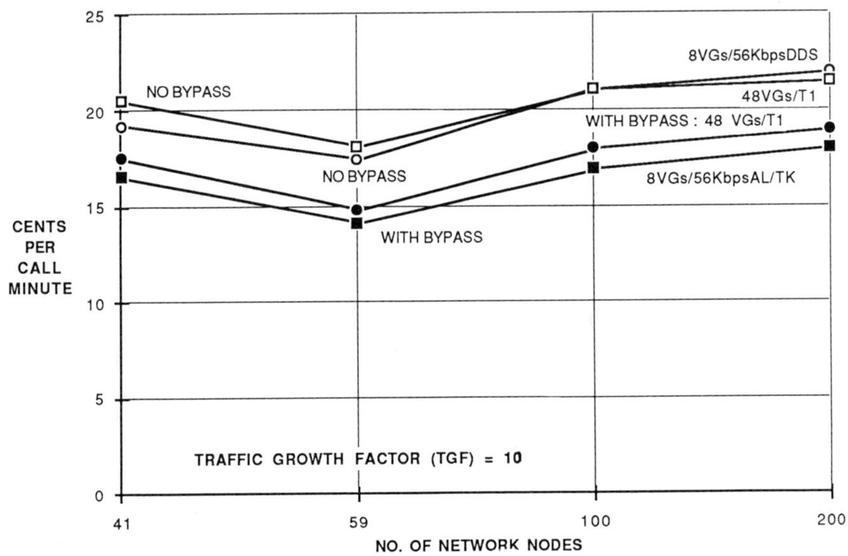

Figure 7.38. Average cost per call minute versus the number of network nodes: TGF = 10, BBTF = 2 for bypass or no-bypass conditions, and two types of access links and trunks.

derive their income primarily through the mechanism of telemarketing. Some other businesses allow homeowners to call sales agents and purchase products being advertised on television. Some large department stores employ an ACD to manage their outgoing calls to their customers.

The so-called 800-Service (or the In-WATS lines) is generally employed to handle the incoming calls from the general public. The outgoing traffic generally requires the use of Wide Area Telecommunications Service (WATS) lines. MegaCom 800 and MegaCom Services provide the same capabilities via T-1 lines. In some special cases, Foreign Exchange (FX) and local business lines may also be used to handle both the incoming and outgoing traffic, and reduce the transmission cost. Most of the OCCs such as AT&T, MCI, and GTE Sprint provide such services, and their tariffs are widely advertised and frequently revised.

The ACD node provides circuit-switched paths between the reservation/sales agents of the business and the terminating trunks (these should be called lines, but most frequently these are called trunks). The ACD switching node also provides a queue for incoming calls whenever all of the agents are busy. A model of the ACD switching node is illustrated in Figure 3.10. See Chapter 3 for a discussion of ACD traffic analysis tools.

The process of traffic data gathering is identical to that for voice networks. Time-consistent traffic loads for each applicable hour must be obtained for computing not only the number of transmission facilities to be leased, but also the number of agents required to handle the incoming calls for all hours. See Figures 7.39(a) and 7.39(b) for the traffic patterns for typical airline reservation and home merchandising businesses. The first pattern implies a predictable level of traffic intensities. Consequently, the task of agent

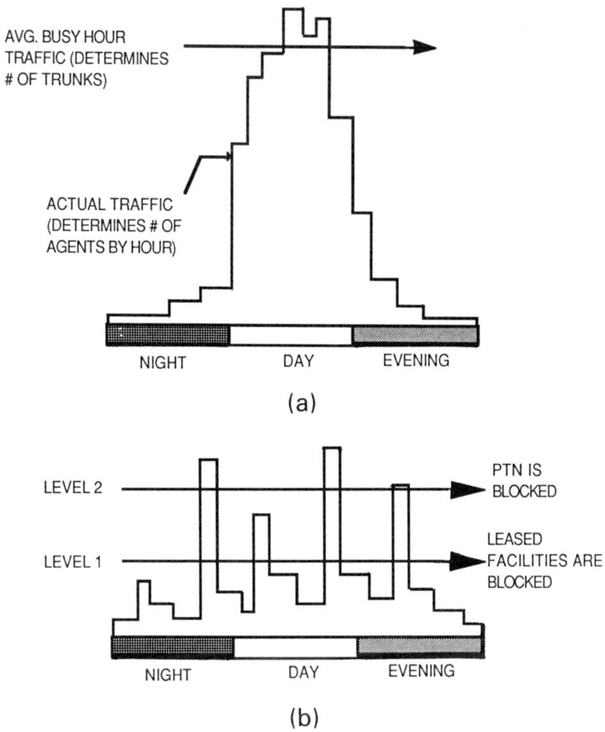

Figure 7.39(a, b). Typical 24-hour traffic patterns for two ACDs employed for airline reservation and home merchandising applications.

scheduling is simplified. The traffic pattern of Figure 7.39(b) will always be tied to the product being advertised on television. If the product has a great appeal to the public and it is attractively priced, the volume of incoming calls may exceed the expectations of even the most astute student of human psychology or behavior. As a result, a large fraction of the traffic may get blocked, not only by the public telephone network of the 800- or WATS-Service provider, but also the private ACD trunk network leased by the customer. Furthermore, the task of agent scheduling becomes quite complex. Some hard decisions need to be made for reaching a compromise between the cost of transmission facilities, sales volumes, and levels of customer satisfaction.

The first step is to locate all of the sources of incoming or outgoing off-net traffic. For the sake of generality, we will construct a hypothetical model based on 49 U.S. metropolitan areas. The call volume from each area is assumed to be proportional to the population. The familiar traffic growth factor (TGF) will be employed to study the effects of call volumes on system costs. Table 7.13 illustrates a data base for the ACDVHD file.

In order to compute the cost of handling off-net traffic, one must first obtain the percent distribution of call volumes for each hour of the day. If such data is not available (e.g.,

Table 7.13 ACD Network Design Data Base (49 States/Regions)

#	-V-	-H-	LOAD (MERLs)	LATA	LINK	NAME
1	8173	1173	526	11	0	TMPFL
2	4422	1249	340	24	0	BOSMA
3	5986	3426	876	15	0	CHIIL
4	5574	2543	506	41	0	CLEOH
5	5510	1575	268	23	0	BALMD
6	8938	3536	358	52	0	HOUTX
7	5251	1458	582	46	0	PHLPA
8	4997	1406	1125	37	0	NYCNY
9	8492	8719	401	5	0	SFOCA
10	9213	7878	923	6	0	LAXCA
11	7954	1031	419	11	0	ORLFL
12	8669	8239	407	5	0	FRSCA
13	7574	7066	449	54	0	SLKUT
14	9250	7810	573	6	0	ANHCA
15	7260	2083	586	12	0	ATLGA
16	5076	2327	486	38	0	BUFNY
17	6263	2680	506	41	0	CINOH
18	5972	2554	468	42	0	COLOH
19	8432	4033	366	51	0	DFWTX
20	5536	2829	537	25	0	DETMI
21	8282	557	459	11	0	FLDFL
22	6272	2992	477	17	0	INDIN
23	7029	4202	497	29	0	KANMO
24	7937	1538	533	11	0	MIAFL
25	5788	3588	505	59	0	MILWI
26	5780	4526	260	27	0	MNPMN
27	4960	1354	321	26	0	NSANY
28	5016	1430	2426	34	0	NWKNJ
29	8486	2638	479	21	0	NORLA
30	9133	6748	519	3	0	PHOAZ
31	5619	2184	613	46	0	PITPA
32	6799	8915	486	44	0	PTLOR
33	9202	7718	526	6	0	RIVCA
34	8303	8581	459	5	0	SACCA
35	6807	3483	293	29	0	STLMO
36	8436	4034	466	52	0	SANTX
37	9462	7632	566	6	0	SNDCA
38	8583	8619	494	5	0	SJSCA
39	6337	8896	533	57	0	SEAWA
40	5623	1578	376	10	0	WASDC
41	6233	1823	419	56	0	SALVA
42	8111	8340	358	32	0	RENNV
43	6206	4167	383	18	0	WTRIA
44	5703	2820	433	41	0	TOLOH
45	6529	2773	444	20	0	LOKKY
46	4913	2195	556	38	0	ROCNY
47	5917	1223	439	39	0	NFKNC
48	4550	1290	444	47	0	PRORI
49	7501	5899	533	7	0	DENCO
TOTAL TRAFFIC LOAD=			25999 MERLs			

for a new ACD system), one must construct such a distribution through the process of consultations with the key players of the business. Such data will not only pinpoint the applicable traffic pattern, but also the number of hours observed during a month for each applicable rate step. Two Daily Traffic Profiles (DTPs) will be considered for analysis. A single DTP file was created to model the reservation-related traffic for every day of the week or month. It should be easy to make a DTP file for each day of the week to reflect a particular traffic pattern. The DTP4 file was created to model the home merchandising traffic. Table 7.14 illustrates these two traffic profiles.

Several special tariff-related tables must be employed to compute the cost of incoming calls. The first table common to all virtual network services defines the Customer Service Area (CSA) Bands, represented by numbers varying from 1 to 6, for each source-destination combination of states/regions (e.g., North Carolina or West Texas). To illustrate, a call coming from North California, and destined to an ACD switch in Memphis,

Tennessee will use a CSA band of 4. Table 7.15 illustrates CSA bands (CSABDS) for all of the combination of states/regions. Each of 60 home state/regions is provided with 62 values of CSA bands, arranged in two rows of values. Two destination regions, namely Virgin Island (VI) and Puerto Rico (PR), are added to the 60 states/regions shown on the leftmost column. A close observation will show that a CSABDS table is not symmetrical in terms of its values.

The second table computes a rate step for a given source state/region and a CSA band. Such a table is called the Rate Step Table (RSTBL), as illustrated in Table 7.16. There are a total of 26 separate rate steps. For each originating home state/region (e.g., North and West Texas), there are six (6) unique rate steps for each of the CSA bands. For example, the rate steps for the home or originating states of Iowa and California are 14 and 18, respectively, considering only the CSA band of 6. The monthly cost of transmission for leasing a certain number of 800-lines could then be computed, if the researcher knew how many hours were utilized for each 800-line, the rate step, and the part of the day (e.g., business hours, evenings, and night-times or weekends). The commonly known Use Rate Table (URTBL), which defines such hourly costs for a leased line, is illustrated in Table 7.17. A subroutine was created for computing the cost of any VPN-based service on a monthly basis. Such a subroutine executes the following tasks:

1. Compute the total number of originating calls handled by each ACD switch during the month;

Table 7.14 DTP and DTP4 Files

Hour	DTP Value	DTP4 Value
1	.01	.041667
2	.01	.041667
3	.01	.041667
4	.01	.041667
5	.02	.041667
6	.02	.041667
7	.02	.041667
8	.03	.041167
9	.04	.041667
10	.06	.041667
11	.08	.041667
12	.12	.041667
13	.10	.041667
14	.12	.041667
15	.10	.041667
16	.08	.041667
17	.04	.041667
18	.03	.041667
19	.02	.041667
20	.02	.041667
21	.02	.041667
22	.02	.041667
23	.01	.041667
24	.01	.041667

Note: DTP and DTP4 columns represent typical airline and home merchandising applications respectively.

Table 7.15 62 CSA Band Values (varying from 1 to 6) for Each State/Region

1 AK	0 5 5 5 5 5 5 5 5 5 5 5 5 5 5 5 5 5 5 5 5 5 5 5 5 5 5 5 5 5 5
	5 5 5 5 5 5 5 5 5 5 5 5 5 5 5 5 5 5 5 5 5 5 5 5 5 5 5 5 5 5 5
2 AL	6 0 5 2 5 5 5 4 4 5 1 1 5 5 3 2 2 3 4 1 1 5 5 5 4 4 4 1 2 5 4
	5 5 4 5 4 4 4 2 5 3 2 3 5 4 4 4 1 5 1 3 3 3 5 5 3 5 3 4 5 5 5
3 AZ	6 4 0 3 2 1 1 5 5 5 5 5 5 2 4 4 4 4 3 4 4 5 5 5 5 5 4 4 4 3 3
	1 5 5 1 5 5 5 5 3 5 5 3 3 5 5 5 5 3 4 3 3 2 1 5 5 3 5 4 2 5 5
4 AR	6 2 5 0 5 5 4 5 4 4 3 3 5 5 2 2 3 3 2 2 1 5 4 5 4 4 4 1 1 5 3
	5 5 5 4 5 5 5 3 4 3 3 1 5 4 4 5 3 4 1 1 3 3 5 5 4 5 3 3 4 5 5
5 NCA	6 4 1 4 0 0 2 5 5 5 5 5 5 1 4 4 4 3 3 4 4 5 5 5 4 4 3 4 3 2 2
	1 5 5 2 5 5 5 5 3 5 5 3 1 5 5 5 5 3 4 3 3 3 1 5 5 1 5 4 2 5 5
6 SCA	6 4 1 4 0 0 2 5 5 5 5 5 5 1 4 4 4 3 3 4 4 5 5 5 4 4 3 4 3 3 3
	1 5 5 1 5 5 5 5 3 5 5 3 1 5 5 5 5 3 4 3 3 2 1 5 5 2 5 4 2 5 5
7 CO	6 5 1 3 3 3 0 5 5 5 5 5 5 2 4 4 4 3 1 5 4 5 5 5 4 4 3 4 3 2 1
	2 5 5 1 5 5 5 5 3 5 5 1 4 5 5 5 5 2 4 3 4 2 1 5 5 4 5 4 1 5 5
8 CT	6 3 5 4 5 5 5 0 2 2 4 3 5 5 3 3 3 4 4 3 4 2 2 1 3 3 4 4 4 5 4
	5 1 1 5 1 1 2 3 4 2 2 4 5 1 2 1 3 4 3 5 5 5 5 1 2 5 2 3 5 5 5
9 DE	6 3 5 4 5 5 5 2 0 1 3 3 5 5 3 3 3 4 4 3 4 3 1 2 3 3 4 4 4 5 4
	5 2 1 5 2 1 2 2 4 3 2 4 5 1 2 2 3 4 3 5 5 5 5 2 1 5 2 4 5 5 5
10 DC	6 3 5 4 5 5 5 2 1 0 3 3 5 5 5 3 3 4 4 3 4 3 1 3 3 3 4 4 4 5 5
	5 3 1 5 2 2 2 2 4 2 2 4 5 1 1 2 3 4 3 5 5 5 5 3 1 5 1 4 5 5 5
11 FL	6 1 5 2 5 5 5 4 2 2 0 1 5 5 3 3 2 4 4 2 1 5 2 4 4 4 5 1 3 5 4
	5 4 3 5 4 4 4 1 5 3 2 3 5 3 3 4 1 5 1 3 3 3 5 4 2 5 2 4 5 4 4
12 GA	6 1 5 2 5 5 5 4 3 3 1 0 5 5 3 2 2 4 4 1 2 5 3 4 3 3 5 1 3 5 4
	5 4 3 5 3 3 3 1 5 3 2 4 5 3 3 4 1 5 1 4 4 4 5 4 2 5 2 4 5 5 5
13 HI	3 5 2 4 1 1 3 5 5 5 5 5 0 2 4 4 5 4 4 5 4 5 5 5 5 5 4 5 4 2 3
	2 5 5 3 5 5 5 5 4 5 5 4 1 5 5 5 5 3 5 4 4 4 2 5 5 2 5 4 3 5 5
14 ID	6 5 3 4 2 2 2 5 5 5 5 5 5 0 4 4 4 3 3 4 4 5 5 5 4 4 3 4 3 1 3
	1 5 5 3 5 5 5 5 3 5 5 3 1 5 5 5 5 3 5 4 4 4 1 5 5 1 5 3 1 5 5
15 NIL	6 3 5 2 5 5 4 5 4 4 5 3 5 5 0 0 1 1 3 2 3 5 4 5 2 1 2 3 1 5 3
	5 5 4 5 4 4 4 3 4 2 2 3 5 3 3 5 3 4 2 4 4 4 5 5 3 5 2 1 5 5 5
16 SIL	6 2 5 2 5 5 4 5 4 4 5 3 5 5 0 0 1 1 3 1 3 5 4 5 3 2 3 2 1 5 3
	5 5 4 5 4 4 4 3 4 2 1 3 5 3 3 5 3 4 1 4 4 4 5 5 3 5 3 2 5 5 5
17 IN	6 3 5 3 5 5 5 4 3 3 4 3 5 5 1 1 0 2 4 1 4 5 3 4 3 1 3 3 2 5 4
	5 4 4 5 3 3 3 3 4 1 1 4 5 3 2 4 3 4 2 5 5 5 5 4 3 5 2 2 5 5 5
18 IA	6 3 5 3 5 5 3 5 5 4 5 4 5 5 1 1 2 0 2 3 3 5 4 5 2 2 1 3 1 4 1
	5 5 5 4 5 5 5 4 3 3 3 2 5 4 4 5 4 1 3 4 4 4 5 5 4 5 3 1 3 5 5
19 KS	6 3 4 2 5 5 1 5 5 5 5 4 5 4 3 2 3 1 0 3 3 5 5 5 4 4 3 3 1 4 1
	4 5 5 2 5 5 5 4 3 4 4 1 5 5 5 5 4 1 3 2 3 2 3 5 4 5 4 3 3 5 5
20 KY	6 2 5 3 5 5 5 4 3 3 4 2 5 5 2 1 1 3 4 0 3 5 3 4 3 3 4 3 1 5 4
	5 4 3 5 4 4 4 2 5 2 1 4 5 3 3 4 2 5 1 4 4 4 5 4 1 5 1 3 5 5 5
21 LA	6 1 5 1 5 5 4 5 4 4 3 2 5 5 3 2 3 3 3 2 0 5 4 5 4 4 4 1 2 5 3
	5 5 4 3 5 5 5 3 5 3 3 1 5 4 4 5 3 4 2 1 1 2 5 5 4 5 3 4 5 5 5
22 ME	6 4 5 4 5 5 5 1 2 2 4 3 5 5 3 3 3 4 4 3 4 0 2 1 3 3 3 4 4 5 4
	5 1 1 5 1 1 2 3 4 2 3 4 5 2 2 1 3 4 3 5 5 5 5 1 2 5 2 3 5 5 5
23 MD	6 3 5 4 5 5 5 2 1 1 3 3 5 5 3 3 3 4 4 3 4 3 0 2 3 3 4 4 4 5 4
	5 3 1 5 2 1 2 2 4 2 2 4 5 1 1 2 3 4 3 5 5 5 5 3 1 5 1 4 5 5 5
24 MA	6 3 5 4 5 5 5 1 2 2 4 3 5 5 3 3 3 3 4 3 4 1 2 0 3 3 4 4 4 5 4
	5 1 1 5 1 1 2 3 4 2 2 4 5 1 2 1 3 4 3 5 5 5 5 1 2 5 2 3 5 5 5
25 NMI	6 4 5 4 5 5 4 3 3 3 4 4 5 5 1 2 2 1 4 2 4 4 3 3 0 0 1 4 2 5 4
	5 3 3 5 3 3 2 3 2 1 2 4 5 3 2 3 4 3 3 5 5 5 5 3 3 5 3 1 5 5 5
26 SMI	6 4 5 4 5 5 4 3 3 2 4 4 5 5 1 2 1 2 4 2 4 4 3 3 0 0 3 4 3 5 4
	5 3 3 5 2 3 1 3 4 1 1 4 5 2 2 3 4 3 3 5 5 5 5 3 3 5 2 1 5 5 5
27 MN	6 4 5 3 5 5 3 5 5 4 5 5 5 4 1 2 2 1 2 3 4 5 4 5 1 2 0 4 2 3 1
	5 5 5 4 4 4 4 5 1 3 3 3 5 3 3 5 5 1 3 4 4 4 4 5 4 5 3 1 3 5 5
28 MS	6 1 5 1 5 5 4 5 4 4 2 1 5 5 3 2 3 3 3 2 1 5 4 5 4 4 4 0 2 5 4
	5 5 4 4 5 5 5 3 5 3 3 2 5 4 4 5 2 4 1 1 3 3 5 5 3 5 3 3 5 5 5
29 MO	6 3 5 1 5 5 4 5 4 4 4 3 5 5 1 1 2 1 1 1 2 5 4 5 3 3 3 2 0 5 1
	5 5 5 4 5 5 5 4 4 3 3 1 5 4 4 5 4 3 1 3 3 3 5 5 4 5 3 2 4 5 5
30 MT	6 5 3 4 2 3 2 5 5 5 5 5 5 1 4 4 4 3 3 4 5 5 5 5 4 4 2 5 3 0 2
	2 5 5 3 5 5 5 5 1 4 4 3 1 5 5 5 5 1 5 4 4 4 1 5 5 1 5 3 1 5 5
31 NE	6 4 4 3 5 5 1 5 5 5 5 4 5 3 2 3 3 1 1 4 4 5 5 5 3 3 2 4 1 3 0
	4 5 5 3 5 5 5 5 2 4 4 2 4 5 5 5 5 1 4 3 3 3 3 5 5 5 4 2 1 5 5
32 NV	6 5 1 4 1 1 2 5 5 5 5 5 5 1 4 4 4 3 3 5 4 5 5 5 4 4 4 4 4 3 3
	0 5 5 3 5 5 5 5 3 5 5 3 1 5 5 5 5 3 4 3 3 3 1 5 5 2 5 4 2 5 5

Table 7.15 (*Continued*)

Region	CSA band values
33 NH	6 4 5 4 5 5 5 1 2 2 4 3 5 5 3 3 3 3 4 3 4 1 2 1 3 3 4 4 4 5 4
	5 0 1 5 1 1 2 3 4 2 2 4 5 2 2 1 3 4 3 5 5 5 5 1 2 5 2 3 5 5 5
34 NJ	6 3 5 4 5 5 5 1 1 1 4 3 5 5 3 3 3 4 4 3 4 3 1 1 3 3 4 4 4 5 4
	5 2 0 5 2 1 2 2 4 2 2 4 5 1 2 1 3 4 3 5 5 5 5 2 2 5 2 3 5 5 5
35 NM	6 4 1 3 3 3 1 5 5 5 5 5 5 3 4 4 4 3 2 4 3 5 5 5 5 5 4 4 3 3 2
	2 5 5 0 5 5 5 5 4 5 5 1 4 5 5 5 5 3 4 2 2 1 1 5 5 4 5 4 2 5 5
36 NENY	6 3 5 4 5 5 5 1 1 2 4 3 5 5 3 3 3 3 4 3 4 2 1 1 3 2 4 4 3 5 4
	5 1 1 5 0 0 0 3 4 2 2 4 5 1 1 1 3 4 3 5 5 5 5 1 2 5 2 3 5 5 5
37 SENY	6 3 5 4 5 5 5 1 1 1 4 3 5 5 3 3 3 3 4 3 4 2 1 1 3 3 4 4 3 5 4
	5 1 1 5 0 0 0 2 4 2 2 4 5 1 2 1 3 4 3 5 5 5 5 1 2 5 2 3 5 5 5
38 WNY	6 3 5 4 5 5 5 2 1 1 4 3 5 5 3 3 3 3 4 3 4 2 1 2 3 2 4 4 3 5 4
	5 2 1 5 0 0 0 3 4 1 2 4 5 1 1 2 3 4 3 5 5 5 5 2 2 5 1 3 5 5 5
39 NC	6 2 5 4 5 5 5 3 1 1 3 1 5 5 3 3 3 4 4 2 4 4 1 3 3 3 4 3 4 5 4
	5 4 2 5 3 3 3 0 5 2 1 4 5 2 2 3 1 5 1 4 4 4 5 3 1 5 1 4 5 5 5
40 ND	6 5 4 3 4 4 2 5 5 5 5 5 5 2 2 3 3 1 2 4 5 5 5 5 2 3 1 4 3 1 1
	3 5 5 3 5 5 5 5 0 4 4 3 3 5 5 5 5 1 4 4 4 4 3 5 5 3 4 2 1 5 5
41 NOH	6 3 5 4 5 5 5 3 2 2 4 3 5 5 2 2 1 3 4 1 4 4 2 3 3 1 4 4 3 5 4
	5 4 3 5 3 3 1 3 4 0 0 4 5 2 1 4 3 4 2 5 5 5 5 3 1 5 1 3 5 5 5
42 SOH	6 3 5 4 5 5 5 3 3 2 4 3 5 5 2 2 1 3 4 1 4 4 2 3 3 1 4 4 3 5 4
	5 4 3 5 3 3 1 2 4 0 0 4 5 2 1 4 3 4 2 5 5 5 5 3 1 5 1 3 5 5 5
43 OK	6 3 4 1 5 5 1 5 5 5 4 4 5 5 3 3 3 3 1 3 2 5 5 5 4 4 3 2 1 4 2
	5 5 5 1 5 5 5 4 4 4 4 0 5 5 5 5 4 3 3 1 2 1 3 5 4 5 4 3 3 5 5
44 OR	6 5 2 4 1 2 3 5 5 5 5 5 5 1 4 4 4 3 3 5 4 5 5 5 4 4 3 4 4 2 3
	1 5 5 3 5 5 5 5 3 5 5 3 0 5 5 5 5 3 5 3 3 3 1 5 5 1 5 4 2 5 5
45 EPA	6 3 5 4 5 5 5 1 1 1 4 3 5 5 3 3 3 4 4 3 4 3 1 2 3 3 4 4 3 5 4
	5 2 1 5 1 1 1 2 4 2 2 4 5 0 0 2 3 4 3 5 5 5 5 2 2 5 2 3 5 5 5
46 WPA	6 3 5 4 5 5 5 2 1 1 4 3 5 5 3 3 3 4 4 3 4 3 1 2 3 2 4 4 3 5 4
	5 3 2 5 1 2 1 2 4 1 1 4 5 0 0 3 3 4 3 5 5 5 5 3 1 5 1 3 5 5 5
47 RI	6 3 5 4 5 5 5 1 2 2 4 3 5 5 3 3 3 4 4 3 4 2 2 1 3 3 4 4 4 5 4
	5 1 1 5 1 1 2 3 4 2 3 4 5 2 2 0 3 4 3 5 5 5 5 1 2 5 2 3 5 4 4
48 SC	6 1 5 3 5 5 5 3 2 2 2 1 5 5 3 3 2 4 4 1 3 4 2 4 4 4 5 3 3 5 5
	5 4 3 5 3 3 3 1 5 2 2 4 5 3 2 4 0 5 1 4 4 4 5 4 1 5 1 4 5 5 5
49 SD	6 4 4 3 5 5 2 5 5 5 5 5 5 3 2 3 3 1 2 4 4 5 5 5 3 3 1 4 2 1 1
	4 5 5 3 5 5 5 5 1 4 4 3 4 5 5 5 5 0 4 3 3 3 3 5 5 4 4 2 1 5 5
50 TN	6 1 5 1 5 5 5 4 4 3 3 1 5 5 2 2 2 3 4 1 3 5 3 5 3 3 4 1 1 5 4
	5 5 4 5 4 4 4 1 5 2 2 3 5 3 3 4 2 5 0 4 4 4 5 5 1 5 2 3 5 5 5
51 ETX	6 2 3 1 5 5 2 5 5 5 3 3 5 4 3 2 3 3 1 3 1 5 5 5 4 4 4 1 1 4 2
	4 5 5 2 5 5 5 4 4 4 4 1 5 5 5 5 4 3 2 0 0 0 3 5 5 5 4 4 3 5 5
52 STX	6 2 2 1 5 5 2 5 5 5 3 3 5 4 3 3 3 3 1 3 1 5 5 5 4 4 4 1 2 4 3
	4 5 5 1 5 5 5 4 4 4 4 1 5 5 5 5 4 3 2 0 0 0 3 5 5 5 4 3 3 5 5
53 WTX	6 3 2 1 5 5 1 5 5 5 3 3 5 4 3 3 3 3 1 3 2 5 5 5 4 4 4 2 2 4 2
	4 5 5 1 5 5 5 4 4 4 4 1 5 5 5 5 4 3 3 0 0 0 2 5 5 5 4 4 3 5 5
54 UT	6 5 1 4 2 2 5 1 5 5 5 5 5 1 4 4 4 3 3 5 4 5 5 5 4 4 4 4 4 2 3
	1 5 5 1 5 5 5 5 3 5 5 3 3 5 5 5 5 3 4 3 3 3 0 5 5 3 5 4 1 5 5
55 VT	6 4 5 4 5 5 5 1 2 2 4 3 5 5 3 3 3 3 4 3 4 1 2 1 3 3 4 4 4 5 4
	5 1 1 5 1 1 2 3 4 2 2 4 5 2 2 1 3 4 3 5 5 5 5 0 2 5 2 3 5 5 5
56 VA	6 3 5 4 5 5 5 3 1 1 3 3 5 5 3 3 3 4 4 1 4 4 1 3 3 3 4 3 4 5 4
	5 3 2 5 3 2 2 1 5 2 1 4 5 2 1 3 2 4 1 5 5 5 5 3 0 5 1 4 5 5 5
57 WA	4 5 3 4 1 2 3 5 5 5 5 5 5 1 4 4 5 3 3 5 5 5 5 5 4 4 3 5 3 1 3
	1 5 5 3 5 5 5 5 2 5 5 3 1 5 5 5 5 3 5 4 4 4 2 5 5 0 5 3 2 5 5
58 WV	6 3 5 4 5 5 5 3 2 1 4 3 5 5 3 3 2 4 4 1 4 4 1 3 3 3 4 4 3 5 4
	5 3 2 5 3 3 2 1 5 1 1 4 5 2 1 3 2 4 2 5 5 5 5 3 1 5 0 3 5 5 5
59 WI	6 4 5 3 5 5 4 4 4 3 5 4 5 5 1 2 2 1 3 3 4 5 3 4 1 1 1 4 2 4 2
	5 4 4 5 3 3 3 4 3 2 2 3 5 3 3 4 4 2 3 5 5 5 5 4 3 5 3 0 4 5 5
60 WY	6 5 2 4 3 3 1 5 5 5 5 5 5 1 4 4 4 3 2 4 4 5 5 5 4 4 3 4 3 1 1
	2 5 5 2 5 5 5 5 2 5 5 3 2 5 5 5 5 1 5 4 4 4 1 5 5 3 5 3 0 5 5

Note: The standard 2-letter state designation terminates the region designation (e.g. North California = NCA or NorthEast New York = NENY). The Number Associated with the state/region is shown on the left. There are 62 CSA band values for each region, i.e. the 60 as shown as the leftmost column plus Virgin Island (VI) and Puerto Rico (PR). The "0" CSA band numbers are for two identical states/regions forming a pair. Examples: The CSA band for the (58-10) state/region pair = 1, for the (55-35) pair = 5 and so on.

Table 7.16 Rate Steps for Each CSA Band and State/Region

	CSA1	CSA2	CSA3	CSA4	CSA5	CSA6
1 AK	4	7	9	11	17	17
2 AL	4	7	9	11	17	17
3 AZ	6	9	12	15	18	18
4 AR	4	7	9	11	15	15
5 NCA	8	12	15	17	18	18
6 SCA	7	11	15	17	18	18
7 CO	7	8	10	12	16	16
8 CT	1	7	10	14	18	18
9 DE	1	5	9	13	18	18
10 DC	1	4	8	12	18	18
11 FL	7	10	12	13	18	18
12 GA	4	7	10	12	18	18
13 HI	23	23	24	25	26	26
14 ID	5	9	13	15	18	18
15 NIL	3	6	8	10	15	15
16 SIL	3	6	8	10	15	15
17 IN	3	6	8	10	16	16
18 IA	4	7	9	11	14	14
19 KS	5	7	9	12	14	14
20 KY	3	5	8	10	17	17
21 LA	5	8	10	13	16	16
22 ME	6	9	12	16	18	18
23 MD	2	5	9	12	18	18
24 MA	2	7	11	14	18	18
25 NMI	5	8	9	12	17	17
26 SMI	4	7	9	12	17	17
27 MN	6	8	10	12	15	15
28 MS	5	7	9	11	16	16
29 MO	5	7	8	10	15	15
30 MT	7	10	12	14	17	17
31 NE	5	8	9	12	14	14
32 NV	5	8	13	16	18	18
33 NH	2	7	11	15	18	18
34 NJ	1	5	9	13	18	18
35 NM	6	8	10	13	17	17
36 NENY	3	7	10	14	18	18
37 SENY	1	7	10	14	18	18
38 WNY	3	5	10	14	18	18
39 NC	4	7	8	12	18	18
40 ND	6	9	11	14	15	15
41 NOH	3	5	7	10	17	17
42 SOH	3	5	8	10	17	17
43 OK	5	7	9	12	15	15
44 OR	5	9	15	17	18	18
45 EPA	1	5	8	12	18	18
46 WPA	3	5	8	12	18	18
47 RI	1	6	11	14	18	18
48 SC	4	7	9	12	18	18
49 SD	5	8	10	12	15	15
50 TN	5	6	8	10	17	17
51 ETX	6	9	11	14	16	16
52 STX	8	11	12	14	16	16
53 WTX	7	9	11	14	16	16
54 UT	6	7	11	14	18	18
55 VT	2	7	11	14	18	18

Table 7.16 (*Continued*)

	CSA1	CSA2	CSA3	CSA4	CSA5	CSA6
56 VA	3	5	8	11	18	18
57 WA	8	11	15	17	18	18
58 WV	2	5	7	11	18	18
59 WI	3	7	9	11	16	16
60 WY	5	9	10	13	16	16

NOTES: CSA = Customer Service Area
CSA1 implies that CSA Band is equal to 1,
CSA2 implies that CSA Band is equal to 2 and so on.
WTX = West Texas, NENY = North East New York and so on

2. Compute the busy-hour erlangs (BHRERL) handled by each ACD switch. This determines the number of virtual circuits;
3. Compute the number of leased line required to handle the traffic for a given AL blocking factor;
4. Using the applicable DTP file, compute the total number of handled erlangs during night or weekend, business, and evening hours of a day for *each leased line*;
5. Using CSABDS, RSTBL, and URTBL, compute the total cost for *each leased line*;
6. Compute the total cost of all leased lines for the entire ACD system;
7. Using the Erlang C formulae (e.g., Equations 3.11 and 3.13 of Chapter 3), compute the number of agents required for each hourly period as a function of the erlangs and the specified GOS (or the probability of delay exceeding a specified delay equal to DREQ, as defined in SDF file);
8. Compute the total cost of agent labor for the entire month;
9. Compute the cost of data trunks required to communicate with the data base; and
10. Compute the total cost of a call-minute.

Figures 7.40 to 7.42 illustrate the ACD network topologies for one, two, and three ACD centers, using the 800-Service for TGF = 1. The corresponding results are tabulated in Tables 7.18 through 7.20. The part that computes the cost of 800-Service for each location inserts a zero for a number of access lines (ALs) and the cost of ALs, since we are dealing with a virtual network. The part that describes the cost of labor lists the traffic intensity handled by each ACD location/site during each hour of the day (in millierlangs), and the number of agents required to handle the traffic while providing the desired GOS for each hour and ACD center. The desired GOS is defined at the start of each printout. The part that analyzes trunking describes each data trunk that connects an ACD center to the ACD data base computer location, as defined by the first switch location as entered. Additional types of ACDs can be analyzed. The type of an ACD system can be defined by the 29th design parameter of the SDF file. Only the star-type topology is employed for trunking. The results are obtained for AT&T's and MegaCom's 800 Services, and WATS tariffs using the design parameter, ACD type, 1, 2, or 3, and either the DTP or DTP4 file. URTBLs for WATS and MegaCom 800 Service are shown in Tables 7.17(a) and 7.17(b). The DTP file represents each day of the month

Table 7.17 The Use-Rate Table for AT&T's 800 Service (1/1/87)

	PER HOUR OF USE,	PER RATE PERIOD,	PER ACCESS LINE				
	First 25 Hours/mo		Next 75 Hours/mo		Over 100 Hours/mo		All Hours Late Night & Weekends
RATE STEP	Business Day	Evening	Business Day	Evening	Business Day	Evening	
1	12.37	9.28	11.75	8.81	11.13	8.35	6.64
2	12.76	9.57	12.11	9.10	11.48	8.61	6.83
3	12.93	9.69	12.28	9.21	11.63	8.73	6.92
4	13.07	9.80	12.41	9.31	11.76	8.82	7.01
5	13.19	9.90	12.53	9.42	11.88	8.90	7.08
6	13.31	9.98	12.65	9.49	11.97	8.98	7.14
7	13.45	10.08	12.77	9.57	12.10	9.08	7.21
8	13.59	10.20	12.92	9.68	12.23	9.18	7.28
9	13.74	10.31	13.06	9.79	12.36	9.27	7.36
10	13.84	10.37	13.14	9.86	12.45	9.34	7.41
11	13.94	10.46	13.24	9.94	12.55	9.42	7.46
12	14.02	10.52	13.31	9.98	12.62	9.46	7.53
13	14.10	10.58	13.39	10.04	12.68	9.52	7.55
14	14.18	10.64	13.46	10.10	12.75	9.56	7.61
15	14.28	10.71	13.56	10.18	12.85	9.63	7.67
16	14.39	10.78	13.66	10.26	12.95	9.72	7.70
17	14.48	10.87	13.77	10.33	13.05	9.78	7.77
18	14.63	10.99	13.91	10.43	13.18	9.89	7.87
19							
20							
21							
22							
23	14.85	11.14	14.12	10.6	13.37	10.02	8.03
24	15.02	11.27	14.28	10.71	13.53	10.15	8.12
25	15.17	11.38	14.42	10.81	13.66	10.26	8.19
26	15.3	11.47	14.53	10.9	13.77	10.33	8.26

NOTE: The Use-Rate Table (URTBL) employs the rate steps obtained from Tables 7.15 and 7.16 for computing the total monthly cost of 800 Service.

for airline-type business. The DTP4 represents each day of the month for televised merchandising businesses. Additional DTP files can be constructed for weekly profiles, with each day handling different daily profiles. The computed results are plotted in Figures 7.43 to 7.48 for several ACD types and TGF values for the airline's daily profile.

The above analyses employ a simplified analytical model only. Additional refinements may be necessary to improve the results. These are discussed next.

7.3.1 IMPROVED MODELS FOR COMPUTING ACD SYSTEM RESOURCES

The usage level of a given number of trunks (T) and agent positions (P) will determine the total monthly cost of any ACD. Consequently, it is necessary to compute the opti-

Table 7.17(a) The Use-Rate Table for AT&T's WATS Service (1/1/87)

	PER HOUR OF USE, PER RATE PERIOD, PER ACCESS LINE						
	First 25 Hours/mo		Next 75 Hours/mo		> 100 Hours/mo		All Hours Late Night & Weekends
RATE STEP	Business Day	Evening	Business Day	Evening	Business Day	Evening	
1	12.06	8.07	11.46	7.68	10.85	7.26	5.41
2	12.68	8.49	12.04	8.06	11.41	7.64	5.69
3	12.99	8.70	12.32	8.26	11.69	7.84	5.83
4	13.21	8.85	12.55	8.40	11.90	7.98	5.92
5	13.41	8.99	12.74	8.53	12.08	8.09	6.00
6	13.57	9.11	12.91	8.64	12.23	8.18	6.07
7	13.81	9.24	13.13	8.79	12.42	8.33	6.19
8	14.07	9.43	13.36	8.96	12.66	8.48	6.31
9	14.28	9.56	13.55	9.09	12.86	8.61	6.39
10	14.45	9.68	13.73	9.19	13.01	8.72	6.51
11	14.62	9.79	13.89	9.31	13.16	8.81	6.56
12	14.78	9.90	14.05	9.41	13.31	8.92	6.62
13	14.90	9.98	14.17	9.48	13.41	8.99	6.69
14	15.05	10.09	14.30	9.57	13.54	9.08	6.76
15	15.22	10.19	14.45	9.68	13.69	9.17	6.83
16	15.41	10.32	14.63	9.80	13.85	9.28	6.90
17	15.56	10.42	14.79	9.91	14.01	9.39	6.96
18	15.86	10.64	15.07	10.10	14.28	9.56	7.11
19							
20							
21							
22							
23	16.24	10.88	15.43	10.34	14.62	9.79	7.33
24	16.54	11.08	15.71	10.52	14.88	9.97	7.45
25	16.80	11.26	15.96	10.69	15.12	10.13	7.59
26	16.99	11.39	16.15	10.82	15.31	10.25	7.68

NOTE: The Use-Rate Table (URTBL) employs the rate steps obtained from Tables 7.15 & 7.16 for computing the total monthly cost of WATS Service. These rates change with time. In case, the increase is uniform over all values, one can use the SDF Parameter#43 to reflect the change.

Table 7.17(b) The Use-Rate Table for AT&T's MegaCom Service (1/1/87)

Rate Schedule	Per Hour of Use, Per Rate Period, Per Access Line		
Service Area	Business Day	Evening	Night / Weekend
1	10.36	8.07	6.74
2	10.93	8.52	7.11
3	11.22	8.75	7.29
4	11.80	9.20	7.66
5	12.07	9.43	7.86
6	13.81	10.76	8.97

NOTE: The Use-Rate Table (URTBL) employs the customer service area value obtained from Table 7.15 for computing the total monthly cost of AT&T's MegaComService. These rates change with time. In case, the increase is uniform over all values, one can use the SDF Parameter#43 to reflect the change.

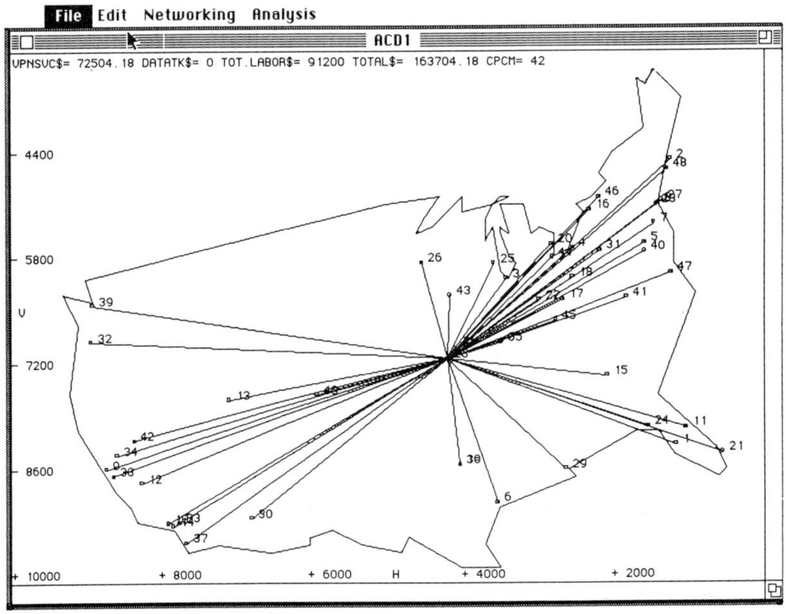

Figure 7.40. An ACD network topology with one switch: 49 regions and AT&T's 800 SVC.

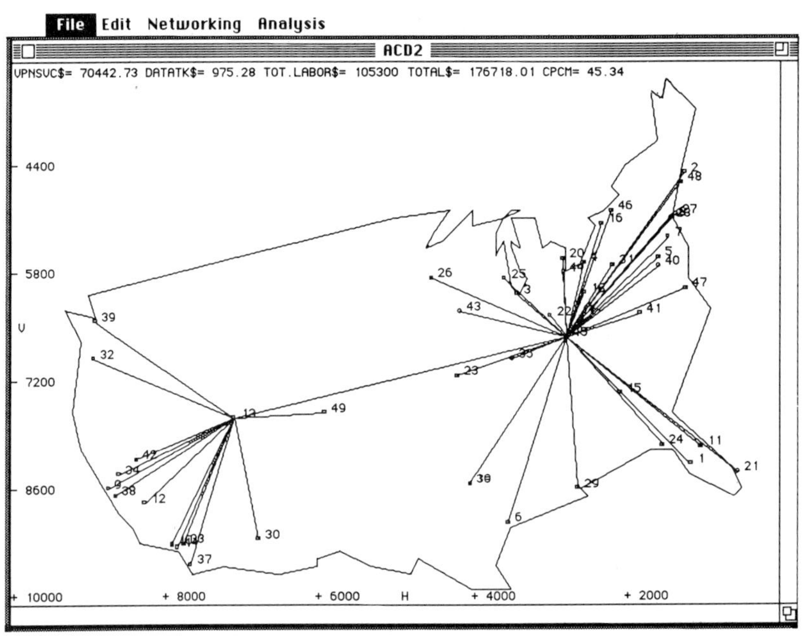

Figure 7.41. An ACD network topology with two switches: 49 regions and AT&T's 800 SVC.

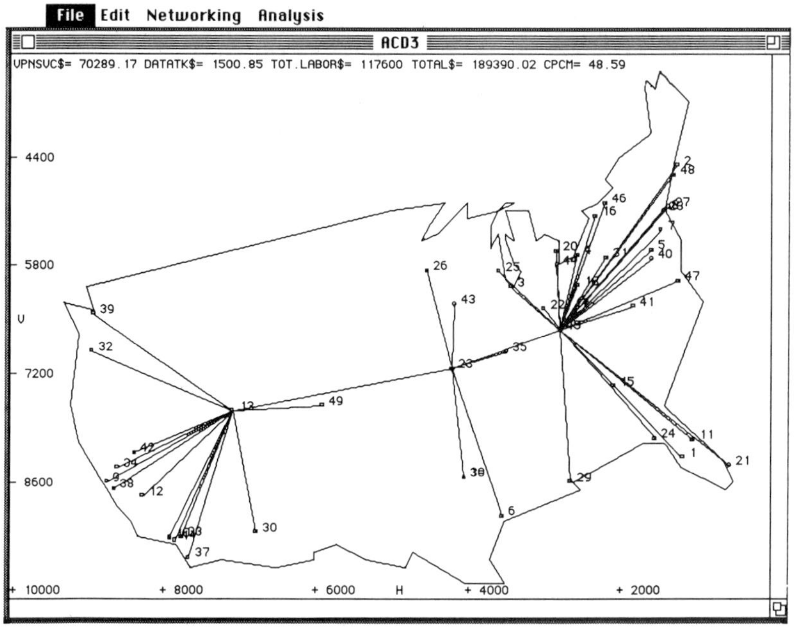

Figure 7.42. An ACD network topology with three switches: 49 regions and AT&T's 800 SVC.

mum values of T and P for a particular GOS specification as correctly as possible. Unfortunately, the state-of-the-art of teletraffic theory does not yield a closed form solution at this time for all practical situations.

The analytical models described earlier assumed two stochastically independent subsystems for computing the number of trunks and agents. The Erlang B and Erlang C formulas used for computing the number of trunks and number of agents, respectively, assume infinite sources and a negative exponential service time distribution. The real situations differ substantially from the ideal conditions assumed for mathematical ease.

The two subsystems are quite interdependent on one another. The number of required agents should consider the fact that the number of calls waiting to be served cannot exceed the number of trunks employed, and it is never infinite. Furthermore, the service time distributions are becoming more and more of the constant-value variety, especially through the use of voice response units (VRUs). In most cases, the upper limit to service time is generally not much larger than the expected call duration (ECD). In any case, closed form solutions do not exist for such general cases. Fortunately, the number of agents required should be less than predicted by the simplified theory. The case of infinite sources for the Erlang B formula may be the only one assumption that remains valid in the light of the results shown in Figure 3.9 when computing the number of trunks. The curves of Figure 3.9 suggest that, if the number of trunks exceed 500, the assumption of infinite sources holds.

The assumption that the call disappears from the system if it is blocked may not apply

Table 7.18 ACD Network Design Data

```
NO. OF SWITCHES= 1                 AT         23
AL BLOCKING FACTOR= .01            AL TRUNK LINK TYPE= 0              #PKHRS/DAY(Fph)= 8.333
ACD-TYPE= 1      DREQ(SEC)= 30     PROB(DELAY>DREQ)= .15                 LABOR COST($/HR)= 10
TRAFFIC GROWTH FACTOR(TGF)= 1      EXP.CALL DURATION-SEC= 300
(A) ACCESS LINE ANALYSIS
```

#	SW	ALs	10*MIs	$/AL	$/ALB
1	23	0	10238	0	1466.87
2	23	0	12456	0	948.16
3	23	0	4110	0	2442.92
4	23	0	6978	0	1411.09
5	23	0	9596	0	747.37
6	23	0	6393	0	998.36
7	23	0	10339	0	1623.04
8	23	0	10930	0	3137.32
9	23	0	15014	0	1118.28
10	23	0	13521	0	2573.99
11	23	0	10445	0	1168.47
12	23	0	13779	0	1135.01
13	23	0	9219	0	1252.13
14	23	0	13397	0	1597.94
15	23	0	6740	0	1634.19
16	23	0	8561	0	1355.32
17	23	0	5388	0	1411.09
18	23	0	6191	0	1305.12
19	23	0	4468	0	1020.67
20	23	0	6414	0	1497.54
21	23	0	12188	0	1280.02
22	23	0	4513	0	1330.22
23	23	0	0	0	1385.99
24	23	0	8900	0	1486.39
25	23	0	4378	0	1408.3
26	23	0	4080	0	725.06
27	23	0	11131	0	895.18
28	23	0	10833	0	6765.45
29	23	0	6759	0	1335.8
30	23	0	10444	0	1447.35
31	23	0	7784	0	1709.49
32	23	0	14921	0	1355.32
33	23	0	13070	0	1466.87
34	23	0	14421	0	1280.02
35	23	0	2379	0	817.09
36	23	0	4480	0	1299.54
37	23	0	13298	0	1578.42
38	23	0	14807	0	1377.63
39	23	0	15004	0	1486.39
40	23	0	9413	0	1048.56
41	23	0	7933	0	1168.47
42	23	0	13525	0	998.36
43	23	0	2604	0	1068.08
44	23	0	6056	0	1207.51
45	23	0	4787	0	1238.19
46	23	0	9222	0	1550.53

to some systems. A percentage of callers always redial immediately after experiencing blocking. Similarly, the assumption that, when a call enters the ACD queue, it will remain there forever until it is answered, may also not hold in some cases. A percentage of callers always abandon the queue after waiting a certain number of seconds and, in some cases, redial again. This will complicate the analysis measurably.

The presence of many input queues dealing with separate trunk bundles (e.g., FX line bundles from different cities, and 800-Service bundles for each mileage-band) also com-

Table 7.18 (*Continued*)

47	23	0	10055	0	1224.25
48	23	0	12093	0	1238.19
49	23	0	5570	0	1486.39

(B) ACCESS LINE ANALYSIS SUMMARY

SW LOC	#VTERMs	BHRERL	NCH/MO	SW COST($)
23	37	25.999	77948	72504.2555034

(C) ACD AGENT POSITION ANALYSIS

SW#	HOUR	#MERLs	#AP REQD	TMNHRS
23	1	2166	5	5
23	2	2166	5	10
23	3	2166	5	15
23	4	2166	5	20
23	5	4333	7	27
23	6	4333	7	34
23	7	4333	7	41
23	8	6499	10	51
23	9	8666	13	64
23	10	12999	17	81
23	11	17332	22	103
23	12	25999	32	135
23	13	21665	27	162
23	14	25999	32	194
23	15	21665	27	221
23	16	17332	22	243
23	17	8666	13	256
23	18	6499	10	266
23	19	4333	7	273
23	20	4333	7	280
23	21	4333	7	287
23	22	4333	7	294
23	23	2166	5	299
23	24	2166	5	304

TOTAL VIRTUAL NETWORK COST ($)= 72504.18   FOR 37 VIRTUAL LINES
TOTAL COST OF LABOR($)= 91200
TOTAL #CALLS PER MO. & BHR ERLANGS = 77948   25.999

plicates the analysis involving a common pool of agents. Whittaker (1975) has analyzed such complexities involved with the analysis of a multiqueue and a shared multiserver system.

One can make the following refinements to improve the quality of the results described earlier:

1. With the availability of the 800-Service, one may prefer to create only one input queue by eliminating different customer service area bands and/or separate FX-line bundles. The results show that costs do not change substantially while simplifying the computations.
2. One should compute the average delay for all handled calls (as expressed by Equation 3.11b or Equation 3.13), and add this to the service time for trunks. The ring time should also be added to the trunk service time in case the system does not provide a delay-recording after the first ring.
3. One may employ an improved value of the traffic offered to the agents (equal to the offered traffic to the trunks minus the blocked traffic). In some case, a certain

Table 7.19 ACD Network Design Data

```
NO. OF SWITCHES= 2                AT        45  13
AL BLOCKING FACTOR= .01           AL TRUNK LINK TYPE= 0            #PKHRS/DAY(Fph)= 8.333
ACD-TYPE= 1      DREQ(SEC)= 30    PROB(DELAY>DREQ)= .15               LABOR COST($/HR)= 10
TRAFFIC GROWTH FACTOR(TGF)= 1     EXP.CALL DURATION-SEC= 300
(A) ACCESS LINE ANALYSIS
```

#	SW	ALs	10*MIs	$/AL	$/ALB
1	45	0	7254	0	1425.16
2	45	0	8223	0	921.2
3	45	0	2685	0	2373.46
4	45	0	3106	0	1370.97
5	45	0	4973	0	726.12
6	45	0	7990	0	969.97
7	45	0	5798	0	1576.89
8	45	0	6492	0	3048.11
9	13	0	5979	0	1086.48
10	13	0	5784	0	2500.81
11	45	0	7117	0	1135.25
12	13	0	5074	0	1102.74
13	13	0	0	0	1216.53
14	13	0	5798	0	1552.5
15	45	0	3178	0	1587.73
16	45	0	4806	0	1316.78
17	45	0	891	0	1370.97
18	45	0	1892	0	1268.01
19	45	0	7217	0	991.65
20	45	0	3145	0	1454.96
21	45	0	8935	0	1243.63
22	45	0	1067	0	1292.4
23	45	0	4787	0	1346.59
24	45	0	5922	0	1444.13
25	45	0	3483	0	1368.26
26	45	0	6028	0	704.45
27	45	0	6689	0	869.73
28	45	0	6397	0	6573.1
29	45	0	6203	0	1297.82
30	13	0	5031	0	1406.19
31	45	0	3427	0	1660.88
32	13	0	6339	0	1316.78
33	13	0	5545	0	1425.16
34	13	0	5316	0	1243.63
35	45	0	2411	0	793.86
36	45	0	7229	0	1262.59
37	13	0	6232	0	1533.54
38	13	0	5856	0	1338.46
39	13	0	6985	0	1444.13
40	45	0	4742	0	1018.74
41	45	0	3146	0	1135.25
42	13	0	4372	0	969.97
43	45	0	4525	0	1037.71
44	45	0	2616	0	1173.18
45	45	0	0	0	1202.99
46	45	0	5427	0	1506.44
47	45	0	5269	0	1189.44
48	45	0	7820	0	1202.99
49	13	0	3697	0	1444.13

percentage of customers may try to redial immediately after experiencing a call blockage. See Chapter 3 for a discussion of the effects of retrials on a pure Loss System.

4. One may employ an improved value of traffic offered to the agents by considering the effects of call abandonments and/or after-call work times. See a published paper by Hoffman (1983) for a technique to consider the effects of call abandonments.

Table 7.19 (*Continued*)

(B) ACCESS LINE ANALYSIS SUMMARY

SW LOC	#VTERMs	BHRERL	NCH/MO	SW COST($)
45	29	18.772	56281	51025.8616289
13	14	7.227	21667	19416.9372638

(C) ACD AGENT POSITION ANALYSIS

SW#	HOUR	#MERLs	#AP REQD	TMNHRS
45	1	1564	4	4
45	2	1564	4	8
45	3	1564	4	12
45	4	1564	4	16
45	5	3128	6	22
45	6	3128	6	28
45	7	3128	6	34
45	8	4693	8	42
45	9	6257	10	52
45	10	9386	13	65
45	11	12514	17	82
45	12	18772	24	106
45	13	15643	20	126
45	14	18772	24	150
45	15	15643	20	170
45	16	12514	17	187
45	17	6257	10	197
45	18	4693	8	205
45	19	3128	6	211
45	20	3128	6	217
45	21	3128	6	223
45	22	3128	6	229
45	23	1564	4	233
45	24	1564	4	237
13	1	602	2	239
13	2	602	2	241
13	3	602	2	243
13	4	602	2	245
13	5	1204	3	248
13	6	1204	3	251
13	7	1204	3	254
13	8	1806	4	258
13	9	2409	5	263
13	10	3613	7	270
13	11	4818	8	278
13	12	7227	11	289
13	13	6022	9	298
13	14	7227	11	309
13	15	6022	9	318
13	16	4818	8	326
13	17	2409	5	331
13	18	1806	4	335
13	19	1204	3	338
13	20	1204	3	341
13	21	1204	3	344
13	22	1204	3	347
13	23	602	2	349
13	24	602	2	351

TOTAL VIRTUAL NETWORK COST ($)= 70442.73 FOR 43 VIRTUAL LINES
TOTAL COST OF LABOR($)= 105300
TOTAL #CALLS PER MO. & BHR ERLANGS = 77948 25.999

A software program based on a complex theory developed by Smith and Smith (1962) was employed to compute an *exact* value of (1) the blocking probability experienced by trunks, and (2) the probability of exceeding a given delay-limit and for any expected call duration for a given number of trunks (T) and agent positions (P) for a composite

Table 7.20 ACD Network Design Data

```
NO. OF SWITCHES= 3                  AT         23  45  13
AL BLOCKING FACTOR= .01             AL TRUNK LINK TYPE= 0             #PKHRS/DAY(Fph)= 8.333
ACD-TYPE= 1      DREQ(SEC)= 30      PROB(DELAY>DREQ)= .15               LABOR COST($/HR)= 10
TRAFFIC GROWTH FACTOR(TGF)= 1       EXP.CALL DURATION-SEC= 300
(A) ACCESS LINE ANALYSIS
```

#	SW	ALs	10*MIs	$/AL	$/ALB
1	45	0	7254	0	1422.05
2	45	0	8223	0	919.2
3	45	0	2685	0	2368.29
4	45	0	3106	0	1367.98
5	45	0	4973	0	724.54
6	23	0	6393	0	967.86
7	45	0	5798	0	1573.45
8	45	0	6492	0	3041.47
9	13	0	5979	0	1084.11
10	13	0	5784	0	2495.36
11	45	0	7117	0	1132.78
12	13	0	5074	0	1100.33
13	13	0	0	0	1213.88
14	13	0	5798	0	1549.12
15	45	0	3178	0	1584.27
16	45	0	4806	0	1313.91
17	45	0	891	0	1367.98
18	45	0	1892	0	1265.25
19	23	0	4468	0	989.49
20	45	0	3145	0	1451.79
21	45	0	8935	0	1240.92
22	45	0	1067	0	1289.58
23	23	0	0	0	1343.65
24	45	0	5922	0	1440.98
25	45	0	3483	0	1365.28
26	23	0	4080	0	702.91
27	45	0	6689	0	867.83
28	45	0	6397	0	6558.77
29	45	0	6203	0	1294.99
30	13	0	5031	0	1403.13
31	45	0	3427	0	1657.26
32	13	0	6339	0	1313.91
33	13	0	5545	0	1422.05
34	13	0	5316	0	1240.92
35	23	0	2379	0	792.13
36	23	0	4480	0	1259.84
37	13	0	6232	0	1530.2
38	13	0	5856	0	1335.54
39	13	0	6985	0	1440.98
40	45	0	4742	0	1016.52
41	45	0	3146	0	1132.78
42	13	0	4372	0	967.86
43	23	0	2604	0	1035.45
44	45	0	2616	0	1170.63
45	45	0	0	0	1200.36
46	45	0	5427	0	1503.16
47	45	0	5269	0	1186.85
48	45	0	7820	0	1200.36
49	13	0	3697	0	1440.98

ACD. The values of P and T were varied using the trial and error approach until the desired GOS was achieved. The results were also computed using a simplified and a refined method. The *simplified* method assumes two independent subsystems with identical negative exponential service time distributions. The *refined* method assumes no retrials or call abandonments but assumes lower traffic for agents and higher service

Table 7.20 (*Continued*)

(B) ACCESS LINE ANALYSIS SUMMARY

SW LOC	#VTERMs	BHRERL	NCH/MO	SW COST($)
23	8	2.623	7862	7119.48042122
45	26	16.149	48419	43752.8521305
13	14	7.227	21667	19416.9372638

(C) ACD AGENT POSITION ANALYSIS

SW#	HOUR	#MERLs	#AP REQD	TMNHRS
23	1	218	2	2
23	2	218	2	4
23	3	218	2	6
23	4	218	2	8
23	5	437	2	10
23	6	437	2	12
23	7	437	2	14
23	8	655	2	16
23	9	874	3	19
23	10	1311	3	22
23	11	1748	4	26
23	12	2623	5	31
23	13	2185	5	36
23	14	2623	5	41
23	15	2185	5	46
23	16	1748	4	50
23	17	874	3	53
23	18	655	2	55
23	19	437	2	57
23	20	437	2	59
23	21	437	2	61
23	22	437	2	63
23	23	218	2	65
23	24	218	2	67
45	1	1345	4	71
45	2	1345	4	75
45	3	1345	4	79
45	4	1345	4	83
45	5	2691	5	88
45	6	2691	5	93
45	7	2691	5	98
45	8	4037	7	105
45	9	5383	9	114
45	10	8074	12	126
45	11	10766	15	141
45	12	16149	21	162
45	13	13457	18	180
45	14	16149	21	201
45	15	13457	18	219
45	16	10766	15	234
45	17	5383	9	243
45	18	4037	7	250
45	19	2691	5	255
45	20	2691	5	260
45	21	2691	5	265
45	22	2691	5	270
45	23	1345	4	274
45	24	1345	4	278
13	1	602	2	280
13	2	602	2	282
13	3	602	2	284
13	4	602	2	286
13	5	1204	3	289
13	6	1204	3	292
13	7	1204	3	295
13	8	1806	4	299
13	9	2409	5	304

Table 7.20 (*Continued*)

13	10	3613	7	311
13	11	4818	8	319
13	12	7227	11	330
13	13	6022	9	339
13	14	7227	11	350
13	15	6022	9	359
13	16	4818	8	367
13	17	2409	5	372
13	18	1806	4	376
13	19	1204	3	379
13	20	1204	3	382
13	21	1204	3	385
13	22	1204	3	388
13	23	602	2	390
13	24	602	2	392

```
          TOTAL VIRTUAL NETWORK COST ($)= 70289.17    FOR 48 VIRTUAL LINES
          TOTAL COST OF LABOR($)= 117600
          TOTAL #CALLS PER MO. & BHR ERLANGS = 77948             25.999

(D) TRUNKING ANALYSIS

        TRUNK LINK/NODE TYPE 0 (MAX LINK CAPACITY@ 9600BPS)
        FDX FACTOR= 1 ,                          BBTF= 2
        MAX TRUNK LOAD = .8 ,                    TKMF=1
        TOTAL INTER/INTRA NODAL TRAFFIC= 80.09 BPS

ORIGINAL TRUNK TRAFFIC MATRIX AS FOLLOW

 0            55.33          24.76
 0            0              0
 0            0              0

MODIFIED TRUNK TRAFFIC MATRIX IS AS FOLLOWS
 0            55.33          24.76
 0            0              0
 0            0              0

MODIFIED TRUNK MATRIX IS AS FOLLOWS

 0            1              1
 0            0              0
 0            0              0

TRUNK COST MATRIX IS AS FOLLOWS

FROM          TO             MILES          COST           TOTALS
 23           45             479            677.57         677.57
 23           13             922            823.28         823.28

        *****SUMMARY  OF SYSTEM  COSTS*****
           VIRTUAL LINE COST= 70289.17
           TOTAL TRUNK COST = 1500.85
           TOTAL SYSTEM COST= 71790.02
```

times for trunks, as discussed earlier. The results are summarized in Table 7.21 for a system handling calls of average duration 300 seconds each, with a trunk GOS of 0.01 blocking probability, and the probability of delay exceeding 30 seconds equalling 0.15.

The results show that the refined method yields a value of T close to the exact solution. The difference between the computed values of P obtained from the refined and exact methods becomes larger with increasing values of erlangs. The values of P obtained

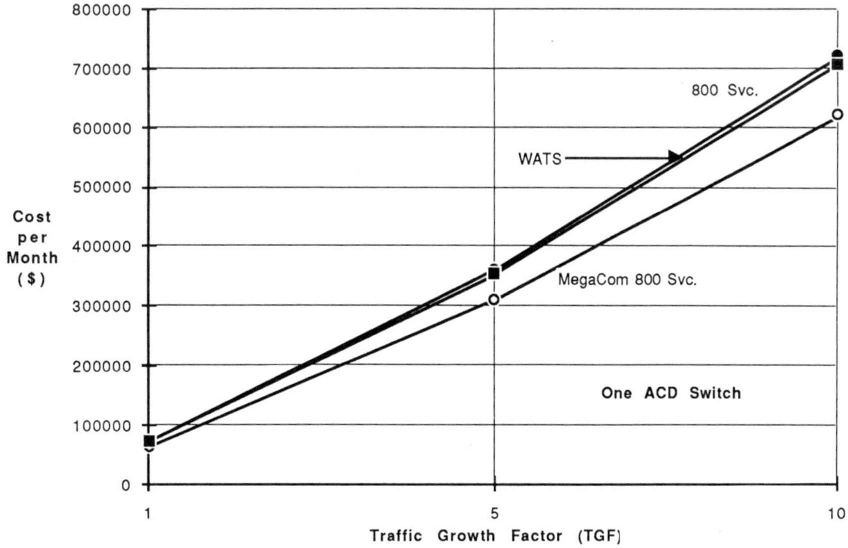

Figure 7.43. Costs per month versus the TGF: 49 regions and AT&T's 800-, MegaCom 800-, and WATS services.

from the exact method are always lower than those obtained from the refined method, for the reasons discussed earlier. By comparing the monthly costs of a trunk with those of an agent, one can compute the penalty of not using the exact method. The exact

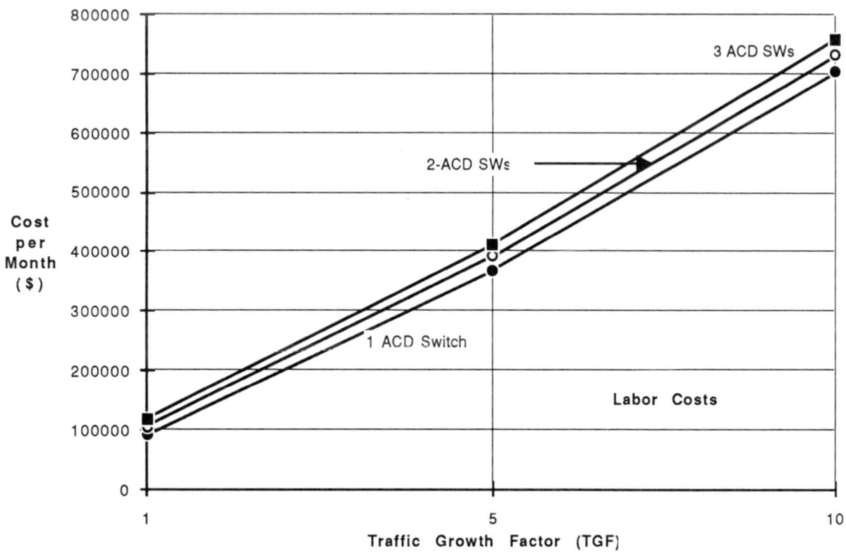

Figure 7.44. Labor costs per month versus the TGF: 49 regions, one to three switches, and any virtual private network service.

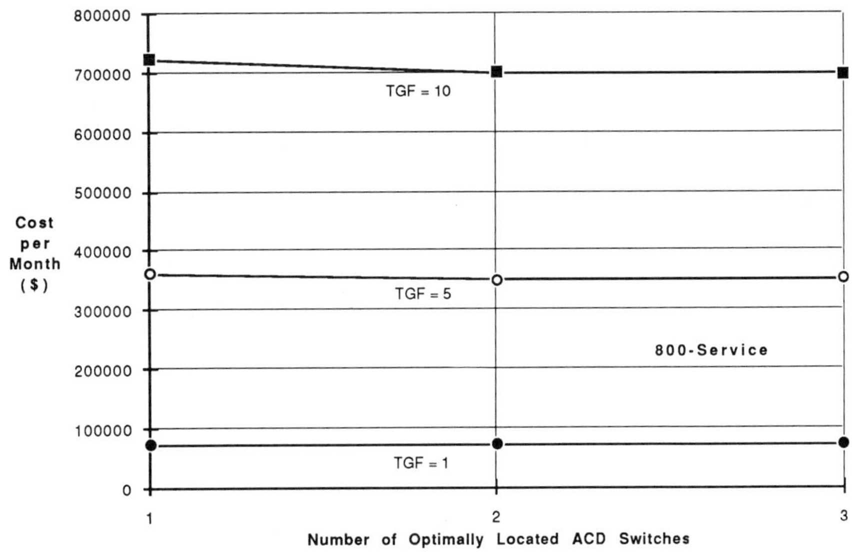

Figure 7.45. Monthly costs of AT&T's 800SVC versus the number of optimally located ACD switches: 49 regions and TGF = 1, 5, or 10.

method employed here does not allow a closed-form solution suitable for a computer. It appears that Whittaker (1975) has developed a closed-form solution, but the analytical complexity of his approach discouraged the author from developing a suitable software

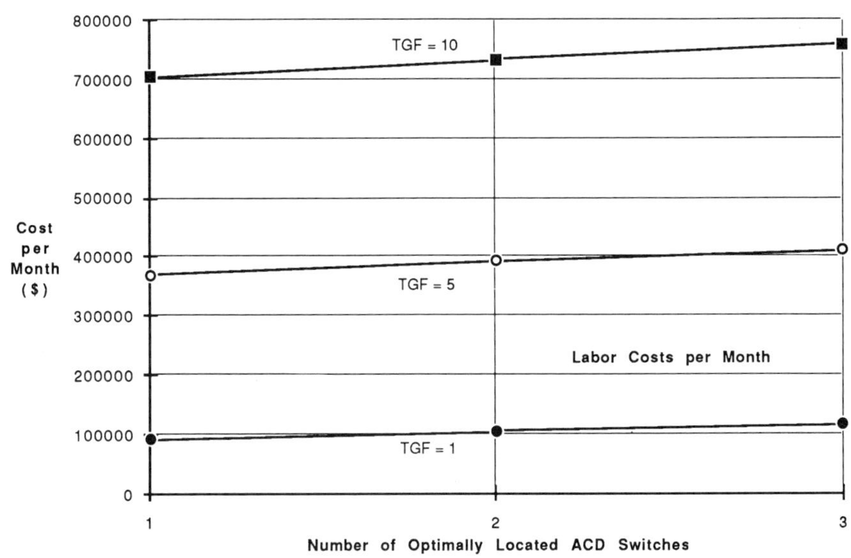

Figure 7.46. Labor costs per month versus the number of optimally located ACD switches: 49 regions and TGF = 1, 5, or 10.

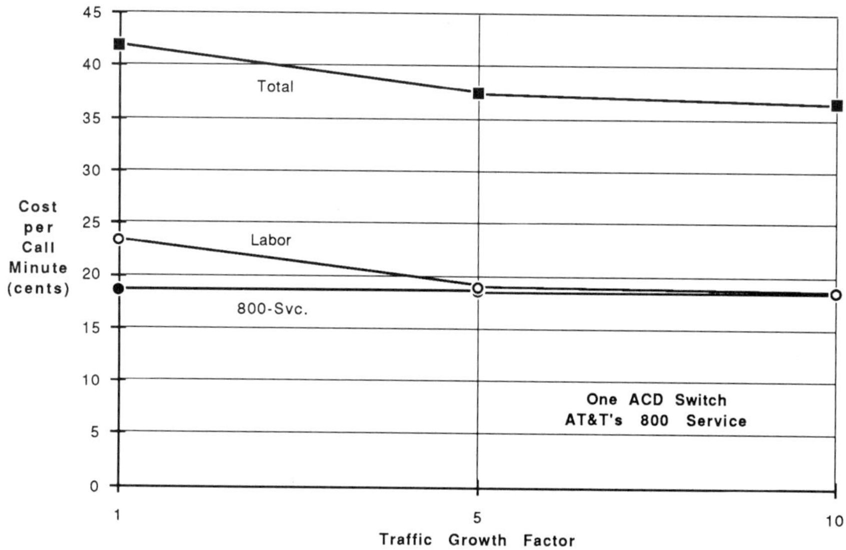

Figure 7.47. Costs per call minute versus the TGF: 49 regions, one ACD switch, and AT&T's 800SVC.

program. Since the exact method assumes a state of statistical equilibrium to derive its answers, and since the number of agents is altered every half-hour or hour which will prevent the achievement of such a state, the results obtained from the exact method will

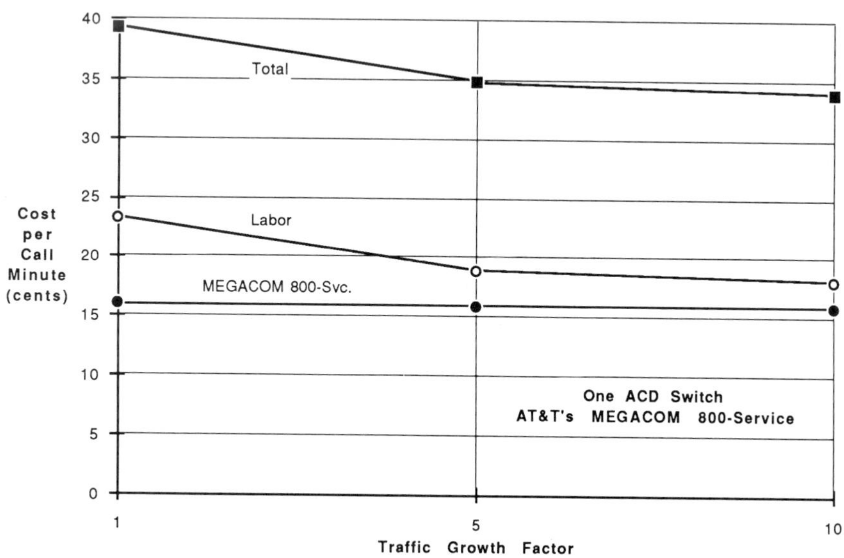

Figure 7.48. Costs per call minute versus TGF: 49 regions, one ACD switch, and AT&T's MegaCom 800SVC.

Table 7.21 Comparison of Various Methods for Computing ACD Performance

	ACD Design Parameters		
	A=3 erlangs	A= 30 erlangs	A= 300 erlangs
Simplified Method:			
Number of Agents(P)	6	36	311
Number of Trunks(T)	8	42	324
Refined Method:			
Number of Agents(P)	6	35	309
Number of Trunks(T)	10	46	338
Exact Method:			
Number of Agents(P)	6	35	304
Number of Trunks(T)	9	45	347

never be quite meaningful. Furthermore, the number of trunks available is always higher than the exact value of trunks required during non-busy hours. Theory shows that such a state will always demand a higher number of agents. This can be explained by higher traffic loads resulting from a lower blocking. Therefore, the refined method should be adequate to give a workable solution. The values of P during busy hours can be reduced carefully if the observed GOS is always much better than what is consistently required. Finally, the refined method is recommended, since it provides a robust system while simplifying the analytical methods.

7.3.2 SUMMARY OF ACD NETWORK SYNTHESES AND CONCLUSIONS

The plots of Figure 7.43 clearly show that the costs for MegaCom 800 Service are approximately 15% lower than those for the pure 800-Services, for all values of TGF. The costs of WATS are approximately 2 percent lower than those for the pure 800-Service, for all values of TGF. Furthermore, one obtains only a 2 percent gain when considering a unit cost per call as the value of TGF is increased from 1 to 10. This may be explained by the fact that the call-hours handled by each line and for each rate step increases only slightly, since the number of VPN lines also increases with TGF. According to the plots of Figure 7.45, the costs of VPN services are slightly reduced (approximately 5–7 percent) as the number of ACD centers is increased from one to three. This results from the use of lower values of customer service bands, as defined by Table 7.15, as the ACD center becomes closer to the traffic sources/destinations.

According to Figure 7.44, costs of agent-hours are not dependent upon the type of VPN services employed. The costs of agent-hours rise slightly as the number of switches is increased. This phenomena can be explained by the fact that the law of large numbers does not apply with an increasing number of ACD centers. The labor costs also rise proportionally with the TGF.

Only fixed labor costs were assumed for all locations in the study. It is possible to locate the switches at places with low labor costs. Another important fact to remember is that the labor costs for a system employing WATS tariffs may not require as many agents as for the ACD handling only incoming calls. The data base generates the outgoing calls as a function of the agent availability. Consequently, the labor cost is then directly determined by the total agent erlangs during the month. Such a quantity is printed in the results. In some cases, the WATS leased lines are employed for handling only the outgoing calls for the employees. For that case, the agent costs can be disregarded.

The results shown in this chapter assume a single bundle of VPN lines. Such a bundle will represent the largest service-area band experienced by the calls. It is possible to lease several VPN line bundles, each reflecting a unique service-area band. But any resulting savings realized from reduced leasing costs will be neutralized by the increase in the total number of required VPN lines. See Figure 3.7 for a justification of this fact when one must maintain the same GOS for each separate bundle.

Figures 7.47 and 7.48 plot costs per call-minute for the VPN service (800-Service and MegaCom 800 Service) and agent labor as a function of TGF. The results are plotted for a single optimally located ACD center. The results show that, whereas MegaCom 800 Service costs about 16 cents per call-minute, the 800-Service costs about 18 cents per call-minute. The labor costs range between 24 and 18 cents per call-minute as TGF is increased from 1 to 10. The labor costs were computed using the assumption that an agent-hour costs $10.00, including overheads dealing with employee benefits and break periods. The results show that one can reduce the costs of an ACD system drastically by employing voice response units (VRUs) instead of live agents. However, this step should only be taken after much planning and study. While cost savings are important, the perception of an indifferent service associated with VRUs must always be kept in mind.

Compared with the costs of a call minute for private voice networks (e.g., see Figures 7.6, 7.17, and 7.29), the VPN service for ACD yields identical costs for large TGF values. For small values of TGF, the cost of VPN services for ACD operations (about 15–21 cents per call minute, as shown in Figures 7.47 and 7.48) is lower than that for the VPN services for plain voice applications (e.g., an ECC of $1.31 divided by 5, or equal to 26 cents per call minute) even for a low monthly call volume of about 78000 (corresponding to TGF = 1). This fact shows the need for selecting the correct VPN / SDN type service for voice networks with small traffic volumes.

It can also be observed that the total cost of a call-minute does not vary with the number of ACD centers. However, the user may decide to realize a multicenter ACD network in order to increase the network availability or survivability. AT&T offers several advanced 800-Services, such as (1) single number service for both intrastate and interstate calls, (2) 800 Area Code Routing for several ACD centers, (3) 800 Call Allocator for distributing the calls according to the capability of an ACD center, (4) 800 Command Routing to handle planned emergencies/call routing to various ACD centers, (5) 800 Time Manager to allow the use of only one (or as few as desired) ACD center during low traffic periods, and so on for achieving greater flexibility and/or survivability of any given multicenter ACD network. Other OCCs offer similar services.

The design of an ACD system to handle only outgoing calls was not discussed in this chapter, due to the fact that delays are directly influenced by the way new calls are launched. In some systems, no new calls are launched during periods characterized by finite queues. In other systems, new calls are launched as a function of the queue lengths. In any case, observed delays are invariably much smaller than those experienced by an ACD designed to handle only incoming calls. Since most ACDs that handle outgoing calls also handle a much greater number of incoming calls, the delay analysis, and the system design, will be primarily influenced by the methodology discussed earlier.

8

Synthesis of Data Network Topologies

8.1 INTRODUCTORY REMARKS

Private data networks exist to provide communications paths between data terminals, PCs, and host computers. About 20 million data communications devices are in use at this time. About 30 percent of these devices are on LANs, and the number is increasing rapidly. The remaining devices are served by either a proprietary network (e.g., IBM's SNA) or a set of point-to-point connections composed of modems and/or multiplexers.

The LAN population is growing at a compound annual rate of about 30 percent. The installed base is mostly composed of CSMA/CD based LANs. However, the token ring LANs are catching up fast. See Chapter 4 and Appendix B for discussions of performance issues and related standards.

The data traffic on the existing LANs is generally packet switched. Some of the low speed data traffic is also handled by existing private voice networks in the form of circuit-switched data calls. A data call is established just like an ordinary voice call. Each established path is generally characterized by short bursts of data separated by long periods of inaction in both directions of traffic flow. Consequently, the use of data calls on a private voice network should be discouraged, in general, and for long-haul connections in particular.

The network topology of a LAN is generally determined by the choice of architecture, and it cannot be optimized further. The choice of a particular LAN is generally dictated by its capability to serve the existing family of PCs and hosts, related applications, ease of use, flexibility related to moves or growth, and the cost of premise wiring/cabling (see Chapter 9 for further discussions) and media interface units.

WANs for data communications can be synthesized in several ways. One can synthesize a WAN simply by interconnecting widely dispersed LANs via bridges, routers, or gateways. Bridges link LANs at the media access control layer, routers link LANs at the network layer, and gateways link dissimilar LANs at higher layers for protocol conversions. See Appendix B for further discussions on related standards. Current technology allows bridges and routers to provide very high speed connections at low costs. Such an approach may seem to be quite attractive considering expediency and departmental control. However, this approach may foster the birth of isolated corporate networks and thus postpone the eventual integration process.

One way to integrate widely dispersed LANs is to employ a backbone network based on the X.25 based packet-switching standard (see Appendix B). This should allow a full sharing of transmission facilities and integrated network management and control.

Another way to synthesize a WAN is through the use of intelligent T1 multiplexers. The bandwidth on shared T1 highways is allocated dynamically to voice or data LANs for internetworking. This approach can be extended to integrate the voice traffic on shared T1 highways of a backbone network. This approach generally provides connections based on the physical layer only. The higher layers are generally vendor dependent.

Another way to synthesize future WANs is through the evolving ISDN-related standards, as described in Appendix B. Future LANs will access a predetermined bandwidth from ISDN COs at the physical layer for the purpose of internetworking. Of course, one can also use the packet-switching capability of the ISDN COs to synthesize a private backbone network based on the X.25 standard.

The question as to which WAN approach will dominate the future marketplace can only be answered by the forces of free enterprise and technology. One should avoid the use of a purely dogmatic approach. No matter what other approach is employed, special network synthesis techniques must be employed to study unique topological and response time constraints imposed by the available packet-switched network architectures. To illustrate, IBM's SNA architecture employs multidrop topology and is generally characterized by low media utilization and large delays resulting from the associated protocols. CCITT's X.25/X.75 architecture employs a mixed star/multidrop topology for the access network and a mesh topology for the backbone network, as illustrated in Figures 1.8 and 1.9. AT&T's ISN architecture is characterized by the use of virtual packets and fast packet switching at the backbone network, but is generally constrained by the star topology at the access line level. The multistar, multicenter topology of Figure 1.8 will generally apply. Doelz's ESR architecture is characterized by protocol-independent, virtual (point-to-point and multipoint) packet switching and high media utilization, but is constrained by the directed-link topology of Figure 1.10.

It should be obvious now that each WAN architecture will require the use of specialized algorithms for computing the relevant response times and optimizing the associated network topologies. In the following pages, we present a number of examples that illustrate the use of well-known network synthesis algorithms, as described in Chapter 5. To limit the scope of this book, only packet-switching techniques will be considered for optimizing data network topologies.

8.2 THE BASIC DESIGN PROCESS FOR A DATA NETWORK

We will synthesize data communications networks using the software package described in Chapter 6. The methodology employed here is similar to that employed for voice networks. To begin with, we will select the appropriate software module to synthesize/analyze the desired network topology. The next step will be to prepare all of the necessary input files, as illustrated in Figure 6.1. The first data file to be prepared is the VHD file. It defines, for each customer location, the associated V and H coordinates and the BHR traffic intensities. Its preparation will be facilitated by the availability of a detailed location/site data file, as illustrated in Figure 6.3. In particular, the knowledge of the street address, its full telephone number, the types of data terminals or hosts located on the premises and any knowledge of associated traffic loads will help to facilitate the network design process. The time-consistent BHR traffic intensity data about each site will be defined in terms of bits per second for data communications, as opposed to millierlangs for voice communications.

There are two ways to define the LINK file for each data site: (1) create a separate LINK file, as defined in Chapter 7, or (2) create a Virtual Circuit Link (VCL) file that defines not only the LINK type for each site, but also up to seven-types of data terminals. Consequently, one can use the VCL file to compute the actual data traffic intensity at each site by specifying a seven-element VC traffic weight vector through SDF parameters 41 to 47. Such a file is quite useful for designing data networks based on some types of Doelz's ESR architecture. The results reported in this book consider only the first approach.

The NAME, NLT, SDF, and the TARIFF files, as described in Chapters 6 and 7, remain unchanged. It may be useful to repeat the fact that the NLT file relates the LINK type number (0, 1, 2, 3 . . .) to the tariff type (1, 2, 3, . . .) and other link attributes, such as link capacity, maximum allowable link rate (equal to Wm), and potential nodal rates. In case a LINK file is not specified, the Link type for each site is assumed to equal "0." The trunk-link type and the maximum allowable utilization for the backbone trunks is specified by the SDF parameters labelled as TKLT and MTKU, respectively. The design parameter F_{fdx} (1 for FDX and others for HDX) specifies whether the trunk can handle FDX data flows. The design parameter BBTF can be used for determining the amount of traffic flowing on the backbone trunks when no From-To files are used. BBTF equal to 1 corresponds to uniformly distributed traffic from every node. BBTF equal to 2 cuts the backbone traffic by a factor of 2. Higher values of BBTF will reduce the cost of trunking even further.

Special design parameters must be specified for computing the response times for each netlink. User port and host port rates are specified by SDF parameters UPR and HPR. SDF parameters IML, RML, NCU, LAMBDA, T_{np}, T_{hm}, BLKL, ICPB, K_{pg}, ATP, and HTT specify the input message length, response (or output) message length (in bytes), number of control units per virtual multipoint circuit (VMPT), average number of messages arriving for a VMPT circuit, nodal packet processing time (ms), half-modem delay (ms), block size (in bytes), information bytes per block/packet, propagation time constant per mile, analysis type (ATP) (1 for ASYNC, 2 for Doelz VMPT, and 3 for SNA multidrop), and host think time respectively. Additional design parameters such as F_{opt} (0 for no optimization, 1 for optimization), FTF (0 for no From-To file, 1 for the use of From-To file), NWF (0 for use of nodal traffic loads, as specified by the VHD file; 1 for using the nodal weights, as specified by the VCL and NLT files). A special design parameter called the F_{dis} is employed to obtain correct distances in miles or kilometers for countries not situated in North America.

In the analysis presented later, each network is analyzed for on-net to on-net traffic flows only. We will obtain the cost of netlinks, backbone trunks, and the entire network for each synthesis job. These values will appear on the associated network graph. The netlinks and the backbone network are completely described in terms of connections and corresponding costs on the printouts that are stored on the disk for later printout if desired. The total number of bits handled by the network during a month will be computed by assuming F_{ph} busy hours per business days and 20 business days per month. We will use this value to derive the cost of transmitting one million bits of user data by the network. This value will also be shown on the associated network graph and printouts. Such a unit cost figure should be quite meaningful for data networks, since most VAN vendors also charge by the number of packets handled. For voice networks, we employed the unit cost of a voice call-minute for network evaluation purposes.

We will start by using the DLNET1.0 and MDNET1.0 modules of the ECONETS package to synthesize networks based on the directed-link, and multidrop topologies, using the algorithms developed by the author during the last two decades. Networks based on the star topology will be designed later. Modules EWNET1.0, EWNET1.1, and EWNET1.2 will be employed to synthesize multidrop topologies based on three popular variations of the Esau-Williams algorithm, as described in Chapter 5. We will also vary the values of TGF, BBTF, Wm (equal to the maximum allowable traffic load on a netlink), and ATP to study the network costs and response times. The maximum and average values of system response times will also be studied along with the total and unit network costs, as discussed above.

8.3 NETWORK SYNTHESIS EXAMPLES

8.3.1 STUDY OF DATA NETWORK TOPOLOGIES WITH A SINGLE SWITCH

We will employ the previously defined VHD41 and VHD59 files, with some changes, to reflect data applications. The nodal traffic load, which represented erlangs for voice, will be multiplied by 300 to represent bits per second (bps) loads during the busy hour. The resulting values will be assumed to reflect a value of TGF = 1. See Tables 8.1 and 8.2 for data base definitions. Furthermore, we will employ two tariffs, the simple 1$/mile tariff and the set of FCC Tariffs 9, 10, and 11, for computing the monthly costs. The simple tariff should also enable the designer to study the total circuit miles. The nodal distribution graphs of Figures 7.1 and 7.12, and the COG analysis graphs of Figures 7.2 and 7.13, still apply. Since the so-called exact solution, as reported by Chandy and Lo (1973), suffers from (1) a lack of design flexibility, and (2) exponential growth of computation time with the number of nodes, only the algorithms based on heuristic principles will be considered. Three popular variations of the Esau-Williams (EW) algorithm and Sharma's DLNET and MDNET algorithms will be compared. The three variations of the EW algorithm can be defined as follows:

1. EWNET1.0 is a simplified version of the EW algorithm that prevents the merging of two components or netlinks, each with more than 1 node. This is supposed to provide good solutions with reduced execution time. An earlier study by Sharma (1983) was based on this variation and the simple 1$/mile tariff.
2. EWNET1.1 is identical to the original EW algorithm, as described in Chapter 5.
3. EWNET1.2 is identical to EWNET1.1 except that the computation of max(T) function is based on distances alone. Consequently, EWNET1.1 and EWNET1.2 provide identical results for the 1$/mile tariff.

Table 8.1 EOSN/Network Design Data Base for the VHD41 Input File

TOTAL NUMBER OF NODES = 41

#	-V-	-H-	LOAD (BPS)	LATA	LINK	NAME
1	7027	4203	2100	524	0	KANMO
2	9213	7878	2100	730	0	LAXCA
3	5015	1430	300	224	0	NWKNJ
4	4687	1373	300	920	0	HFDCT
5	6263	2679	300	922	0	CINOH
6	5972	2555	300	324	0	COLOH
7	4997	1406	2700	132	0	NYCNY
8	8492	8719	1500	722	0	SFOCA
9	7501	5899	300	656	0	DENCO
10	7489	4520	600	532	0	WICKA
11	7947	4373	300	536	0	OKLOK
12	8436	4034	300	552	0	DALTX
13	8266	5076	300	546	0	AMLTX
14	8938	3536	300	560	0	HOUTX
15	8483	2638	300	490	0	NOLLA
16	8549	5887	900	664	0	ALBNM
17	6807	3482	1800	520	0	STLMO
18	7010	2710	300	470	0	NASTN
19	8173	1147	300	952	0	TMPFL
20	8351	527	600	460	0	MIAFL
21	7260	2083	600	438	0	ATLGA
22	9345	6485	300	668	0	TCSAR
23	9135	6748	900	666	0	PHXAR
24	8486	8695	300	722	0	OAKCA
25	8665	7411	300	721	0	LASNE
26	5986	3426	1500	358	0	CHIIL
27	7707	4173	300	538	0	TULOK
28	5536	2828	300	340	0	DETMI
29	4422	1249	900	128	0	BOSMA
30	5574	2543	300	320	0	CLEOH
31	6261	4021	600	635	0	CDRIA
32	6272	2992	300	336	0	INDIN
33	6529	2772	300	462	0	LOUKY
34	6657	1698	300	422	0	CHTNC
35	6113	2705	300	328	0	DATOH
36	5251	1458	1500	228	0	PHLPA
37	5622	1583	900	236	0	WASDC
38	5510	1575	300	238	0	BALMA
39	5621	2185	1800	234	0	PITPA
40	5363	1733	300	226	0	HARPA
41	5166	1585	300	228	0	ALNPA
TOTAL TRAFFIC LOAD=			Σ 28500 BPS			

See Figures 8.1 to 8.12 for the plots that relate monthly costs to the number of EXCHANGE passes for the two VHD files, two previously described tariffs, TGF = 1, and three capacity constraints (or Wm): 3900 bps, 5100 bps, and 6300 bps. A maximum line capacity (or C) of 9600 bps is assumed for all cases. The EXCHANGE algorithm is identical to that described in Chapter 5, and it is always required for Sharma's MDNET algorithm. The three popular variations of EWNET never employ the exchange algorithm for further cost reductions. The cost values for the zero EXCHANGE pass should, therefore, reflect the traditional EW algorithms. The results described in this chapter show that additional savings are possible through the use of Sharma's exchange algo-

(Text continued on pg. 229)

Table 8.2 EOSN/Network Design Data Base for VHD59

TOTAL NUMBER OF NODES = 59

#	-V-	-H-	LOAD (BPS)	LATA	LINK	NAME
1	7260	2083	3600	438	0	ATLGA
2	7089	1674	600	442	0	AGSGA
3	6749	2001	300	420	0	AVLNC
4	5510	1575	300	238	0	BALMD
5	7518	2446	600	476	0	BHMAL
6	8806	3298	300	562	0	BPTTX
7	8476	2874	300	492	0	BTRLA
8	6901	1589	600	434	0	CAESC
9	7098	2366	300	472	0	CHATN
10	5986	3426	1800	358	0	CHIIL
11	7021	1281	300	436	0	CHSWV
12	6261	4021	1200	635	0	CIDIA
13	6657	1698	600	422	0	CLTNC
14	5972	2555	300	324	0	CMHOH
15	7556	2045	600	438	0	CSGGA
16	6263	2679	1200	922	0	CVGOH
17	8436	4034	1800	552	0	DALTX
18	6113	2705	300	328	0	DAYOH
19	5622	1583	900	236	0	WASDC
20	7501	5899	900	656	0	DENCO
21	5536	2828	600	340	0	DETMI
22	8409	3168	300	486	0	ESFLA
23	6729	3019	300	330	0	EVVIN
24	5942	2982	300	334	0	FWAIN
25	8479	4122	1800	552	0	GSWTX
26	7827	3554	300	528	0	HOTAR
27	8938	3536	900	560	0	HOUTX
28	6272	2992	600	336	0	INDIN
29	8035	2880	600	482	0	JANMS
30	7649	1276	300	452	0	JAXFL
31	8665	7411	300	721	0	LASNV
32	9213	7878	600	730	0	LAXCA
33	6459	2562	300	466	0	LEXKY
34	7721	3451	300	528	0	LITAR
35	7364	1865	300	446	0	MCNGA
36	7899	2639	300	482	0	MEIMS
37	7471	3125	1500	468	0	MEMTN
38	7692	2247	300	478	0	MGMAL
39	8351	527	3300	460	0	MIAFL
40	7027	4203	600	524	0	MKCMO
41	8148	3218	300	486	0	MLOLA
42	5781	4525	600	628	0	MSPMN
43	8483	2638	1500	490	0	MSYLA
44	4997	1406	2400	132	0	NYCNY
45	7954	1031	300	458	0	ORLFL
46	8165	606	300	460	0	PBIFL
47	5251	1458	300	228	0	PHLPA
48	6982	3088	300	468	0	PUKTN
49	7266	1379	600	440	0	SAVGA
50	6529	2772	300	462	0	SDFKY
51	9468	7629	300	732	0	SDOCA
52	8492	8719	600	722	0	SFOCA
53	7310	3836	300	522	0	SGFMO
54	8272	3495	600	486	0	SHVLA
55	7496	1340	300	440	0	SSIGA
56	6807	3482	900	520	0	STLMO
57	5704	2820	300	326	0	TOLOH
58	8173	1147	600	952	0	TPAFL
59	6801	2251	600	474	0	TYSTN
TOTAL TRAFFIC LOAD=			Σ 42000 BPS			

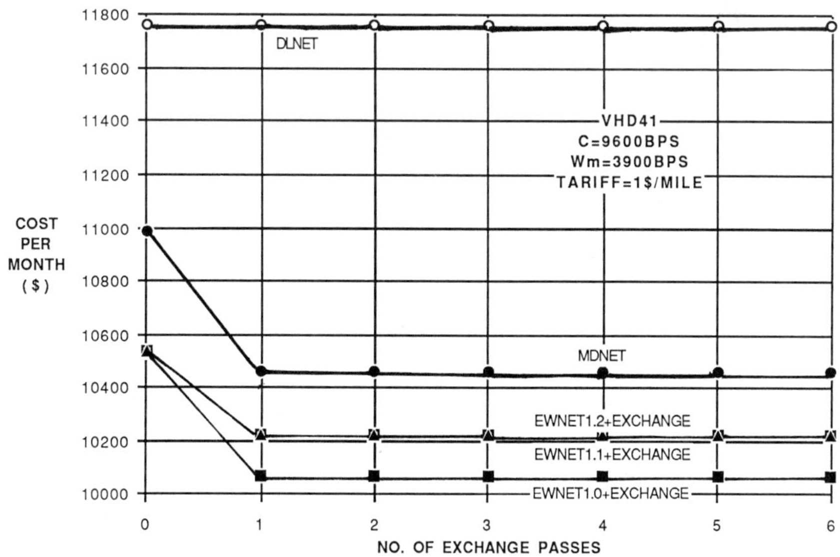

Figure 8.1. Costs per month versus the number of exchange passes for VHD41, 1$ per mile tariff, link capacity (C) = 9600bps, maximum allowable link rate (W_m) = 3900bps, and for directed path (DL), three variations of Esau-Williams (EW), and multidrop (MD) network topologies.

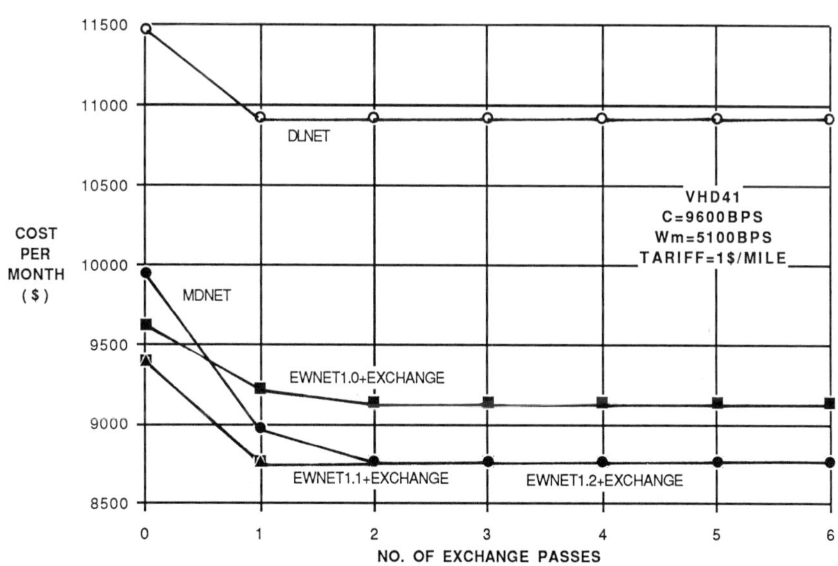

Figure 8.2. Costs per month versus the number of exchange passes for VHD41, 1$ per mile tariff, link capacity (C) = 9600bps, maximum allowable link rate (W_m) = 5100bps, and for directed path (DL), three variations of Esau-Williams (EW), and multidrop (MD) network topologies.

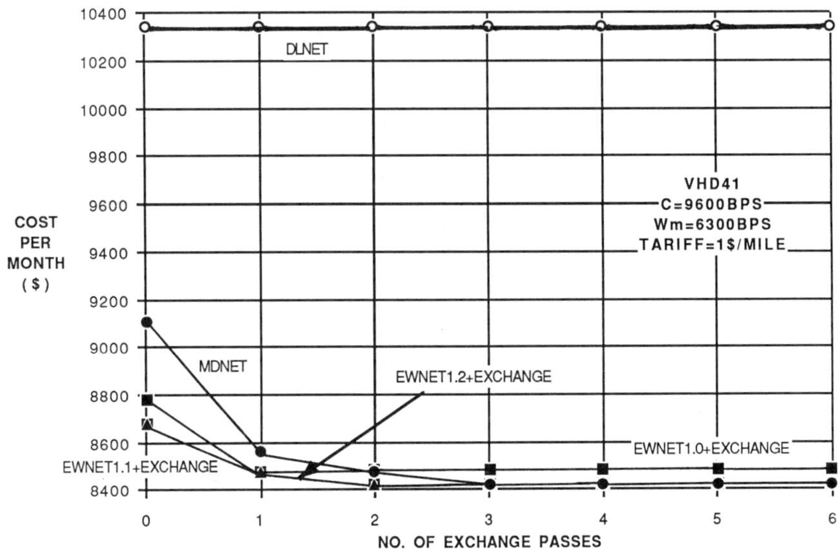

Figure 8.3. Costs per month versus the number of exchange passes for VHD41, 1$ per mile tariff, link capacity (C) = 9600bps, maximum allowable link rate (W_m) = 6300bps, and for directed path (DL), three variations of Esau-Williams (EW), and multidrop (MD) network topologies.

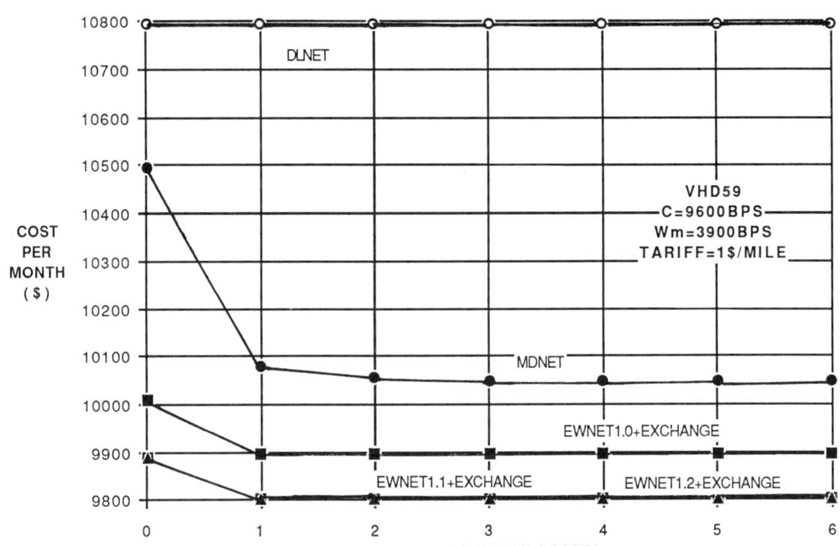

Figure 8.4. Costs per month versus the number of exchange passes for VHD59, 1$ per mile tariff, link capacity (C) = 9600bps, maximum allowable link rate (W_m) = 3900bps, and for directed path (DL), three variations of Esau-Williams (EW), and multidrop (MD) network topologies.

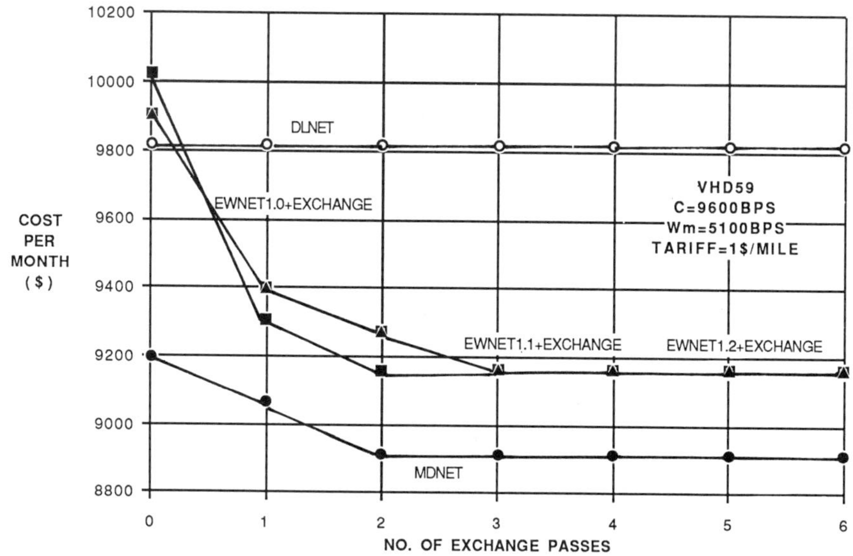

Figure 8.5. Costs per month versus the number of exchange passes for VHD59, 1$ per mile tariff, link capacity (C) = 9600bps, maximum allowable link rate (W_m) = 5100bps, and for directed path (DL), three variations of Esau-Williams (EW), and multidrop (MD) network topologies.

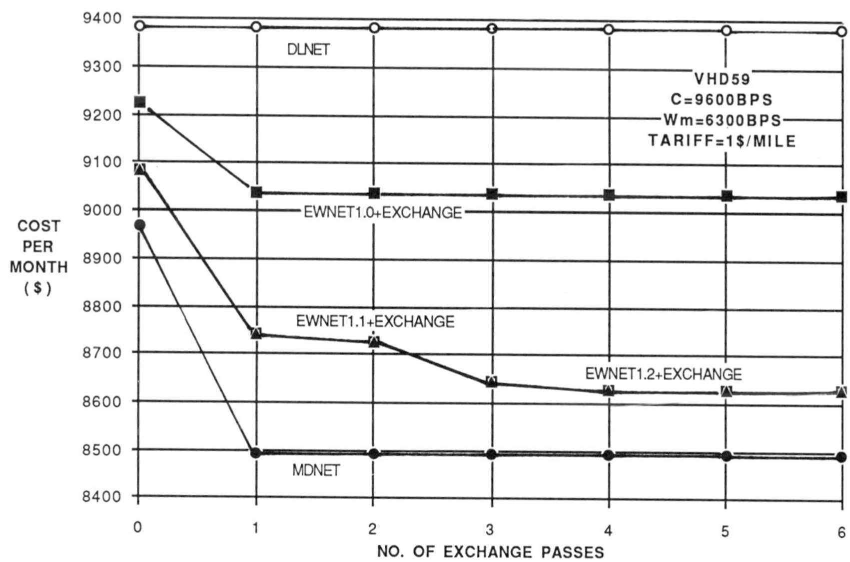

Figure 8.6. Costs per month versus the number of exchange passes for VHD59, 1$ per mile tariff, link capacity (C) = 9600bps, maximum allowable link rate (W_m) = 6300bps, and for directed path (DL), three variations of Esau-Williams (EW), and multidrop (MD) network topologies.

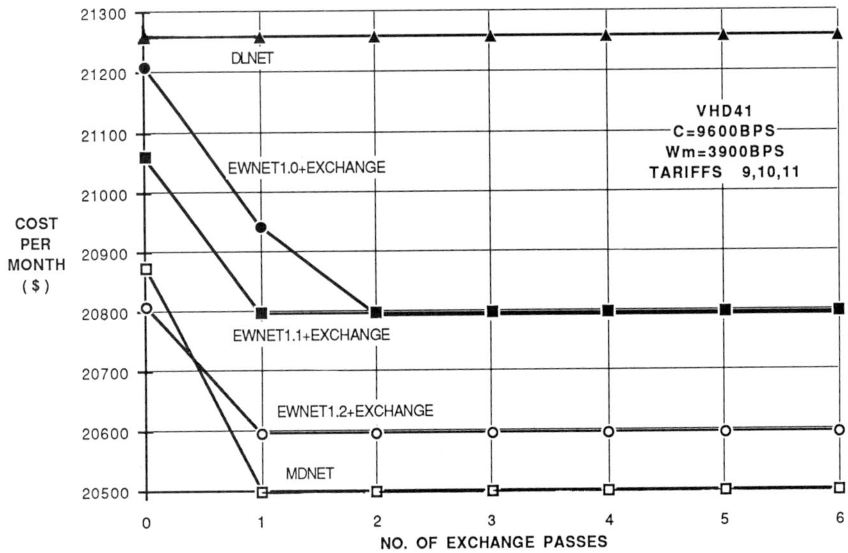

Figure 8.7. Costs per month versus the number of exchange passes for VHD41, Tariffs 9, 10, and 11, link capacity (C) = 9600bps, maximum allowable link rate (W_m) = 3900 bps, and for directed path (DL), three variations of Esau-Williams (EW), and multidrop (MD) network topologies.

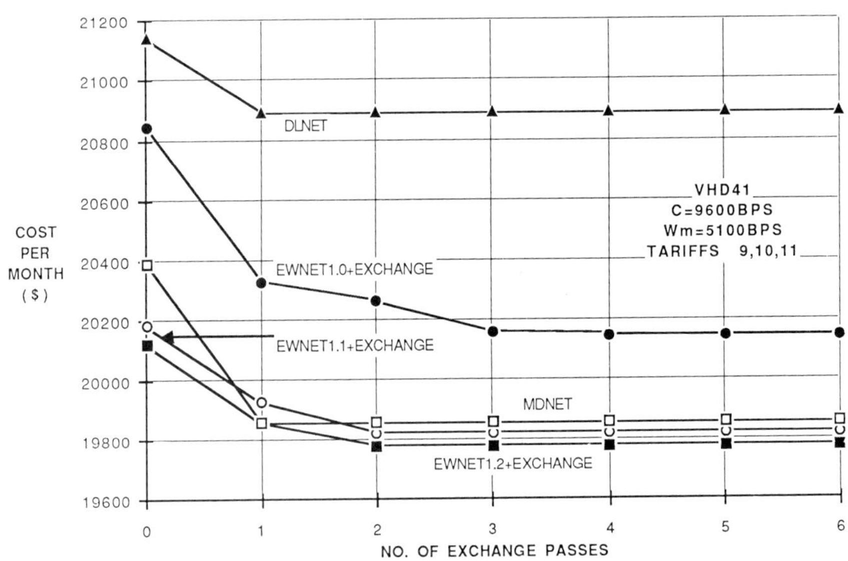

Figure 8.8. Costs per month versus the number of exchange passes for VHD41, Tariffs 9, 10, and 11, link capacity (C) = 9600bps, maximum allowable link rate (W_m) = 5100bps, and for directed path (DL), three variations of Esau-Williams (EW), and multidrop (MD) network topologies.

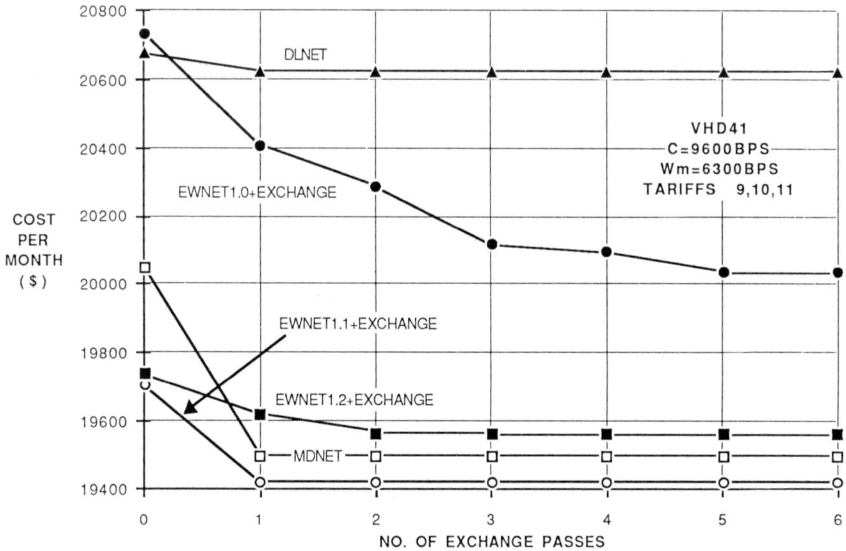

Figure 8.9. Costs per month versus the number of exchange passes for VHD41, Tariffs 9, 10, and 11, link capacity (C) = 9600bps, maximum allowable link rate (W_m) = 6300bps, and for directed path (DL), three variations of Esau-Williams (EW), and multidrop (MD) network topologies.

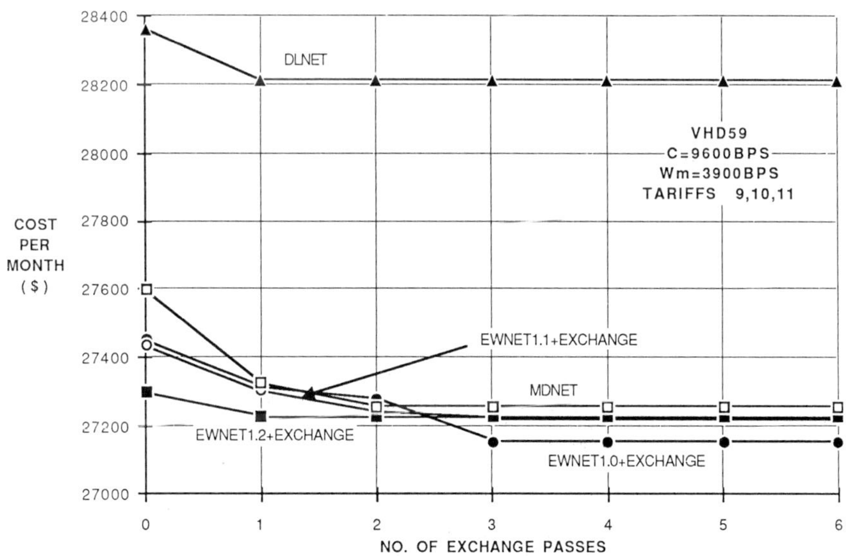

Figure 8.10. Costs per month versus the number of exchange passes for VHD59, Tariffs 9, 10, and 11, link capacity (C) = 9600bps, maximum allowable link rate (W_m) = 3900bps, and for directed path (DL), three variations of Esau-Williams (EW), and multidrop (MD) network topologies.

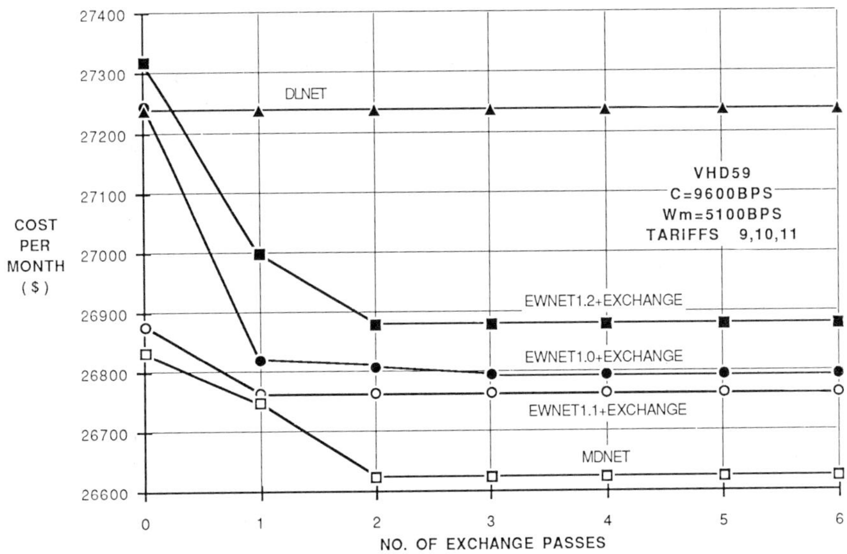

Figure 8.11. Costs per month versus number of exchange passes for VHD59, Tariffs 9, 10, and 11, link capacity (C) = 9600bps, maximum allowable link rate (W_m) = 5100bps, and for directed path (DL), three variations of Esau-Williams (EW), and multidrop (MD) network topologies.

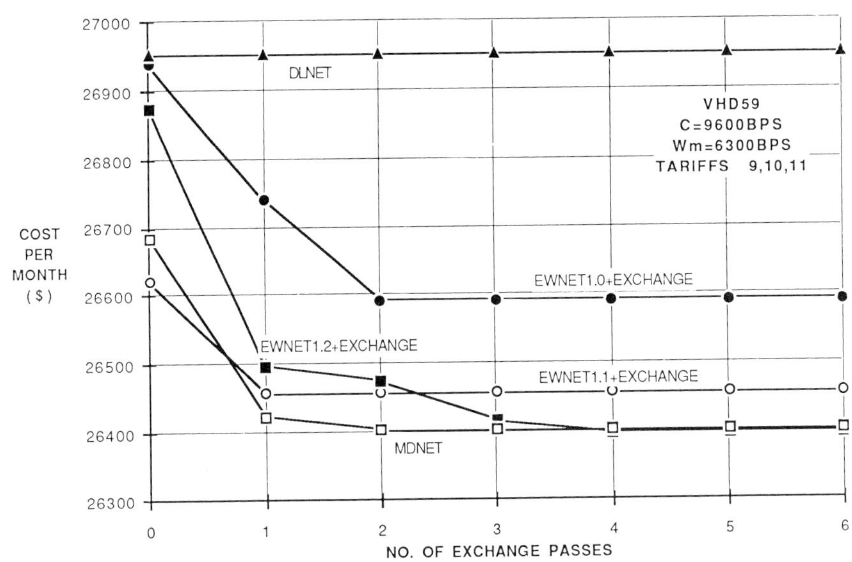

Figure 8.12. Costs per month versus number of exchange passes for VHD59, Tariffs 9, 10, and 11, link capacity (C) = 9600bps, maximum allowable link rate (W_m) = 6300bps, and for directed path (DL), three variations of Esau-Williams (EW), and multidrop (MD) network topologies.

rithm. The results also reflect both the actual and the simple tariff of 1$/mile. Kerschenbaum and Chu (1974) have wrongly stated that the exchange algorithm costs an enormous amount of computation time. The fact is that each exchange pass costs only one to three times the computation time required for the normal EWNET1.1 algorithm. This is much less than the 70-fold increase in the computation time experienced when working with the algorithm developed by Karnaugh (1972) and based on elaborate line cost tradeoffs performed at every stage of the EW algorithm. Karnaugh (1972) achieved a cost improvement of only about 1 percent. Here, the exchange algorithm is shown to yield better cost improvements over Karnaugh's algorithm with very little increase in computational times.

The cost improvements, in percent, are drastically different for the two tariffs. Whereas the cost improvements for the simple tariff (1$/mile) can range up to 10 percent, the cost improvements for the realistic set of tariffs (FCC Tariffs 9, 10, 11) are less than 2 percent. This points to a need for utmost care when considering claims made by several authors. The results presented here clearly show that no heuristic algorithm can claim to be the least costly in all cases. One can only recommend one solution over another on a statistical basis. A quick study of Figures 8.1 to 8.12 clearly shows that the MDNET algorithm yields a lower cost network after 1 exchange pass compared to a network based on the EWNET1.1 algorithm with no exchange in almost all cases. The MDNET algorithm is generally more efficient than the EWNET1.1 algorithm even when the exchange algorithm is applied to the EW algorithm. The EWNET1.0 algorithm is generally inferior to EWNET1.1 and MDNET algorithms. But in some rare cases (see Figure 8.10), it is shown to be better than the EWNET1.1 algorithm. In general, the EWNET1.2 algorithm is only slightly inferior to the EWNET1.1 algorithm. The DLNET algorithm is generally inferior to other algorithms, but it yields much lower response times, as shown later. Only in one case (see Figure 8.5) was it found to be better than all variations of the EW algorithm without the exchange technique. That was quite a surprise. A close observation of the DL topology will show that some netlinks are characterized by loops. Although the human eye can eliminate such loops and hence reduce the network cost somewhat, it is difficult to design an equivalent computer algorithm. This should be a challenge to a keen student of network science.

One can also study the network topologies yielded by the various algorithms. Figures 8.13 and 8.14 represent the network topologies achieved through EWNET1.0 and EWNET1.1/MDNET when one lets Wm go to infinity. It is shown that, while EWNET1.1 and MDNET yield a single minimal spanning tree without any capacity constraint, EWNET1.0 yields an inferior minimal spanning tree due to design constraints, as described earlier. Figures 8.15 to 8.35 represent some typical network topologies for the two tariffs and a capacity constraint of 5100 bps for the two previously defined VHD41 and VHD59 data files, using the EWNET1.1, MDNET and DLNET algorithms with and without the exchange only. *It is startling to observe that network topologies practically remain unchanged with varying tariffs* (i.e., 1$/mile or Tariffs 9, 10, and 11).

Tables 8.3 to 8.11 define some selected network topologies in terms of the nodes on each connection, the mileage and cost of each connection, and running totals of mileage

(*Text continued on pg. 241*)

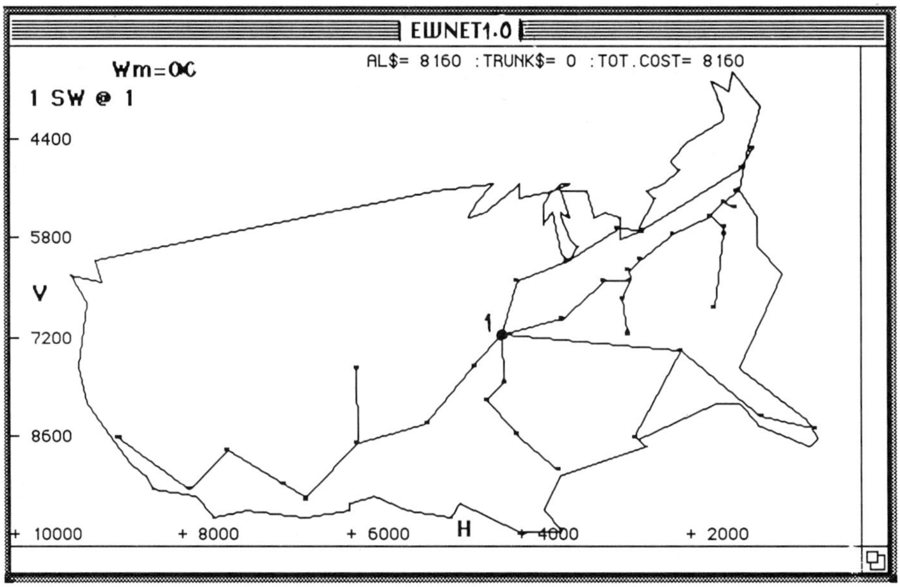

Figure 8.13. Esau-Williams (EW) network topology for VHD41, 1$ per mile tariff, and for the case W_m is infinite.

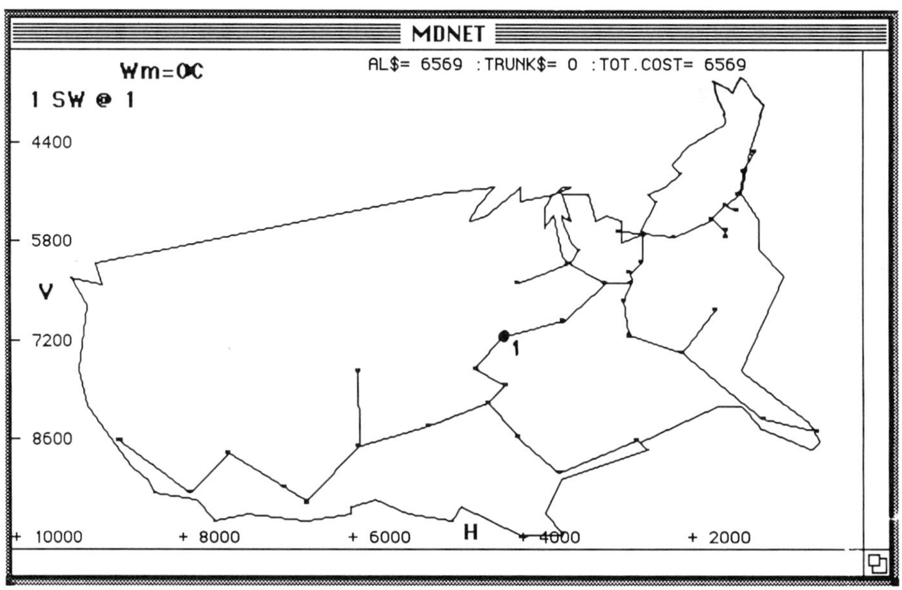

Figure 8.14. Multidrop (MD) network topology for VHD41, 1$ per mile tariff, and for the case where W_m is infinite.

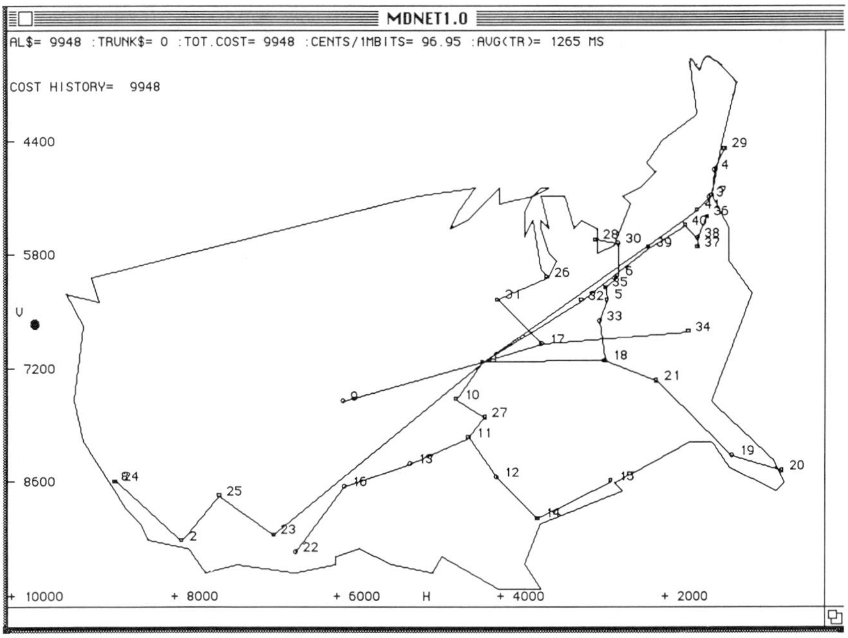

Figure 8.15. Unoptimized multidrop (MD) network topology for VHD41, 1$ per mile tariff, link capacity (C) = 9600bps, and maximum allowable link rate (W_m) = 5100bps.

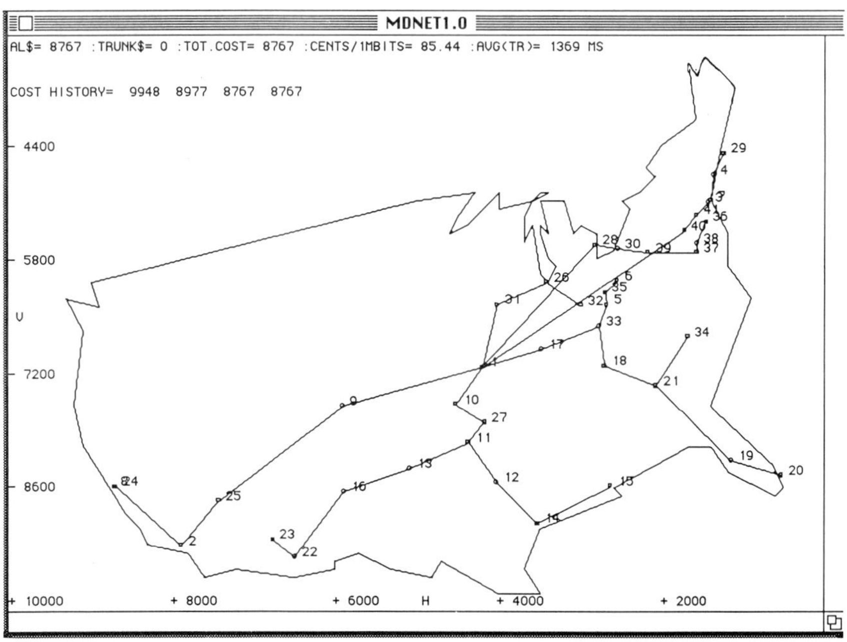

Figure 8.16. Optimized multidrop (MD) network topology for VHD41, 1$ per mile tariff, link capacity (C) = 9600bps, and maximum allowable link rate (W_m) = 5100bps.

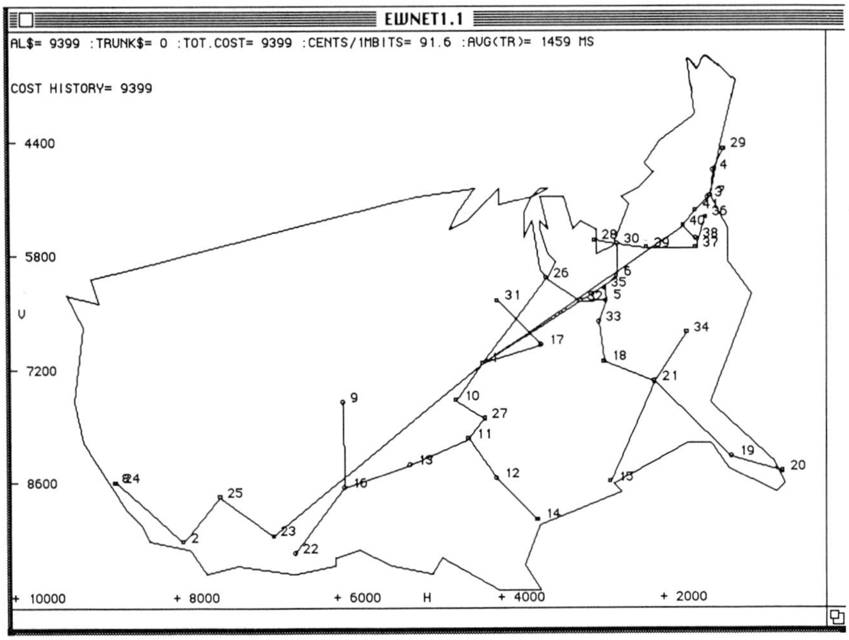

Figure 8.17. Unoptimized Esau-Williams (EW) network topology for VHD41, 1$ per mile tariff, link capacity (C) = 9600bps, and maximum allowable link rate (W_m) = 5100bps.

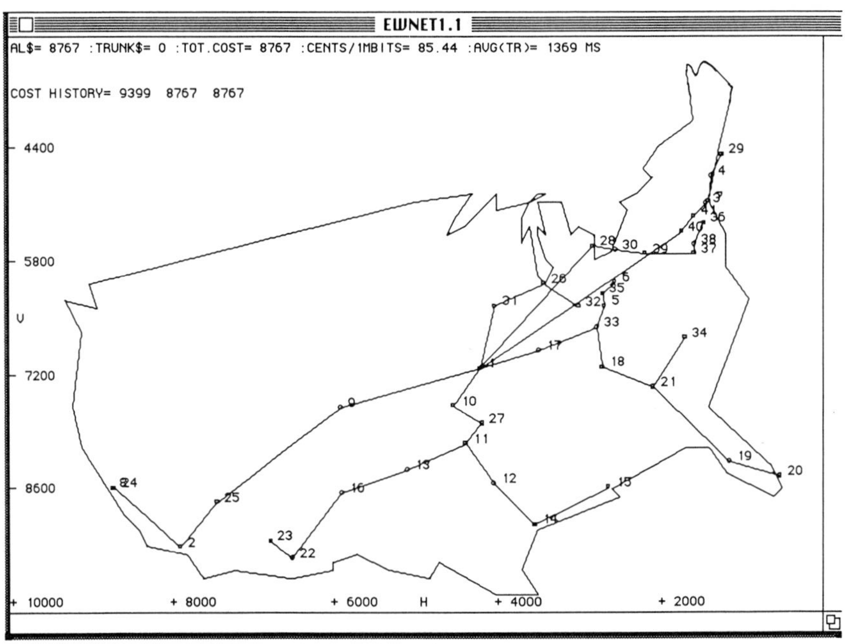

Figure 8.18. Optimized Esau-Williams (EW) network topology for VHD41, 1$ per mile tariff, link capacity (C) = 9600bps, and maximum allowable link rate (W_m) = 5100bps.

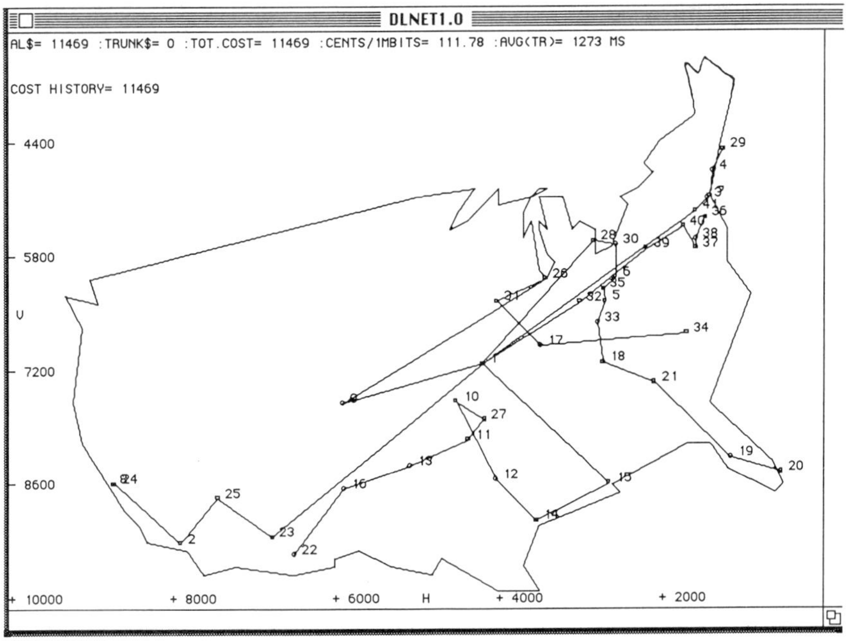

Figure 8.19. Unoptimized directed link (DL) network topology for VHD41, 1$ per mile tariff, link capacity (C) = 9600bps, and maximum allowable link rate (W_m) = 5100bps.

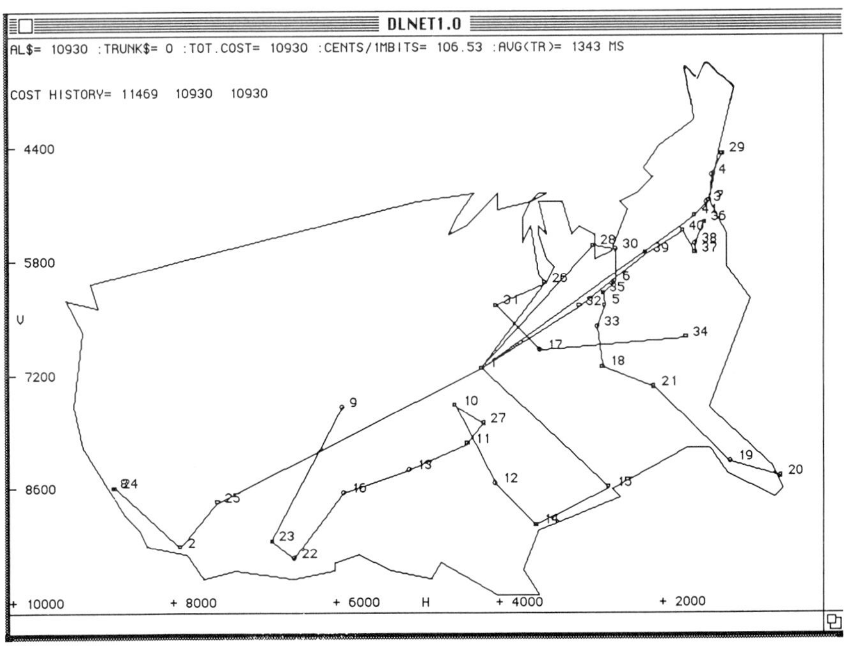

Figure 8.20. Optimized directed link (DL) network topology for VHD41, 1$ per mile tariff, link capacity (C) = 9600bps, and maximum allowable link rate (W_m) = 5100bps.

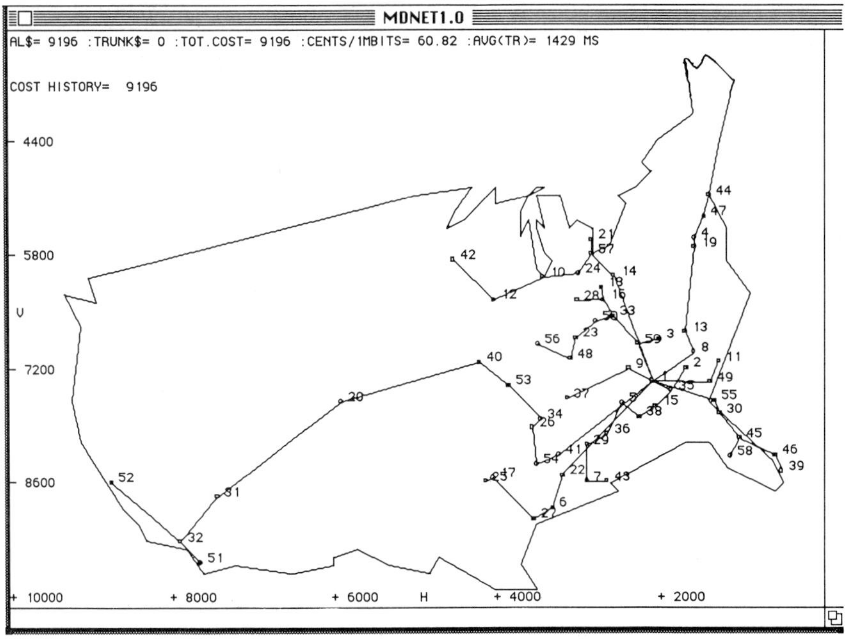

Figure 8.21. Unoptimized multidrop (MD) network topology for VHD59, 1$ per mile tariff, link capacity (C) = 9600bps, and maximum allowable link rate (W_m) = 5100bps.

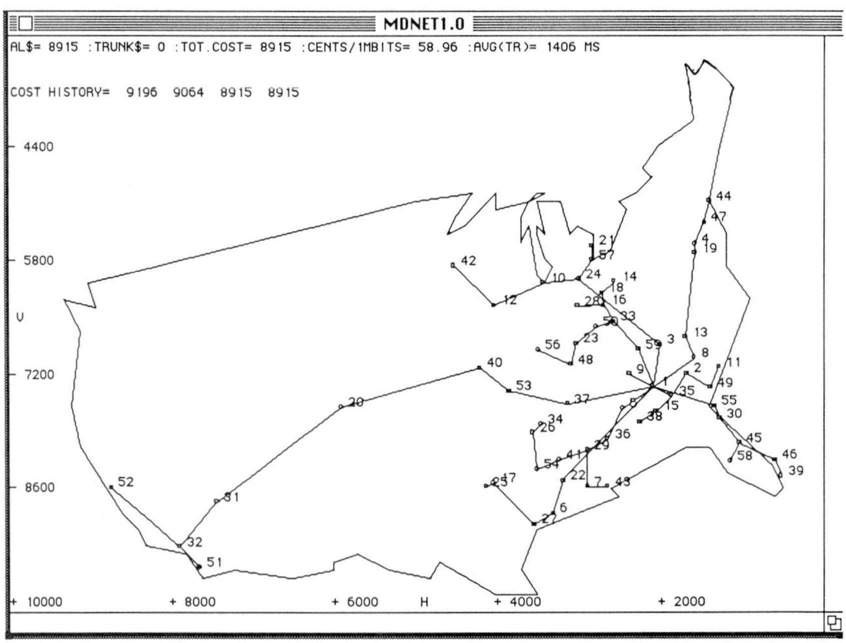

Figure 8.22. Optimized multidrop (MD) network topology for VHD59, 1$ per mile tariff, link capacity (C) = 9600bps, and maximum allowable link rate (W_m) = 5100bps.

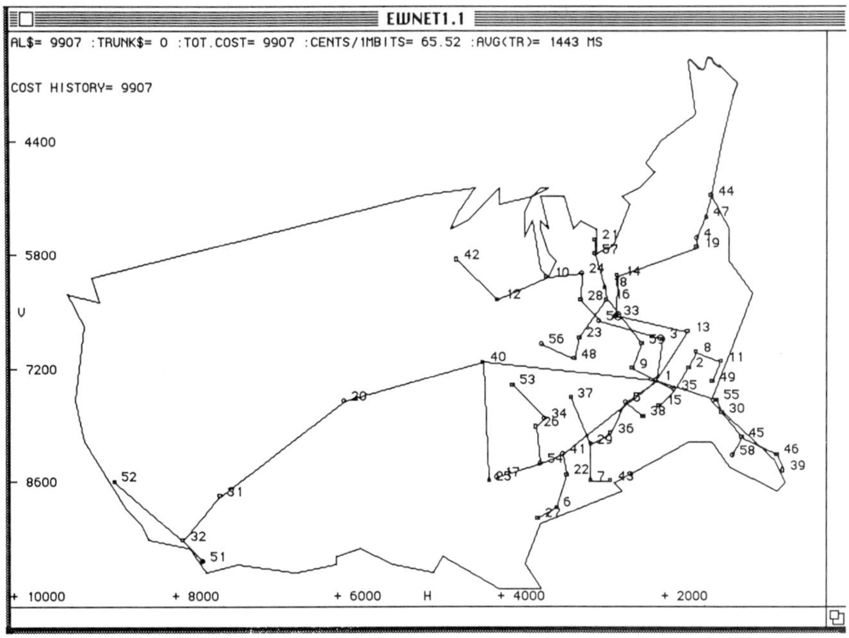

Figure 8.23. Unoptimized Esau-Williams (EW) network topology for VHD59, 1\$ per mile tariff, link capacity (C) = 9600bps, and maximum allowable link rate (W_m) = 5100bps.

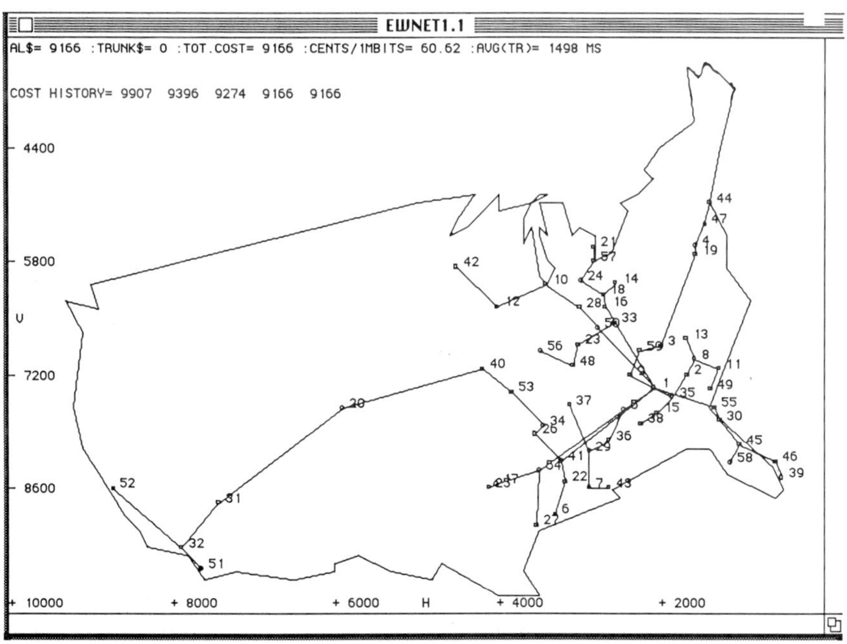

Figure 8.24. Optimized Esau-Williams (EW) network topology for VHD59, 1\$ per mile tariff, link capacity (C) = 9600bps, and maximum allowable link rate (W_m) = 5100bps.

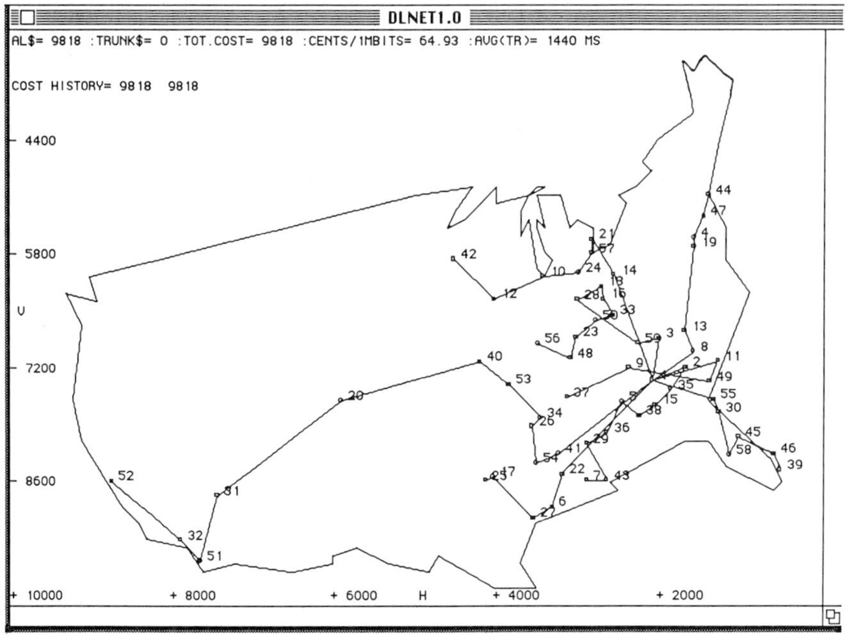

Figure 8.25. Optimized directed link (DL) network topology for VHD59, 1$ per mile tariff, link capacity (C) = 9600bps, and maximum allowable link rate (W_m) = 5100bps.

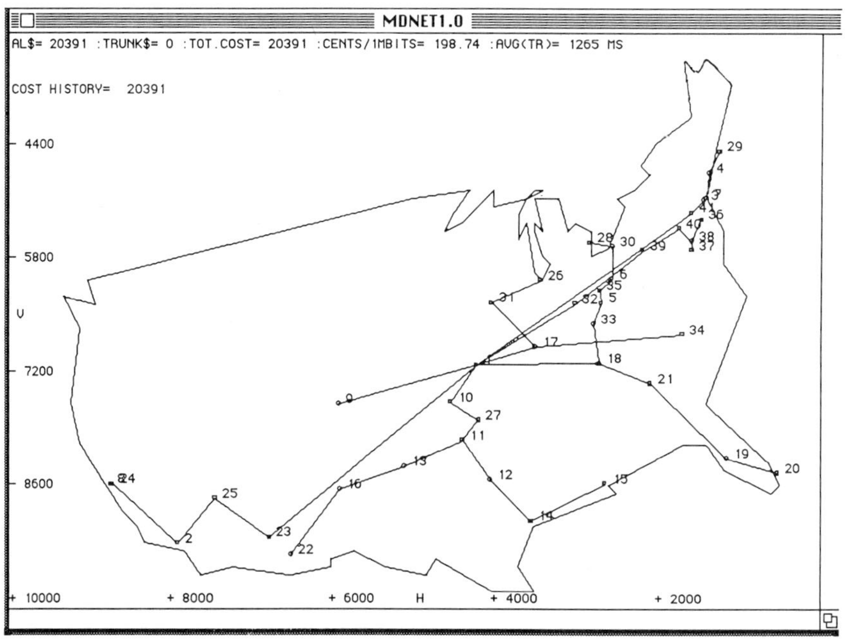

Figure 8.26. Unoptimized multidrop (MD) network topology for VHD41, Tariffs 9, 10, and 11, link capacity (C) = 9600bps, and maximum allowable link rate (W_m) = 5100bps.

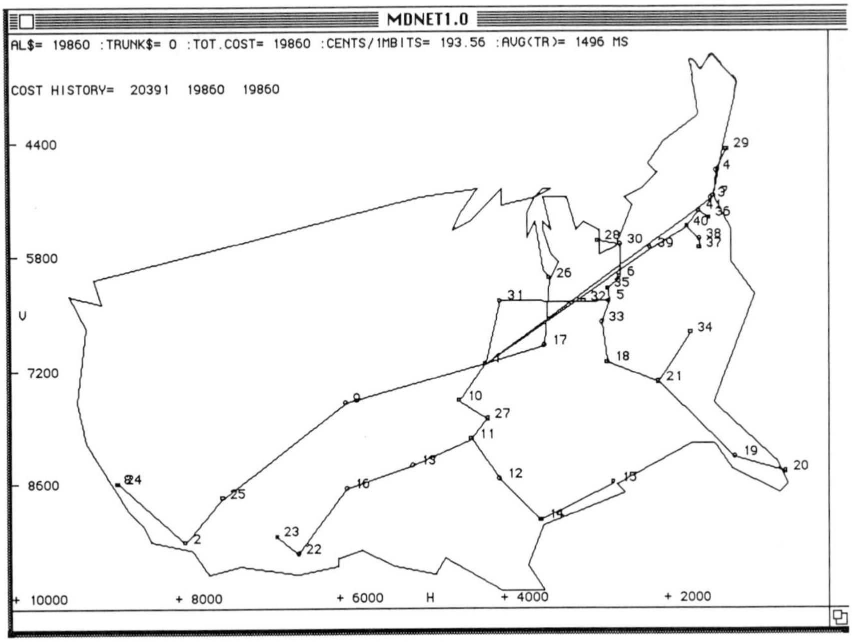

Figure 8.27. Optimized multidrop (MD) network topology for VHD41, Tariffs 9, 10, and 11, link capacity (C) = 9600bps, and maximum allowable link rate (W_m) = 5100bps.

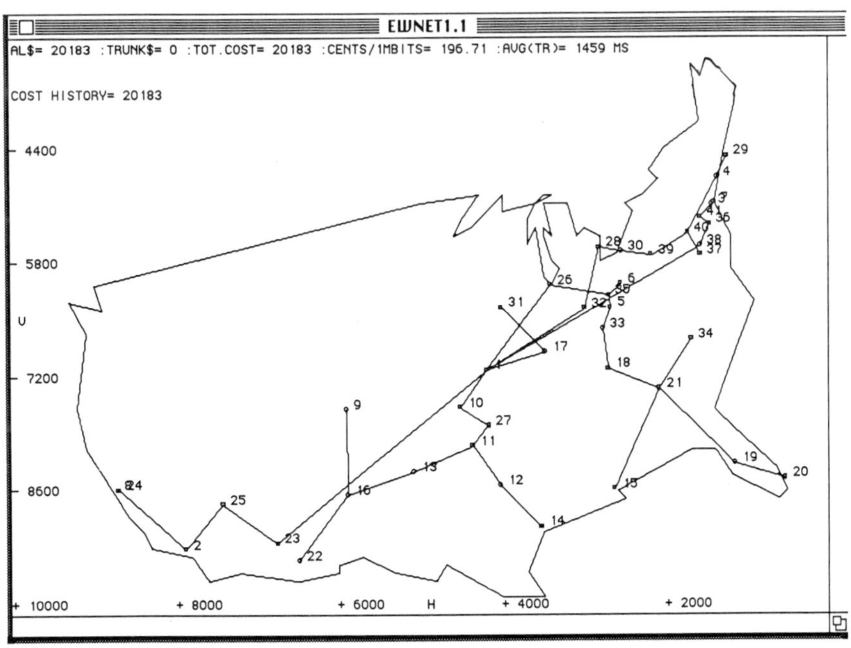

Figure 8.28. Unoptimized Esau-Williams (EW) network topology for VHD41, Tariffs 9, 10, 11, and capacity (C) = 9600bps, and maximum allowable link rate (W_m) = 5100bps.

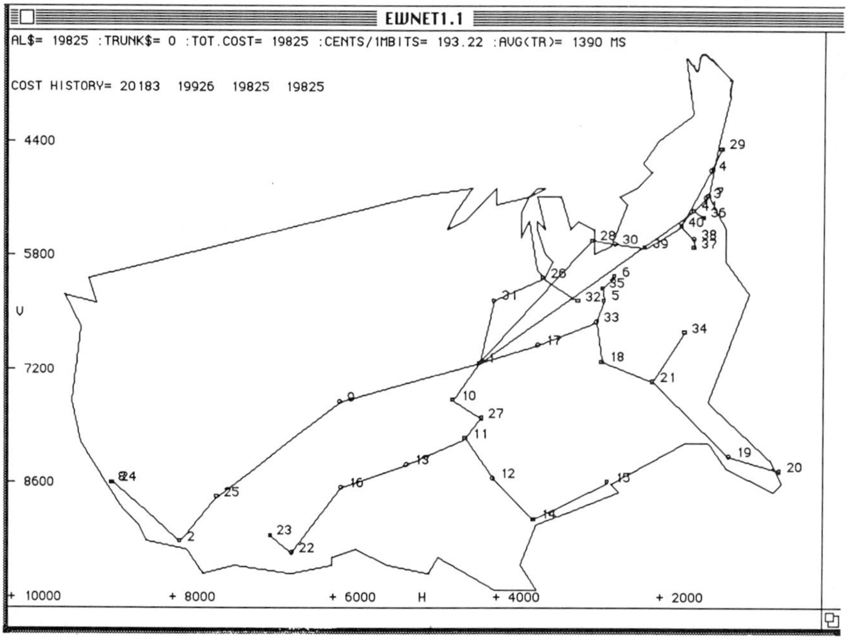

Figure 8.29. Optimized Esau-Williams (EW) network topology for VHD41, Tariffs 9, 10, and 11, link capacity (C) = 9600bps, and maximum allowable link rate (W_m) = 5100bps.

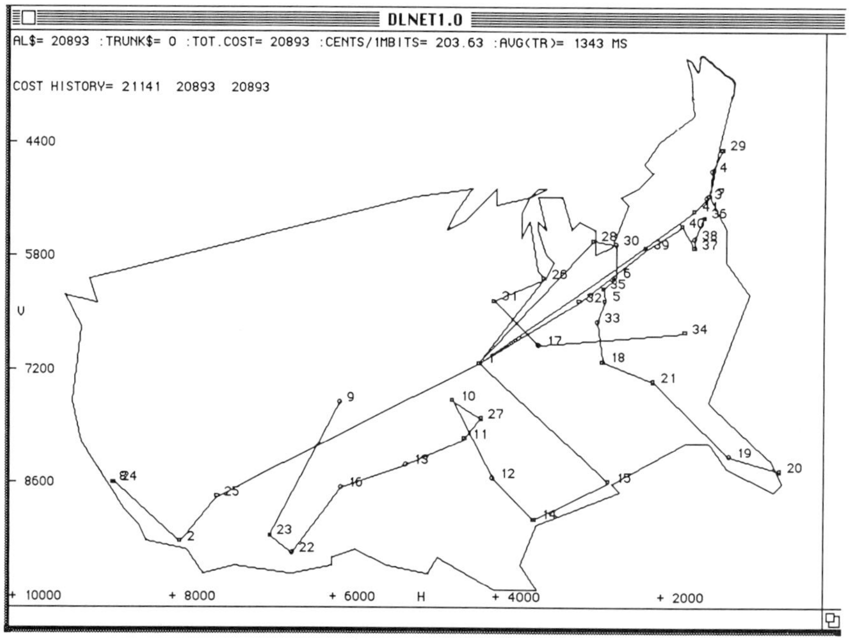

Figure 8.30. Optimized directed link (DL) network topology for VHD41, Tariffs 9, 10, and 11, link capacity (C) = 9600bps, and maximum allowable link rate (W_m) = 5100bps.

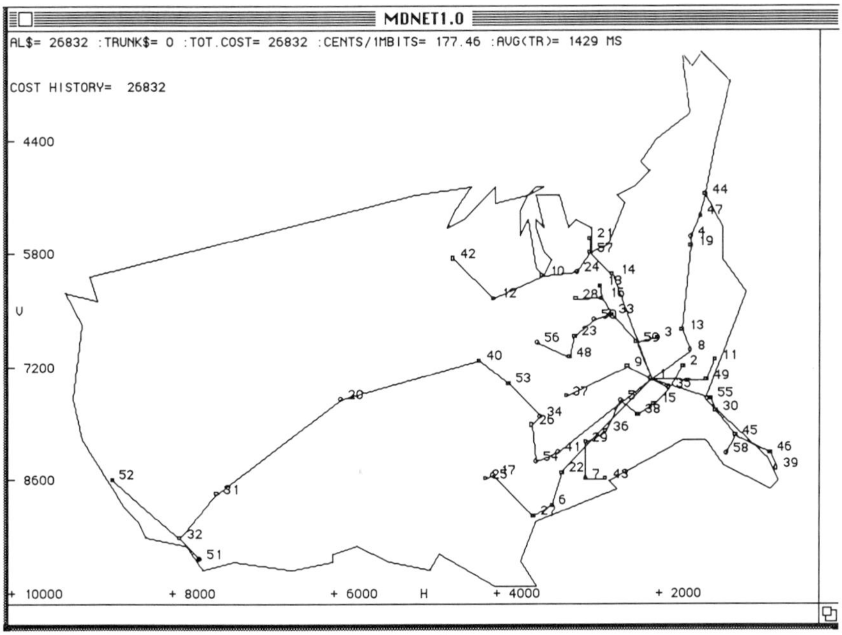

Figure 8.31. Unoptimized multidrop (MD) network topology for VHD59, Tariffs 9, 10, and 11, link capacity (C) = 9600bps, and maximum allowable link rate (W_m) = 5100bps.

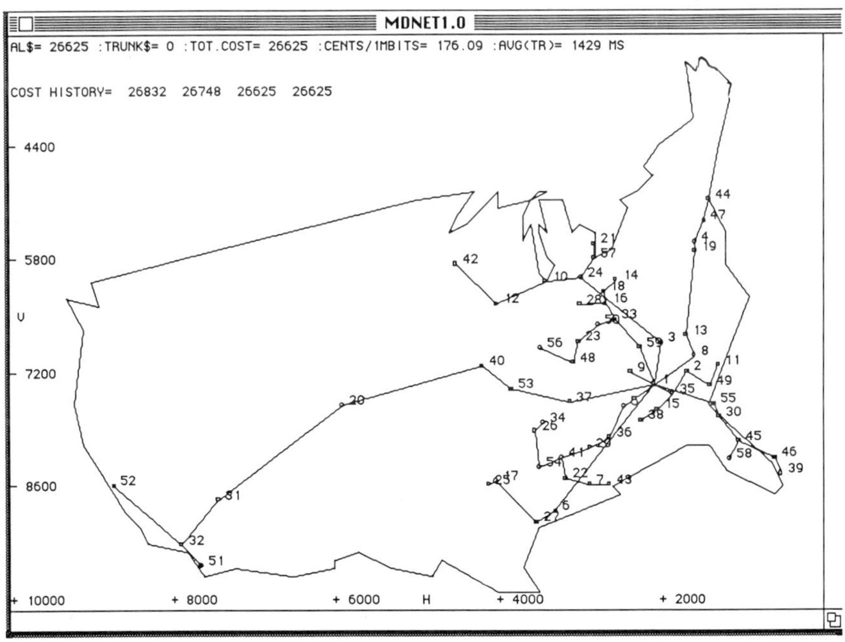

Figure 8.32. Optimized multidrop (MD) network topology for VHD59, Tariffs 9, 10, and 11, link capacity (C) = 9600bps, and maximum allowable link rate (W_m) = 5100bps.

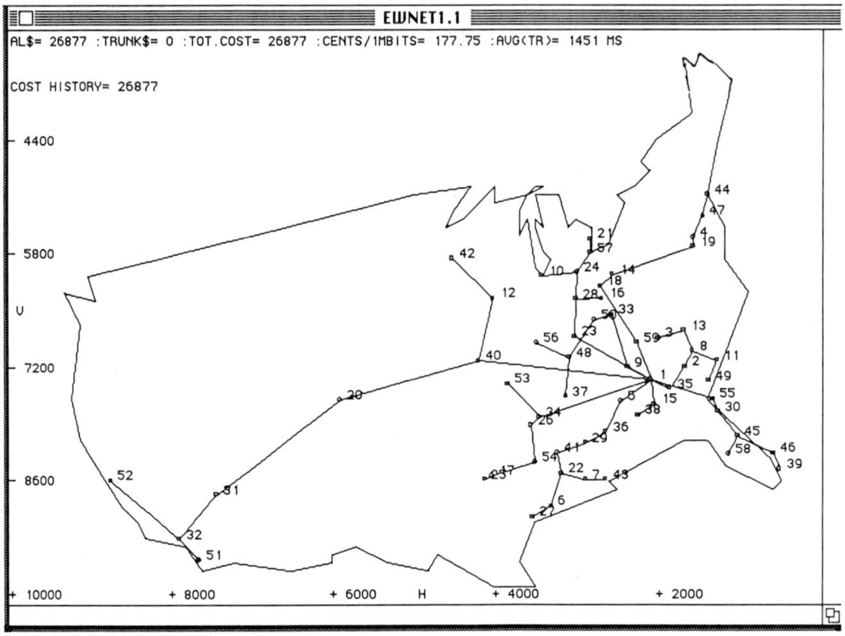

Figure 8.33. Unoptimized Esau-Williams (EW) network topology for VHD59, Tariffs 9, 10, and 11, link capacity (C) = 9600bps, and maximum allowable link rate (W_m) = 5100bps.

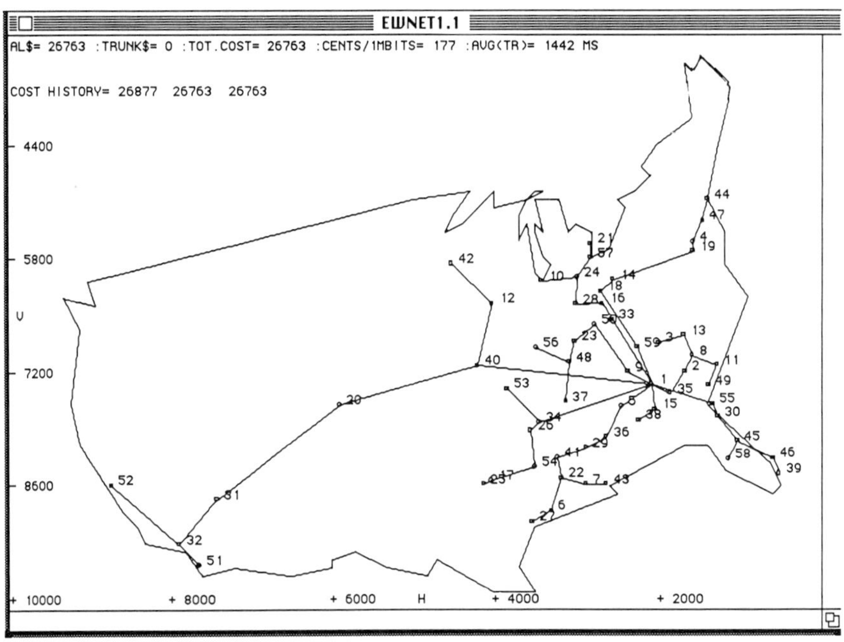

Figure 8.34. Optimized Esau-Williams (EW) network topology for VHD59, Tariffs 9, 10, and 11, link capacity (C) = 9600bps, and maximum allowable link rate (W_m) = 5100bps.

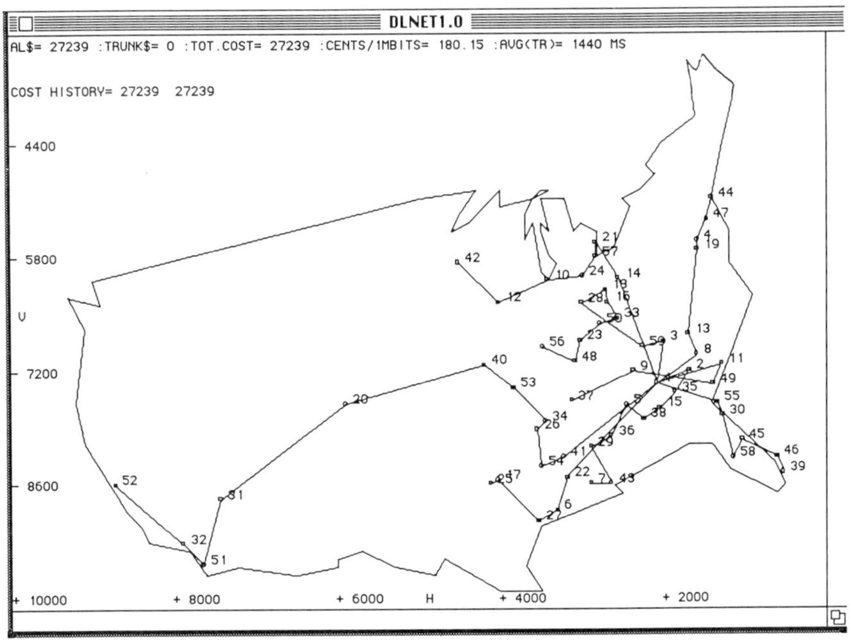

Figure 8.35. Optimized directed link (DL) network topology for VHD59, Tariffs 9, 10, and 11, link capacity (C) = 9600bps, and maximum allowable link rate (W_m) = 5100bps.

and costs. The reader should ignore the slight discrepancy between the mileage and cost of a link for the 1$/mile tariff. It is caused by computational round-offs. The data of Tables 8.3 to 8.11 should be combined with the corresponding topology graphs to study the underlying differences between MDNET and EWNET algorithms. EWNET1.1 is always sensitive to the order in which the various nodes are selected for integration into the network. For the MDNET or DLNET algorithms, only the farthest unused node to switch is considered at the beginning of a netlink design process. It then considers only the closest unused node to the previously chosen node of a netlink. All of the algorithms considered here apply the capacity constraint at each decision point. Whereas the MDNET and EWNET algorithms apply the minimal spanning tree algorithm to the nodes of each netlink, the DLNET does not alter the order of netlink nodes. For further clarification, study the algorithms as defined in Chapter 5.

The traffic growth factor (or TGF) can now be increased to study the cost of transmitting 1Mbits. The computations were based on the assumptions that there are 5 peak hours per business day and there are 20 business days in a typical month. The results are shown in Figures 8.36 and 8.37 for TGF values of 1, 5, 10, and 20 for both network types and the three values of Wm. The costs per 1Mbit for VHD41 are somewhat higher than those for VHD59, since the nodes are scattered further away from the switch location. Such unit network costs will be used to study other networks later.

The star topology is also frequently used to synthesize data networks based on AT&T's ISN/Datakit and CCITT's X.25 architectures. In particular, the star topology is an ideal

(*Text continued on pg. 254*)

Table 8.3 DATANET Link Analysis for VHD41 and DLNET with No Optimization

```
NO. OF SWITCHES= 1                    AT          1
MAXIMUM ALLOWED LINK RATE (Wm)= 5100 BPS

SW#LK #LKWT AQT ATAT E(TR)                     NODES
 1  1  5100  6  497  1063 *** 8  24  2  25  23  1
 1  2  4500  5  452  884  *** 29  4  7  3  41  1
 1  3  3600  3  872  1789 *** 20  19  21  18  33  5  35  6  30  28  1
 1  4  5100  6  513  1144 *** 36  38  37  40  39  32  1
 1  5  3600  3  860  1713 *** 22  16  13  11  27  10  12  14  15  1
 1  6  4500  5  536  1049 *** 34  17  31  26  9  1

MAXTR= 1789      AVG. TR= 1273

OVERALL SYSTEM AVERAGE(TR)-MS = 1273
```

FROM	TO	MILES	COST$	TOT$	TMILES
8	24	7	8	8	7
24	2	345	346	354	353
2	25	227	227	581	581
25	23	256	257	838	838
23	1	1045	1045	1883	1883
29	4	92	92	92	92
4	7	98	98	190	191
7	3	9	9	199	200
3	41	68	68	267	268
41	1	1015	1015	1282	1284
20	19	203	204	204	203
19	21	413	413	617	617
21	18	213	213	830	830
18	33	153	153	983	984
33	5	89	89	1072	1073
5	35	48	48	1120	1121
35	6	65	65	1185	1186
6	30	125	126	1311	1312
30	28	90	91	1402	1403
28	1	641	641	2043	2044
36	38	89	90	90	89
38	37	35	35	125	125
37	40	94	94	219	219
40	39	164	164	383	384
39	32	327	328	711	712
32	1	451	451	1162	1163
22	16	314	315	315	314
16	13	271	271	586	586
13	11	244	244	830	830
11	27	98	98	928	929
27	10	129	129	1057	1058
10	12	336	336	1393	1395
12	14	223	223	1616	1618
14	15	318	318	1934	1937
15	1	675	676	2610	2613
34	17	566	566	566	566
17	31	242	242	808	808
31	26	207	207	1015	1015
26	9	917	917	1932	1933
9	1	556	557	2489	2489

```
            TOTAL AL COST= 11469
       TOTAL AL,SL MILES=  11469,     0
     TOTAL AL AND SL COSTS= 11469,    0
  TOTAL NUMBER OF SPECIAL LINKS=      0
```

Table 8.4 DATANET Link Analysis for VHD41 and DLNET with Optimization

```
NO. OF SWITCHES= 1                      AT            1
MAXIMUM ALLOWED LINK RATE (Wm)= 5100 BPS

SW# LK# LKWT AQT ATAT E(TR)                         NODES
         BPS  MS  MS    MS
 1  1  4200  4  409  735    *** 8  24  2  25  1
 1  2  4500  5  452  884    *** 29  4  7  3  41  1
 1  3  3600  3  872  1789   *** 20  19  21  18  33  5  35  6  30  28  1
 1  4  5100  6  513  1144   *** 36  38  37  40  39  32  1
 1  5  4800  5  1122 2807   *** 9  23  22  16  13  11  27  10  12  14  15  1
 1  6  4200  4  392  703    *** 34  17  31  26  1

MAXTR= 2807     AVG. TR= 1343

OVERALL SYSTEM AVERAGE(TR)-MS = 1343
```

FROM	TO	MILES	COST$	TOT$	TMILES
8	24	7	8	8	7
24	2	345	346	354	353
2	25	227	227	581	581
25	1	1139	1139	1720	1720
29	4	92	92	92	92
4	7	98	98	190	191
7	3	9	9	199	200
3	41	68	68	267	268
41	1	1015	1015	1282	1284
20	19	203	204	204	203
19	21	413	413	617	617
21	18	213	213	830	830
18	33	153	153	983	984
33	5	89	89	1072	1073
5	35	48	48	1120	1121
35	6	65	65	1185	1186
6	30	125	126	1311	1312
30	28	90	91	1402	1403
28	1	641	641	2043	2044
36	38	89	90	90	89
38	37	35	35	125	125
37	40	94	94	219	219
40	39	164	164	383	384
39	32	327	328	711	712
32	1	451	451	1162	1163
9	23	582	582	582	582
23	22	106	106	688	688
22	16	314	315	1003	1003
16	13	271	271	1274	1275
13	11	244	244	1518	1519
11	27	98	98	1616	1617
27	10	129	129	1745	1747
10	12	336	336	2081	2084
12	14	223	223	2304	2307
14	15	318	318	2622	2625
15	1	675	676	3298	3301
34	17	566	566	566	566
17	31	242	242	808	808
31	26	207	207	1015	1015
26	1	410	410	1425	1426

```
          TOTAL AL COST = 10930
      TOTAL AL,SL MILES =  10930,   0
   TOTAL AL AND SL COSTS= 10930,    0
 TOTAL NUMBER OF SPECIAL LINKS=     0
```

Table 8.5 Netlink Analysis for VHD41 and EWNET1.1 with No Optimization

NO. OF SWITCHES= 1 AT 1
MAX ALLOWED LINK RATE= 5100 BPS

SW#	LK#	LKWT	AQT MS	ATAT MS	E(TR) MS		NODES
1	1	5100	6	497	1057	***	8 24 2 25 23 1
1	2	3600	3	897	1772	***	12 14 9 22 16 13 11 27 10 1
1	3	2400	2	213	283	***	17 31 1
1	4	5100	6	1171	3048	***	18 35 15 19 20 21 34 5 33 32 26 1
1	5	5100	6	524	1162	***	36 37 39 30 28 6 1
1	6	5100	6	623	1432	***	41 3 4 29 7 40 38 1

MAXTR= 3048 AVG. TR= 1459

OVERALL SYSTEM AVERAGE(TR)-MS = 1459

FROM	TO	MILES	COST$	TOT$	TMILES
8	24	7	8	8	7
2	25	227	227	235	235
23	25	256	257	492	492
1	23	1045	1045	1537	1537
2	24	345	346	1883	1883
12	11	188	188	188	188
14	12	223	223	411	411
9	16	331	331	742	743
22	16	314	315	1057	1057
13	11	244	244	1301	1302
27	11	98	98	1399	1400
10	27	129	129	1528	1530
1	10	177	177	1705	1707
16	13	271	271	1976	1978
17	1	238	238	238	238
31	17	242	242	480	480
18	33	153	153	153	153
35	5	48	48	201	201
15	21	424	424	625	626
19	20	203	204	829	830
34	21	226	226	1055	1056
32	5	99	99	1154	1155
26	32	164	164	1318	1319
1	26	410	410	1728	1730
19	21	413	413	2141	2143
21	18	213	213	2354	2357
5	33	89	89	2443	2446
36	37	123	124	124	123
39	30	114	114	238	237
28	30	90	91	329	328
6	30	125	126	455	454
1	6	618	618	1073	1073
39	37	190	190	1263	1263
41	3	68	68	68	68
4	29	92	92	160	160
7	3	9	9	169	170
40	38	68	68	237	238
1	40	941	941	1178	1180
40	41	77	78	1256	1258
4	7	98	98	1354	1356

TOTAL AL COST = 9399
TOTAL AL & SL MILES = 9399, 0
TOTAL AL AND SL COSTS= 9399, 0
TOTAL NUMBER OF SPECIAL LINKS= 0

Table 8.6 Netlink Analysis for VHD41 and MDNET with No Optimization

NO. OF SWITCHES= 1 AT 1
MAX LINK RATE= 5100 BPS

SW#	LK#	LKWT	AQT MS	ATAT MS	E(TR) MS		NODES
1	1	5100	6	497	1057	***	8 24 2 25 23 1
1	2	4500	5	452	879	***	29 4 7 3 41 1
1	3	3600	3	872	1774	***	20 19 21 18 33 5 35 6 30 28 1
1	4	5100	6	513	1137	***	36 38 37 40 39 32 1
1	5	3600	3	860	1700	***	22 16 13 11 27 10 12 14 15 1
1	6	4500	5	536	1043	***	34 17 31 26 9 1

MAXTR= 1774 AVG.TR (FOR MAIN & SPEC. LINKS)= 1265

OVERALL SYSTEM AVERAGE(TR)-MS = 1265

FROM	TO	MILES	COST$	TOT$	TMILES
8	24	7	8	8	7
2	25	227	227	235	235
23	25	256	257	492	492
1	23	1045	1045	1537	1537
2	24	345	346	1883	1883
29	4	92	92	92	92
7	3	9	9	101	101
41	3	68	68	169	170
1	41	1015	1015	1184	1186
7	4	98	98	1282	1284
20	19	203	204	204	203
21	18	213	213	417	417
33	5	89	89	506	506
35	5	48	48	554	554
6	35	65	65	619	619
30	28	90	91	710	710
1	18	472	472	1182	1182
30	6	125	126	1308	1308
33	18	153	153	1461	1461
21	19	413	413	1874	1875
36	38	89	90	90	89
37	38	35	35	125	125
40	38	68	68	193	193
39	40	164	164	357	358
32	39	327	328	685	685
1	32	451	451	1136	1137
22	16	314	315	315	314
13	11	244	244	559	558
27	11	98	98	657	657
10	27	129	129	786	787
12	11	188	188	974	975
14	12	223	223	1197	1198
15	14	318	318	1515	1517
1	10	177	177	1692	1694
13	16	271	271	1963	1965
34	17	566	566	566	566
31	26	207	207	773	773
9	1	556	557	1330	1330
1	17	238	238	1568	1568
31	17	242	242	1810	1811

TOTAL AL COST = 9948
TOTAL AL & SL MILES = 9948, 0
TOTAL AL & SL COSTS= 9948, 0
TOTAL NUMBER OF SPECIAL LINKS= 0

Table 8.7 Netlink Analysis for VHD41 and MDNET/EWNET1.1 with Optimization

```
NO. OF SWITCHES= 1                    AT          1
MAX LINK RATE= 5100 BPS

SW#LK# LKWT AQT ATAT E(TR)                        NODES
            MS  MS   MS
 1  1  4500  5  484  941    *** 8  24  2  25  9  1
 1  2  4800  5  521  1102   *** 29  4  7  3  41  40  1
 1  3  5100  6  908  2296   *** 20  19  34  21  6  35  5  33  18  17  1
 1  4  5100  6  513  1137   *** 36  38  37  39  30  28  1
 1  5  4500  5  1022  2353  *** 23  22  16  15  14  13  12  11  27  10  1
 1  6  2400  2  278  388    *** 32  26  31  1

MAXTR= 2353      AVG.TR (FOR MAIN & SPEC. LINKS)= 1369

OVERALL SYSTEM AVERAGE(TR)-MS = 1369
```

FROM	TO	MILES	COST$	TOT$	TMILES
8	24	7	8	8	7
2	25	227	227	235	235
9	1	556	557	792	792
9	25	603	603	1395	1395
2	24	345	346	1741	1741
29	4	92	92	92	92
7	3	9	9	101	101
41	3	68	68	169	170
40	41	77	78	247	248
1	40	941	941	1188	1189
7	4	98	98	1286	1288
20	19	203	204	204	203
34	21	226	226	430	430
6	35	65	65	495	495
5	35	48	48	543	543
33	5	89	89	632	632
18	33	153	153	785	785
17	1	238	238	1023	1024
17	33	241	241	1264	1265
18	21	213	213	1477	1478
21	19	413	413	1890	1891
36	38	89	90	90	89
37	38	35	35	125	125
39	30	114	114	239	239
28	30	90	91	330	330
1	28	641	641	971	971
39	37	190	190	1161	1161
23	22	106	106	106	106
16	13	271	271	377	378
15	14	318	318	695	696
12	11	188	188	883	884
27	11	98	98	981	983
10	27	129	129	1110	1112
1	10	177	177	1287	1289
12	14	223	223	1510	1513
11	13	244	244	1754	1757
16	22	314	315	2069	2072
32	26	164	164	164	164
31	26	207	207	371	371
1	31	248	249	620	620

```
        TOTAL AL COST =  8767
   TOTAL AL & SL MILES=  8767,   0
    TOTAL AL & SL COSTS= 8767,   0
 TOTAL NUMBER OF SPECIAL LINKS=  0
```

Table 8.8 DATANET Link Analysis for VHD59 and DLNET with No Optimization

NO. OF SWITCHES= 1 AT 1
MAXIMUM ALLOWABLE LINK RATE= 5100 BPS

SW#	LK#	LKWT BPS	AQT MS	ATAT MS	E(TR) MS		NODES
1	1	5100	6	1080	2836	***	52 32 51 31 20 40 53 34 26 54 41 1
1	2	5100	6	593	1373	***	42 12 10 24 57 21 14 1
1	3	5100	6	433	927	***	25 17 27 6 22 1
1	4	5100	6	485	1082	***	44 47 4 19 13 8 1
1	5	5100	6	481	1073	***	39 46 45 58 30 55 1
1	6	5100	6	778	1986	***	56 48 23 50 33 16 18 28 59 3 1
1	7	5100	6	690	1708	***	7 43 29 36 5 38 15 35 2 1
1	8	2700	2	357	538	***	37 9 49 11 1

MAXTR= 2836 AVG. TR= 1440

OVERALL SYSTEM AVERAGE(TR)-MS = 1440

FROM	TO	MILES	COST$	TOT$	TMILES
52	32	350	350	350	350
32	51	112	112	462	463
51	31	263	263	725	726
31	20	603	603	1328	1329
20	40	556	557	1885	1886
40	53	146	146	2031	2032
53	34	178	178	2209	2210
34	26	46	46	2255	2257
26	54	141	142	2397	2399
54	41	95	96	2493	2495
41	1	455	455	2948	2951
42	12	220	220	220	220
12	10	207	207	427	427
10	24	141	141	568	568
24	57	91	91	659	659
57	21	53	53	712	712
21	14	162	162	874	874
14	1	433	433	1307	1308
25	17	30	31	31	30
17	27	223	223	254	254
27	6	86	86	340	340
6	22	132	132	472	472
22	1	499	499	971	972
44	47	81	82	82	81
47	4	89	90	172	171
4	19	35	35	207	207
19	13	329	329	536	536
13	8	84	84	620	621
8	1	193	193	813	814
39	46	63	64	64	63
46	45	150	150	214	213
45	58	78	78	292	292
58	30	170	170	462	462
30	55	52	52	514	515
55	1	246	246	760	761
56	48	136	136	136	136
48	23	82	83	219	219
23	50	100	100	319	319
50	33	70	70	389	389
33	16	72	72	461	461
16	18	48	48	509	509
18	28	103	103	612	613
28	59	287	288	900	901

Table 8.8 (*Continued*)

FROM	TO	MILES	COST$	TOT$	TMILES
59	3	80	80	980	982
3	1	163	163	1143	1145
7	43	74	74	74	74
43	29	161	161	235	235
29	36	87	87	322	323
36	5	135	135	457	458
5	38	83	83	540	541
38	15	77	77	617	618
15	35	83	83	700	701
35	2	105	106	806	807
2	1	140	140	946	947
37	9	267	267	267	267
9	49	316	316	583	584
49	11	83	83	666	667
11	1	264	264	930	932

```
            TOTAL AL COST  = 9818
         TOTAL AL,SL MILES   = 9818      0
         TOTAL AL AND SL COSTS= 9818      0
        TOTAL NUMBER OF SPECIAL LINKS=     0

        *******SYSTEM DESIGN PARAMETERS*******
USER PORT SPEED(BPS)= 9600               HOST PORT SPEED(BPS)= 9600
INP MSG LENGTH(BYTES)= 8                 OUT MSG LENGTH(BYTES)= 64
AVG # CUs/VMPTC= 8                       AVG # MSGS/SEC/VMPTC= 2.78E-03
HOST THINK TIME(SEC)= 1E-03              SHARED MEDIA SPEED(BPS)= 9600
NODAL PKT PROC TIME(MS)= 10              HALF MODEM DELAY(MS)= 4
PACKET SIZE(BYTES)= 64                   INF BYTES/PACKET= 56
PROPAGATION CONSTANT(MS/MI)= .01         ANALYSIS TYPE= 3
MAXIMUM ALLOWABLE LINK RATE= 5100        TRAFFIC GROWTH FACTOR= 1
```

Table 8.9 Netlink Analysis for VHD59 and EWNET1.1 with No Optimization

```
NO. OF SWITCHES= 1                    AT          1

MAX ALLOWED LINK RATE (Wm) = 5100 BPS

SW# LK# LKWT AQT ATAT E(TR)                          NODES
        BPS  MS  MS   MS
 1  1   5100  6  596  1371 *** 12  42  10  24  28  50  3  1
 1  2   3000  3  467  780  *** 15  8  11  49  2  35  1
 1  3   5100  6  824  1893 *** 25  51  32  52  31  20  40  1
 1  4   5100  6  585  1344 *** 38  37  7  43  29  36  5  1
 1  5   5100  6  485  1076 *** 46  39  45  58  30  55  1
 1  6   5100  6  590  1357 *** 47  44  4  19  14  33  13  1
 1  7   5100  6  768  1886 *** 53  34  26  17  54  6  27  22  41  1
 1  8   4800  5  788  1844 *** 57  59  21  18  16  23  56  48  9  1

MAXTR= 1893      AVG. TR= 1443

OVERALL SYSTEM AVERAGE(TR)-MS = 1443
```

FROM	TO	MILES	COST$	TOT$	TMILES
12	10	207	207	207	207
42	12	220	220	427	427
24	28	104	104	531	531
50	28	106	107	638	638
3	1	163	163	801	802
3	50	253	253	1054	1055
24	10	141	141	1195	1196
15	35	83	83	83	83
8	2	65	65	148	148
11	49	83	83	231	231

Table 8.9 *(Continued)*

	TO	MILES	COST$	TOT$	TMILES
1	35	76	76	307	308
11	8	104	104	411	412
2	35	105	106	517	518
25	40	459	460	460	459
51	32	112	112	572	572
52	32	350	350	922	922
31	32	227	227	1149	1150
20	40	556	557	1706	1707
1	40	674	674	2380	2381
31	20	603	603	2983	2985
38	5	83	83	83	83
37	29	194	194	277	277
7	43	74	74	351	352
36	29	87	87	438	440
1	5	140	141	579	580
7	29	139	139	718	720
36	5	135	135	853	855
46	39	63	64	64	63
45	58	78	78	142	142
30	55	52	52	194	194
1	55	246	246	440	441
30	45	123	123	563	564
45	46	150	150	713	714
47	44	81	82	82	81
4	19	35	35	117	117
14	33	154	154	271	271
13	1	226	226	497	497
13	33	280	280	777	777
14	19	326	326	1103	1104
4	47	89	90	1193	1194
53	34	178	178	178	178
26	34	46	46	224	224
17	54	178	178	402	402
6	27	86	86	488	488
22	41	84	84	572	572
1	41	455	455	1027	1028
41	54	95	96	1123	1124
6	22	132	132	1255	1256
54	26	141	142	1397	1398
57	21	53	53	53	53
59	9	100	100	153	153
18	16	48	48	201	201
23	48	82	83	284	284
56	48	136	136	420	421
1	9	103	103	523	524
23	16	182	182	705	706
18	57	134	134	839	840
59	16	217	217	1056	1058

```
TOTAL AL COST  = 9907
TOTAL AL & SL  MILES = 9907,       0
TOTAL AL AND SL COSTS= 9907,       0
TOTAL NUMBER OF SPECIAL LINKS=     0
```

```
              *******SYSTEM DESIGN PARAMETERS*******
USER PORT SPEED(BPS)= 9600             HOST PORT SPEED(BPS)= 9600
INP MSG LENGTH(BYTES)= 8               OUT MSG LENGTH(BYTES)= 64
AVG # CUs/VMPTC= 8                     AVG # MSGS/SEC/VMPTC= 2.7E-03
NODAL PKT PROC TIME(MS)= 10            HALF MODEM DELAY(MS)= 4
PACKET SIZE(BYTES)= 64                 INF BYTES/PACKET= 56
PROPAGATION CONSTANT(MS/MI)= .01       ANALYSIS TYPE= 3
HOST THINK TIME(SEC)= 1E-03            TRAFFIC GROWTH FACTOR= 1
```

Table 8.10 Netlink Analysis for VHD59 and MDNET with Optimization

```
NO. OF SWITCHES= 1                          AT          1
MAX ALLOWED LINK RATE(Wm)= 5100 BPS

SW#LK#LKWT  AQT ATAT E(TR)                             NODES
             MS  MS   MS
 1  1  5100  6  791  1880 *** 52  32  51  31  20  40  53  37  1
 1  2  5100  6  634  1458 *** 42  12  21  10  57  24  3  1
 1  3  5100  6  433   922 *** 25  17  27  6  22  1
 1  4  5100  6  485  1076 *** 44  47  4  19  13  8  1
 1  5  5100  6  481  1066 *** 39  46  45  58  30  55  1
 1  6  5100  6  824  2084 *** 56  14  28  18  16  23  48  50  33  59  1
 1  7  4800  5  762  1783 *** 54  26  7  34  41  43  29  36  5  1
 1  8  3000  3  571   985 *** 11  49  38  2  9  15  35  1

MAXTR= 2084       AVG.TR (FOR MAIN & SPEC. LINKS)= 1406

OVERALL SYSTEM AVERAGE(TR)-MS = 1406
```

FROM	TO	MILES	COST$	TOT$	TMILES
52	32	350	350	350	350
51	32	112	112	462	463
31	32	227	227	689	690
20	40	556	557	1246	1247
53	40	146	146	1392	1393
37	53	230	230	1622	1624
1	37	336	336	1958	1960
20	31	603	603	2561	2563
42	12	220	220	220	220
21	57	53	53	273	273
10	24	141	141	414	414
3	1	163	163	577	577
3	24	401	401	978	979
24	57	91	91	1069	1070
10	12	207	207	1276	1277
25	17	30	31	31	30
27	6	86	86	117	116
22	6	132	132	249	249
1	22	499	499	748	748
27	17	223	223	971	972
44	47	81	82	82	81
4	19	35	35	117	117
13	8	84	84	201	201
1	8	193	193	394	395
13	19	329	329	723	724
4	47	89	90	813	814
39	46	63	64	64	63
45	58	78	78	142	142
30	55	52	52	194	194
1	55	246	246	440	441
30	45	123	123	563	564
45	46	150	150	713	714
56	48	136	136	136	136
14	18	65	65	201	201
28	16	99	99	300	300
23	48	82	83	383	383
50	33	70	70	453	453
59	33	146	146	599	599
1	59	154	154	753	753
33	16	72	72	825	826
16	18	48	48	873	874
50	23	100	100	973	974
54	41	95	96	96	95
26	34	46	46	142	142
7	43	74	74	216	217
29	36	87	87	303	304
5	36	135	135	438	439
1	5	140	141	579	580

Table 8.10 (*Continued*)

FROM	TO	MILES	COST$	TOT$	TMILES
29	41	112	112	691	693
7	29	139	139	830	832
26	54	141	142	972	974
11	49	83	83	83	83
38	15	77	77	160	160
2	35	105	106	266	266
9	1	103	103	369	369
1	35	76	76	445	445
35	15	83	83	528	528
2	49	108	108	636	637

```
          TOTAL AL COST  = 8915
        TOTAL AL & SL MILES = 8915,     0
       TOTAL AL & SL COSTS= 8915,       0
      TOTAL NUMBER OF SPECIAL LINKS=    0
```

```
                 *******SYSTEM DESIGN PARAMETERS*******
USER PORT SPEED(BPS)= 9600                  HOST PORT SPEED(BPS)= 9600
INP MSG LENGTH(BYTES)= 8                    OUT MSG LENGTH(BYTES)= 64
AVG # CUs/VMPTC= 8                          AVG # MSGS/SEC/VMPTC= 2.7E-03
NODAL PKT PROC TIME(MS)= 10                 HALF MODEM DELAY(MS)= 4
PACKET SIZE(BYTES)= 64                      INF BYTES/PACKET= 56
PROPAGATION CONSTANT(MS/MI)= .01            ANALYSIS TYPE= 3 (SNA/SDLC)
HOST THINK TIME(SEC)= 1E-03                 TRAFFIC GROWTH FACTOR= 1
```

Table 8.11 Netlink Analysis for VHD59 and MDNET with Optimization (Tariffs 9, 10, 11)

```
NO. OF SWITCHES= 1                    AT            1
MAX LINK RATE= 5100 BPS

SW# LK# LKWT AQT ATAT E(TR)                          NODES
               MS  MS   MS
 1  1  5100  6  791  1880 *** 52  32  51  31  20  40  53  37  1
 1  2  5100  6  634  1458 *** 42  12  21  10  57  24  3  1
 1  3  4800  5  366  711  *** 25  17  27  6  1
 1  4  5100  6  485  1076 *** 44  47  4  19  13  8  1
 1  5  5100  6  481  1066 *** 39  46  45  58  30  55  1
 1  6  5100  6  824  2084 *** 56  14  28  18  16  23  48  50  33  59  1
 1  7  5100  6  859  2174 *** 54  22  26  7  34  41  43  29  36  5  1
 1  8  3000  3  571  985  *** 11  49  38  2  9  15  35  1

MAXTR= 2174      AVG.TR (FOR MAIN & SPEC. LINKS)= 1429

OVERALL SYSTEM AVERAGE(TR)-MS = 1429
```

FROM	TO	MILES	COST$	TOT$	TMILES
52	32	350	592	592	350
51	32	112	440	1032	463
31	32	227	513	1545	690
20	40	556	706	2251	1247
53	40	146	462	2713	1393
37	53	230	515	3228	1624
1	37	336	583	3811	1960
20	31	603	721	4532	2563
42	12	220	509	509	220
21	57	53	382	891	273
10	24	141	458	1349	414

Table 8.11 (*Continued*)

FROM	TO	MILES	COST$	TOT$	TMILES
3	1	163	472	1821	577
3	24	401	625	2446	979
24	57	91	422	2868	1070
10	12	207	500	3368	1277
25	17	30	334	334	30
27	6	86	417	751	116
1	6	621	727	1478	738
27	17	223	511	1989	962
44	47	81	413	413	81
4	19	35	345	758	117
13	8	84	415	1173	201
1	8	193	491	1664	395
13	19	329	578	2242	724
4	47	89	421	2663	814
39	46	63	393	393	63
45	58	78	409	802	142
30	55	52	381	1183	194
1	55	246	526	1709	441
30	45	123	447	2156	564
45	46	150	464	2620	714
56	48	136	455	455	136
14	18	65	395	850	201
28	16	99	431	1281	300
23	48	82	414	1695	383
50	33	70	400	2095	453
59	33	146	461	2556	599
1	59	154	467	3023	753
33	16	72	402	3425	826
16	18	48	374	3799	874
50	23	100	432	4231	974
54	41	95	427	427	95
22	41	84	415	842	179
26	34	46	371	1213	226
7	43	74	405	1618	301
29	36	87	418	2036	388
5	36	135	454	2490	523
1	5	140	458	2948	664
29	41	112	440	3388	777
7	22	95	427	3815	872
26	54	141	459	4274	1014
11	49	83	414	414	83
38	15	77	407	821	160
2	35	105	435	1256	266
9	1	103	434	1690	369
1	35	76	407	2097	445
35	15	83	414	2511	528
2	49	108	437	2948	637

```
     TOTAL AL COST  =  26625
  TOTAL AL & SL MILES =  8959,      0
  TOTAL AL & SL COSTS=  26625,      0
  TOTAL NUMBER OF SPECIAL LINKS=    0
```

```
              *******SYSTEM DESIGN PARAMETERS*******
USER PORT SPEED(BPS)= 9600                 HOST PORT SPEED(BPS)= 9600
INP MSG LENGTH(BYTES)= 8                   OUT MSG LENGTH(BYTES)= 64
AVG # CUs/VMPTC= 8                         AVG # MSGS/SEC/VMPTC= 2.7E-03
NODAL PKT PROC TIME(MS)= 10                HALF MODEM DELAY(MS)= 4
PACKET SIZE(BYTES)= 64                     INF BYTES/PACKET= 56
PROPAGATION CONSTANT(MS/MI)= .01           ANALYSIS TYPE= 3
HOST THINK TIME(SEC)= 1E-03                TRAFFIC GROWTH FACTOR= 1
```

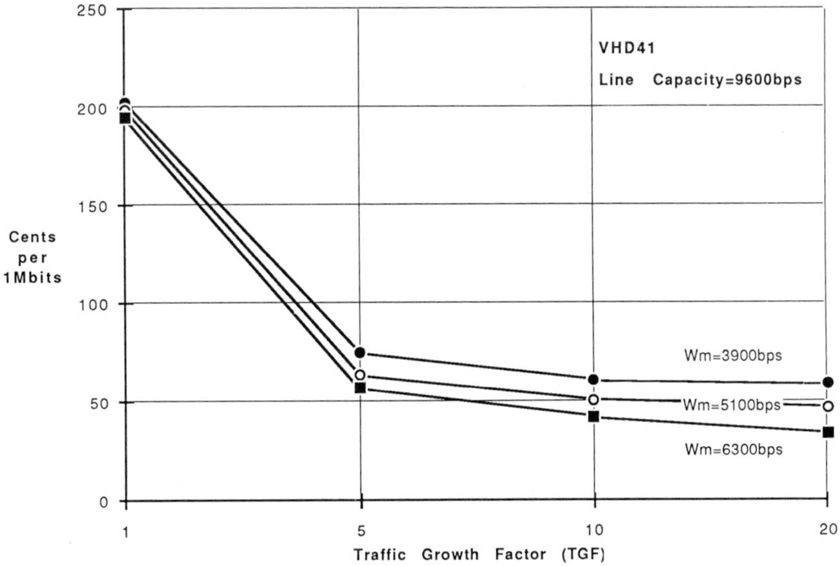

Figure 8.36. Cents per 1Mbits versus TGF for VHD41, Tariffs 9, 10, and 11, link capacity (C) = 9600bps, and maximum allowable link rates (W_m) = 3900bps, 5100bps, and 6300bps.

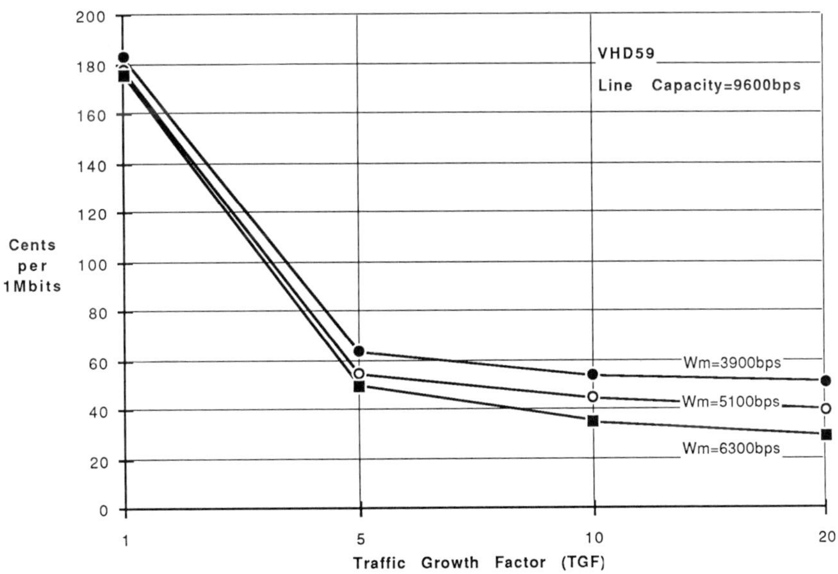

Figure 8.37. Cents per 1Mbits versus TGF for VHD59, Tariffs 9, 10, and 11, link capacity (C) = 9600bps, and maximum allowable link rates (W_m) = 3900 bps, 5100 bps, and 6300bps.

solution for interconnecting LANs distributed over a campus or a metropolitan environment. The network topologies obtained by using the STARNET item under the NETWORKING menu with properly modified input files of VHD41, VHD59, VHD100, and VHD200 will appear identical to those for the voice networks, as described in Chapter 7. Each netlink will serve exactly two nodes—one at the switch location and the other at one of the remaining locations. If a nonswitch location node has a traffic load (TL_i) greater than W_m bps, then exactly $[\text{int}(TL_i/W_m) + 1]$ netlinks will be formed, with $[\text{int}(TL_i/W_m)]$ netlinks with loads of W_m and the last netlink with a load equal to the Remainder = TL_i MOD W_m. The cost of a data network based on the star topology will be generally higher than that based on the multidrop (DL or MD) topology, especially for small values of TGF. Figure 8.38 shows a comparison of costs for the MDNET and STARNET as a function of TGF, and for an VHD41 input file and two values of W_m = 3900 and 6300 bps. The results show that the cost difference disappears as the value of TGF becomes larger than 5. This is due to the fact that the special links of an MDNET are identical to the netlinks of a STARNET as TGF increases.

The response times can be studied for DLNET, MDNET, and STARNET topologies, using the ATP values of 1, 2, and 3 respectively. The ATP values of 1, 2, and 3 represent the ASYNC, Doelz's virtual multipoint, and the SNA multipoint protocols, respectively. The results using the EWNET module should be similar to those for the MDNET module. The average bid-to-start (BST) delays or Average (T_r) are presented in Figures 8.39 through 8.41 as a function of TGF for the VHD41 input file. The value of BST is identical to the average access time as defined by Schwartz (1987). Four interesting observations are cited in the following paragraphs.

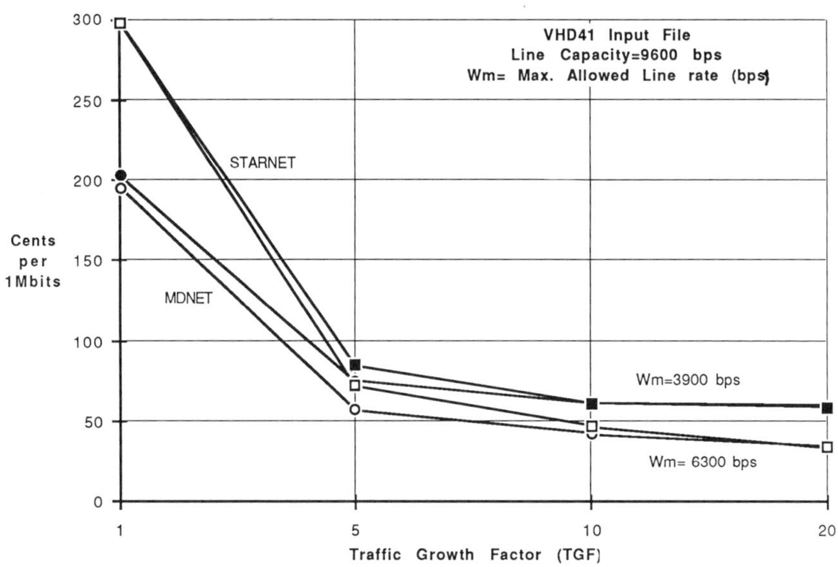

Figure 8.38. Cents per 1 Mbits versus TGF for VHD41, Tariffs 9, 10, and 11, link capacity (C) = 9600bps, maximum allowable link rates (W_m) = 3900bps, 5100bps, and 6300bps, and MDNET/STARNET topologies.

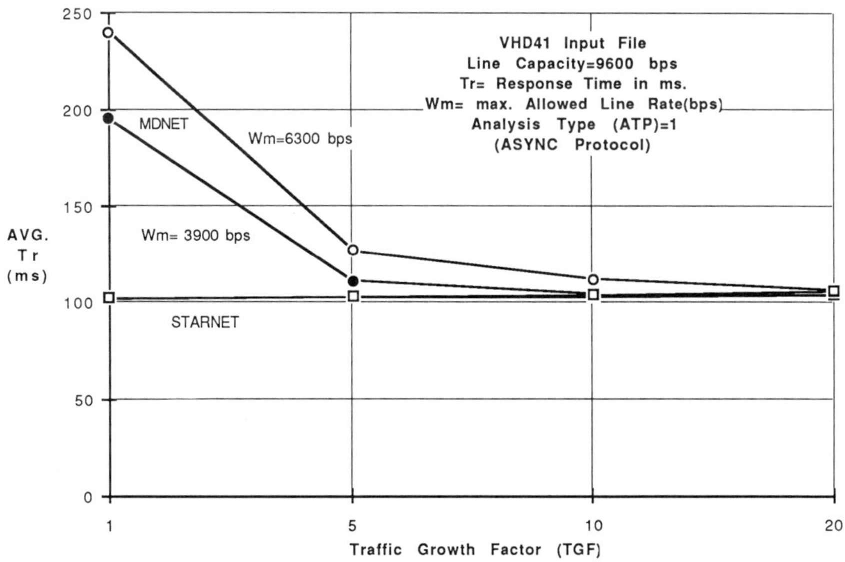

Figure 8.39. Average response time (T_r) versus TGF for VHD41, Tariffs 9, 10, and 11, link capacity (C) = 9600bps, maximum allowable link rates (W_m) = 3900bps, 5100bps, and 6300bps, and ASYNC traffic.

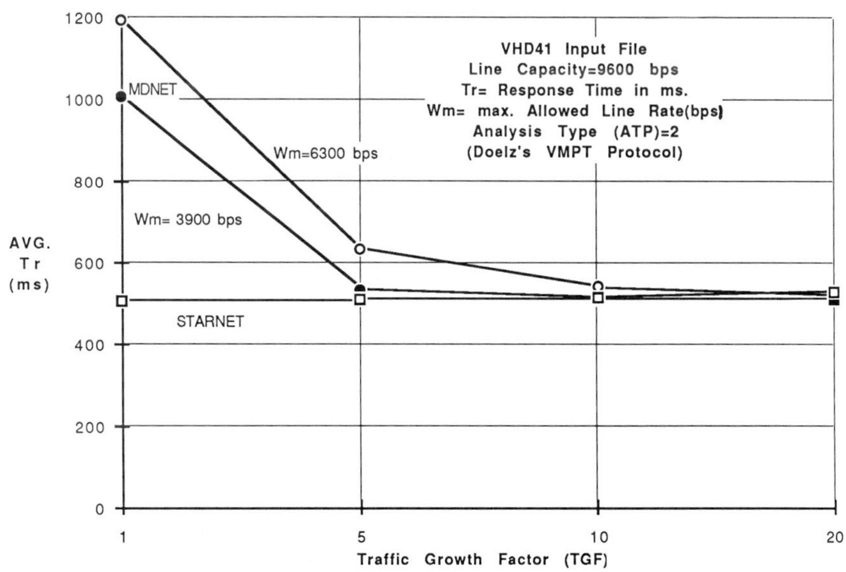

Figure 8.40. Average response time (T_r) versus TGF for VHD41, Tariffs 9, 10, and 11, link capacity (C) = 9600bps, maximum allowable link rates (W_m) = 3900bps, 5100bps, and 6300bps, and Doelz architecture.

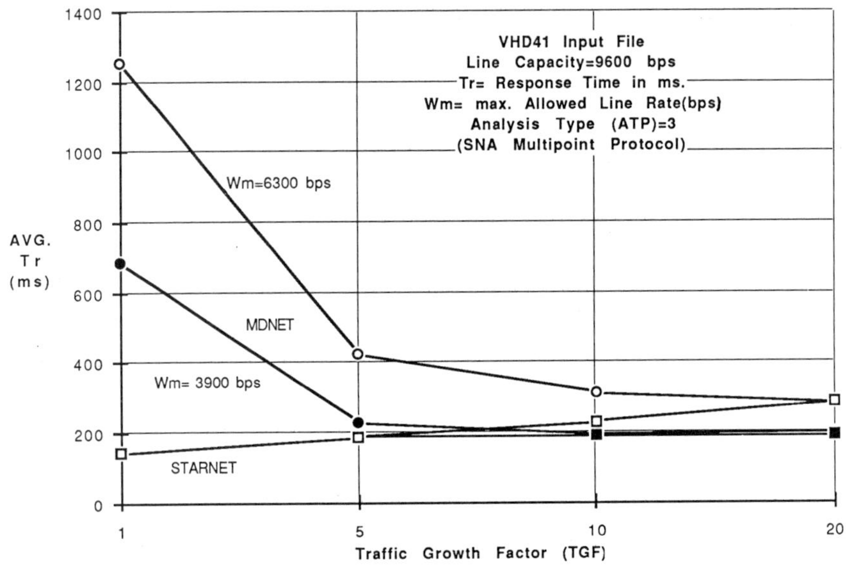

Figure 8.41. Average response time (T_r) versus TGF for VHD41, Tariffs 9, 10, and 11, link capacity (C) = 9600bps, maximum allowable link rates (W_m) = 3900bps, 5100bps, and 6300bps, and SNA architecture.

1. The value of BST increases as W_m increases. According to the queuing theory, as described in Chapter 4, the end-to-end queuing delays will tend to go to infinity as the value of Wm nears the value of circuit capacity, C. The average values of BST, as plotted, are valid for the actual number of controllers (N) for each synthesized multipoint netlink, and N_{cu} = 4 for each virtual multipoint circuit on the ESR. The value of N_{cu} is assumed to equal N, if N is less than 4. The Doelz network is less sensitive to heavy traffic load than a conventional MPT network. However, the Doelz network exhibits slightly higher access delays for low traffic loads, as expected. See Chapter 4 for an explanation based on the virtual packet-switching technique employed for resource sharing.

2. The response times are significantly reduced with TGF greater than 5. This can be explained by the fact that the number of nodes (N or N_{cu}) per netlink are reduced as the nodal traffic loads are increased. This is especially true for the increase in the number of special links, each consisting of only two nodes, as is the case with a STARNET.

3. The value of BST for the virtual point-to-point ASYNC operation is always much lower than the BST value for virtual multipoint or conventional multipoint operations. This can be attributed to the elimination of polling, which is employed only to achieve centralized control through the mainframe or host. The ASYNC operation employs a distributed control. Its main disadvantage is related to the loss of stability at very high traffic loads. Since an ESR-based network achieves traffic flow control through the use of a virtual circuit concept, such an instability is prevented. Since an ESR-based network also achieves relatively low queuing delays even at very heavy traffic loads, it becomes an ideal vehicle for ASYNC operation.

4. The response times for a STARNET are always less than any other type of mul-

tidrop network and for any given protocol. Only at high values of TGF does the response time of a STARNET become equivalent to those of multidrop network topologies for any given protocol.

In the analysis presented above, we maintained a fixed line capacity of 9600 bps. It should be quite instructive to study the network costs as a function of line capacities. See Figures 8.42 and 8.43 for the plots of costs versus TGF and three line capacities. The effects of bypass on the 56 Kbps and 1.544 Mbps links are also shown. The plots show that the use of a 9600 bps line results in the cheapest network. The results also reveal that the use of a 56 Kbps line would have been justified only for TGF > 20. A VHD 41 network for TGF = 1 would have consisted of only one 56 Kbps netlink serving all nodes, according to a true minimal spanning tree topology. A network designed for TGF = 5 will consist of four or more netlinks. The plots suggest that the use of a 1.544 Mbps line capacity will always result in the most expensive network for the range of TGF values considered. The use of a T-1 line should result in the cheapest network for TGF values much greater than 40. The lesson of Figures 8.42 and 8.43 should now be quite obvious. One must always study the effect of line capacity on network costs before studying the network topologies in detail. This can be done by simply changing the tariff number for a given link type on the NLT input file. See Chapter 6 for directions.

8.3.2 STUDY OF DATA NETWORK TOPOLOGIES WITH MULTIPLE SWITCHES

In order to synthesize an optimum network topology, one must vary the number of optimally located switches in the network, as shown in the previous chapter dealing with

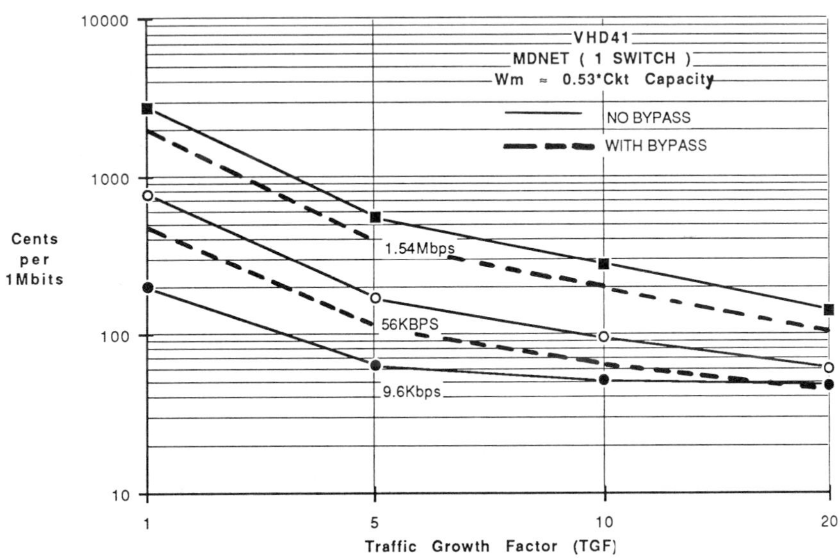

Figure 8.42. Cents per 1Mbits versus the TGF for VHD41, Tariffs 9, 10, and 11, maximum allowable link rates (W_m) = 0.53 * C, and link capacities (C) = 9600bps, 56kbps, and 1.544mbps.

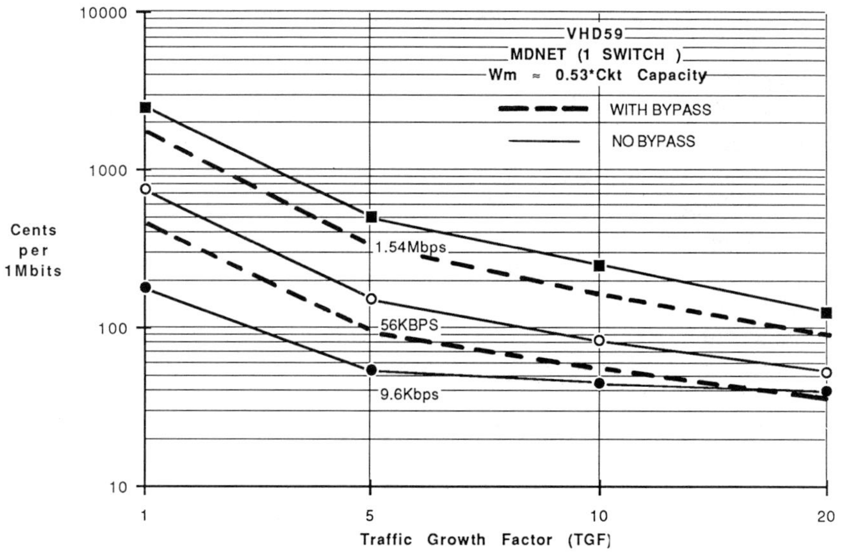

Figure 8.43. Cents per 1Mbits versus the TGF for VHD59, Tariffs 9, 10, and 11, maximum allowable link rates (W_m) = 0.53 * C, and link capacities (C) = 9600bps, 56kbps, and 1.544mbps.

voice applications. The COG approach was shown to be quite effective for selecting any desired number of switch locations.

The design process can be illustrated by selecting the VHD41 data file and increasing the number of switches from one to five in the network employing 9600 bps circuits. Table 8.12 illustrates the trunking analysis summarizing the design assumptions, and the total link and trunk costs for two data switches located at nodes 16 and 39. The table shows that 9600 bps links were used to implement FDX trunks, and that the maximum utilization of the trunk was 0.8. The table also shows that a backbone trunking factor (BBTF) of 2 was employed for determining the traffic flows on the backbone network, as defined by the Original Trunk Traffic matrix and for TGF = 1.

The 2 × 2 matrix shows that the intranodal traffic flows of 7446 bps and 14346 bps occur at switch nodes 1 and 2, respectively. The 2 × 2 matrix also shows that 3354 bps flow in both directions of the trunk connecting nodes 1 and 2. Such a symmetric flow results from the nature of the simplified mathematical model, as discussed in Subsection 2.4.7. Only the use of a From-To traffic file (see Chapter 9 for an example) can provide actual traffic flows between the nodes of a backbone. The 2 × 2 Modified Trunk Traffic matrix shows only the effective traffic flow on the backbone network for computing the number of trunks needed. The 2 × 2 Modified Trunk matrix shows that only one trunk is needed to connect the two switch nodes. The final Trunk Cost matrix defines each trunk bundle, its mileage, unit trunk cost, and the total cost of the trunk bundle.

The cost summaries of Table 8.12 and other similar tables were used to draw curves as plotted in Figure 8.44 for TGF values of 1, 5, 10, and 20. Although most of the curves are rather flat, it is possible to see that a network topology with three switches

Table 8.12 Trunking Analysis for a 2-Switch Data Communication Network

```
                                   TRUNK LINK/NODE TYPE 0 (@9600bps)

                                   FDX FACTOR = 1        :      BBTF=2

                                   MAX TRUNK LOAD = .8   :       TGF=1

                                   TOTAL INTER/INTRA NODAL TRAFFIC= 28499.98

                                   TARIFFS 9,10,11

                                   NO. OF SWITCHES= 2     AT     16     39

ORIGINAL TRUNK TRAFFIC MATRIX AS FOLLOW

 7446.31          3353.68

 3353.68          14346.31

MODIFIED TRUNK TRAFFIC MATRIX IS AS FOLLOWS

 0                3353

 0                0

MODIFIED TRUNK MATRIX IS AS FOLLOWS

 0                1

 0                0

TRUNK COST MATRIX IS AS FOLLOWS

FROM            TO              MILES           COST            TOTALS
 16              39              1493            1006            1006

                    *****SUMMARY OF SYSTEM COSTS*****
                          ACCESS LINK COST= 18697
                          TOTAL TRUNK COST= 1006
                          TOTAL SYSTEM COST= 19703
                          CENTS PER 1MBITS= 192.03
```

generally produces a least-cost network for higher values of TGF. Two sample network topologies are shown in Figures 8.45 and 8.46 for the 9600 bps network with two and three optimally located switches. Similar results can be obtained for the VHD59 and other data files.

The results clearly show that unit network costs can be reduced significantly by increasing the value of TGF and/or selecting the appropriate line capacity (56 Kbps versus 9600 bps). The optimum choice of switches does not influence the network costs as much in comparison. This observation is similar to that for the voice networks, where increasing the TGF and the use of multiplexed lines also decreased the unit network costs significantly.

8.3.3 STUDY OF DATA NETWORK TOPOLOGIES WITH SATELLITES

Recent technological developments have made data communications networks based on very small aperture terminals (VSATs) quite attractive economically. High-powered Ku-

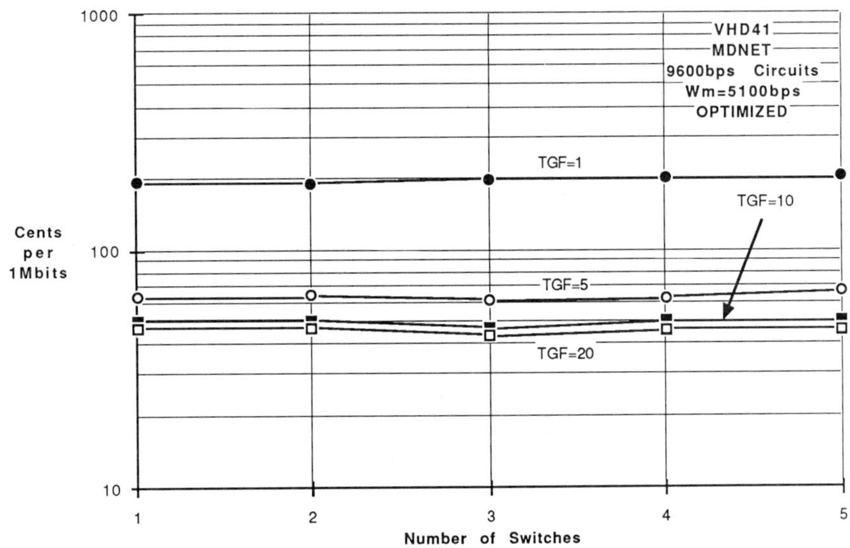

Figure 8.44. Cents per 1Mbits versus the number of optimally located switches for VHD41, Tariffs 9, 10, and 11, link capacity (C) = 9600bps, and TGF = 1, 5, 10, and 20.

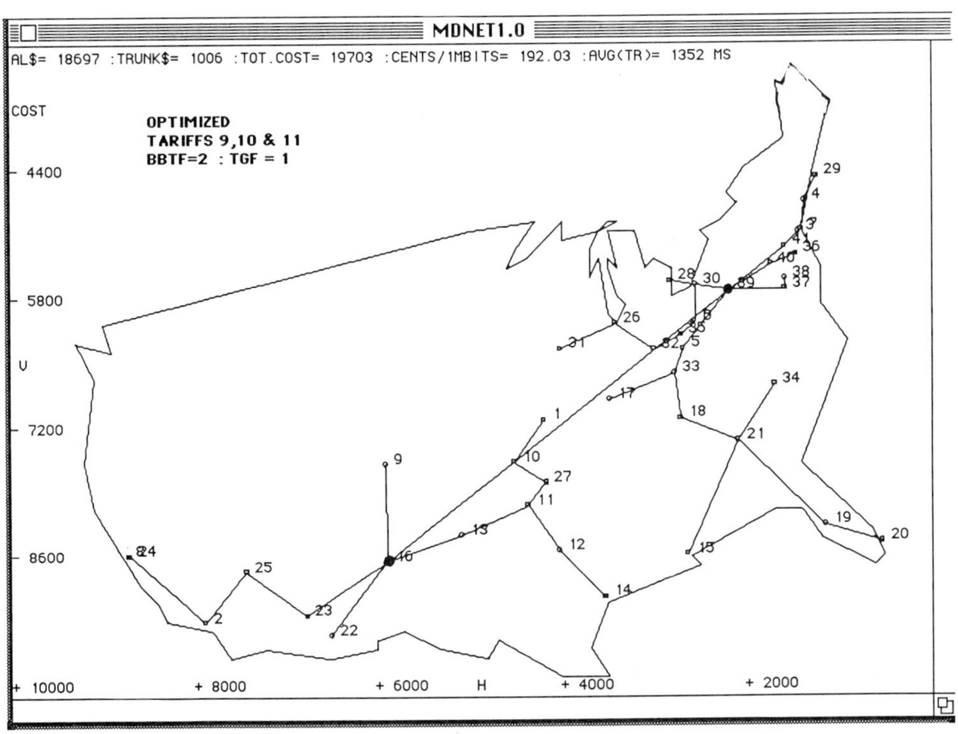

Figure 8.45. Optimized multidrop (MD) network topology for VHD41, Tariffs 9, 10, and 11, two optimally located switches, TGF = 1, BBTF = 2, link capacity (C) = 9600bps, and maximum allowable link rate (W_m) = 5100bps.

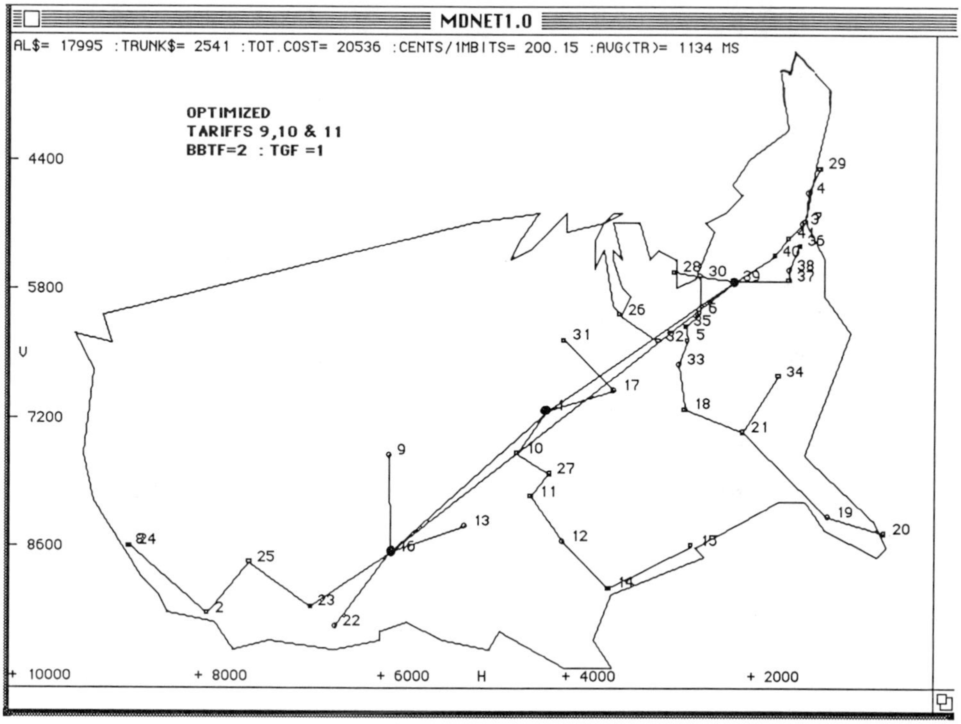

Figure 8.46. Optimized multidrop (MD) network topology for VHD41, Tariffs 9, 10, and 11, three optimally located switches, TGF = 1, BBTF = 2, link capacity (C) = 9600bps, and maximum allowable link rate (W_m) = 5100bps.

band satellites, coupled with the advancements in the design of a master ground station (MGS), enable the use of a large number of VSATs on customer premises. The use of VSATs makes it possible to economically bypass both the LECs and IECs for low data rate communications applications involving a large number of widely dispersed customer locations. Examples of such applications are the IBM SNA/SDLC, CCITT's X.25/X.75, and DB management/sharing. Applications requiring very heavy traffic loads from every location will tend to justify a terrestrial network.

The major cost of a VSAT network deals with the hardware. The cost of a VSAT node will vary between \$6000 and \$20,000, depending upon the number of ports served and performance requirements dealing with traffic, protocol, and reliability. The cost of an MGS will vary between \$1 million and \$2 million, depending upon the traffic, protocol, and reliability requirements. The monthly cost of the space segment (or the leased transponder bandwidth) per VSAT ranges from \$30 to \$150 per month, depending upon the bandwidth and nodal intelligence. More intelligence incorporated into the VSAT design implies a lower cost of the space segment. In contrast, almost all of the cost of a terrestrial data communications network is associated with the leased private LEC and IEC lines. The costs of switching hardware on a monthly basis is generally negligible.

To study the applicability of a VSAT network, one can use the plots from Figure 8.47. Four networks, consisting of 41, 59, 100, and 200 nodes, were considered. Their

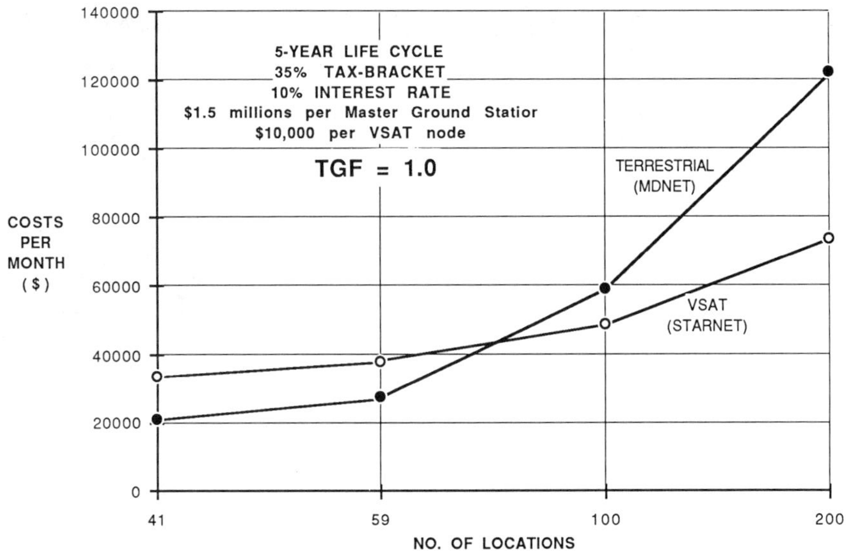

Figure 8.47. Costs per month versus the number of locations for a terrestrial network with MD topology and a satellite network with a single hub.

VHD files have been defined in this and the previous chapters. The costs of the terrestrial networks were computed using the MDNET module with the correct number of switches. The costs of VSAT networks were computed using the following assumptions:

1. Each VSAT node costs $10,000, including the electronics and 1-meter dish;
2. Each master ground station costs $1,500,000;
3. A cost of $100 per VSAT per space segment;
4. A five-year life cycle with 10 percent interest rate; and
5. A 35 percent corporate tax rate and no income tax credit.

Accordingly to the return-on-investment (ROI) analysis of Chapter 2, the one-time hardware cost should be divided by a factor of 47.065 to compute the monthly cost due to hardware. The result show that a VSAT network appears very attractive for over 100 locations. Of course, other design considerations, such as response times and security/privacy, must not be ignored while making the selection.

8.4 SUMMARY OF DATA NETWORK SYNTHESIS AND CONCLUSIONS

Several algorithms were employed to synthesize capacitated minimal spanning trees (MSTs), directed link, and star topology networks. Two techniques, MDNET and three variations of the EWNET algorithm, were compared for the synthesis of capacitated MSTs. It was shown that the MDNET algorithm was more efficient than the EWNET algorithm for synthesizing cheaper networks. It was also shown that one can employ the exchange algorithm to further reduce the cost of any network synthesized using the EWNET technique. However, it was interesting to observe that such costs were usually higher than those associated with the MDNET-derived networks. Another important observation deals with the fact that network topologies remain intact when only the tariffs are changed (e.g., when the 1$/mile tariff was changed to AT&T's Tariffs 9, 10, and 11). A slight change in a given tariff should have practically no effect on the network topology. These facts should allow one to accelerate the network design process significantly by using only simplified models of tariffs. The exact models of tariffs involving large data bases should only be applied to finalized network topologies.

Unit network costs were shown to be closely related to the value of TGF and the capacity of the circuits employed for synthesizing capacitated MSTs. The selection of an optimum number of switches did not reduce the network costs significantly. The use of concentrators can also be studied for networks with heavy traffic loads. In many cases, the unit cost of transporting 1Mbits can be reduced by increasing the number of network layers (or by using concentrators and switches). See Chapter 6 for directions related to the creation of special LINK files that specify the locations of concentrators at any level of the network hierarchy.

Data communications networks using VSATs were also studied, and their costs were compared with those obtained with terrestrial networks and multidrop topologies. It was shown that a VSAT network appears very attractive when the number of locations exceeds 100, and when relatively low traffic volumes are involved. Other design considerations, such as response times and privacy/security must also be considered while making the selection. A detailed analysis is specially recommended when high traffic volumes are involved.

There may be a situation where most of the data terminals are clustered in buildings scattered over a campus or a metropolitan area or a single LATA. A data network based on the star topology (resulting from the STARNET option under the NETWORKING menu) is recommended when several such campuses or metropolitan areas are interconnected to form a corporate data network. For this case, most of the links are leased from the local exchange carrier for connecting buildings together using the star topology. No OCCs are involved. In order to compute the correct costs, one must define the LATA number for each location through the LATA file. In the analyses discussed above, all of the links were assumed to be of the IEC variety used for WANs. The analysis presented here clearly shows that data networks based on the star topology always provide superior response time performance and equivalent cost performance at high traffic loads (as exemplified by large values of TGF). This is quite a revealing fact favoring the star topology, especially in view of the recent increases in the cost of leasing multidrop line segments.

9

Synthesis of Integrated Voice/Data Network Topologies

9.1 INTRODUCTORY REMARKS

Techniques for synthesizing network topologies for either voice or data applications have been described in preceding chapters. It should be stressed that such techniques are effective for designing not only WANs but also LANs. This chapter will discuss several design considerations for implementing integrated voice and data (IVD) networks for both WAN and LAN applications. First we will deal with the process of selecting the proper switching nodes for voice, data, and IVD applications. Next we will describe the process of selecting a suitable premise wiring system (PWS) for voice, data, and IVD applications. Finally, we will employ the principle of superimposition for realizing integrated voice and data networks. The reader will discover that each of these considerations has a unique influence on the resulting LAN and WAN topologies.

9.2 NODAL DESIGN CONSIDERATIONS

9.2.1 SWITCHING NODES FOR VOICE NETWORKS

The most common type of switching node used for voice LAN applications is a PABX. The switching network topology, digital/analog signal interfaces, and switching network control must be known in detail. The nodal architecture thus determines whether one is considering an analog or a digital PABX. With the availability of cost-effective digital transmission facilities, such as 56000 bps DDS and 1.544 Mbps T-1, one can eliminate the cost of A-to-D (analog-to-digital) conversion hardware by selecting a digital PBX. In some nodal designs, the A-to-D hardware costs as much 30 percent of the entire PABX cost.

The second feature of major interest deals with the call handling capacity of the PABX. It is measured in terms of busy hour erlangs. A nonblocking switching network within the PABX will allow a connection between two idle ports, even when all other ports are busy. Some other PABX designs allow a certain amount of call blocking during busy hours. This feature allows a higher termination capability on the PABX. Other nodal characteristics also deal with the main CPU power, the available random access memory, and the number of microprocessors available for handling peripheral tasks, such as line/trunk interfaces and call processing features such as redial, call conferencing, and ACD. Such nodal characteristics will determine capabilities such as the allowable number of peak period call attempts and ACD call intensity. To illustrate, some PABXs with integrated ACD capabilities allow a much lower peak hour ACD call intensity and, as a result, may reflect a very low cost per termination. In comparison, pure ACD systems generally allow much higher peak hour intensities and even a nonblocking switch. As a consequence, these pure ACDs may cost even less than the integrated PABXs for applications requiring very high peak hour traffic intensities.

The third feature of importance deals with the networking features. Some PABXs can also act as tandem voice switches, in addition to handling voice LAN tasks. Such a capability allows the synthesis of a corporate voice WAN. The techniques discussed in earlier chapters are valid for synthesizing both LAN and WAN topologies. Again, one must be very careful in assessing the usefulness of these integrated PABX/tandem nodes, considering the future growth of traffic. One special area deals with interswitch signaling and the ability to integrate it with the existing network or emerging networks based on CCITT approved Signaling System 7 (SS7).

Additional features dealing with such items as power consumption, equipment footprint, air conditioning, and universal power supply requirements may also play important parts in the selection of a PABX. Considerations of such features are out of the scope of this book.

9.2.2 SWITCHING NODES FOR DATA NETWORKS

Most of the LANs employ CSMA/CD, Token Ring, and Token Bus access techniques for interchanging high-speed data between hosts, workstations, and data terminals on fast PS connections. Some LANs employ the voice PABXs to provide data exchange between hosts and data terminals when data rates do not exceed 64 Kbps and CS connection delays are not very critical. As a result, PABXs are sometimes used for both voice and data applications in the LAN environments.

Most of the LANs deal with interconnecting PCs to one another and to LAN servers that allow sharing of data bases and laser printers. Many LANs also allow PCs to communicate with different types of mainframes. This is accomplished through the use of hardware modules (which emulate terminals or communications interfaces) and software modules. Bridges, routers, and gateways are employed to interconnect LANs. Bridges are generally used for extending the local area environ at the media access control layer. But their use is increasing for wide area applications. Routers are generally used to interconnect widely dispersed LANs based on a common protocol at the network layer. Gateways are generally employed to get an access to a WAN based on a different protocol. Chapters 7 and 8 primarily provide techniques for synthesizing economical WANs by interconnecting LANs. In that sense, a data LAN appears the same as a voice PABX does, to the voice WAN.

Most of the WANs employ packet switches as network nodes. PABXs are seldom used for high-speed WAN data applications because of excessive connection delays and misuse of long-haul transmission facilities. There are PABXs that provide an integral packet switch for handling WAN applications. One can also employ specialized data switches that employ fast circuit switching, allowing protocol-free permanent connections between hosts and data terminals with no incremental delays. Such CS connections can also handle very high data rates.

The nodal capabilities can be described in terms of busy hour throughputs measured in packets per second, protocols handled, and gateway availability.

9.2.3 NETWORK NODES FOR INTEGRATED VOICE/DATA NETWORKS (IVDNs)

The early version of a network node for IVD application came in the form of a simple T1 multiplexer that allocated 56 Kbps streams to either data or digitized voice on a backbone network route consisting of one or more T1 circuits. A fixed network configuration was created or updated manually on a network console unit each time the demand profile changed. The modern version of an IVD network node can be called a dynamic bandwidth allocator (DBA), which provides not only the required bandwidth on demand, but also fully integrated network management and control capabilities. This implies many features, such as subrate multiplexing of low-bit data channels, multiple DS0 channel multiplexing, network security, network-wide diagnostics, fault recovery, and automatic alternate path routing. See Figure 9.1 for an illustration of a backbone network based on the use of DBAs. As shown, each DBA serves different network switches such as PABXs, X.25 packet switches, data circuit switches (DCSs), video compression units (VCUs), hosts, PCs, consoles, and data terminals. Each of these devices have requirements for varying amounts of bandwidth for wide-area networking. Since the use of T1 circuits results in great savings in monthly transmission costs, when compared to the

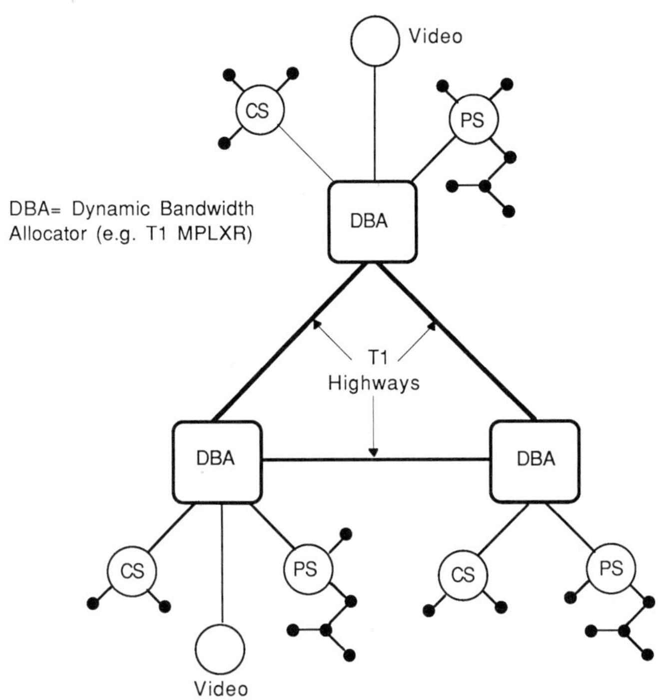

Figure 9.1. Topology of an IVDN employing dynamic bandwidth allocators.

costs of individual voice and 56 Kbps data networks, the DBA is a very cost-effective vehicle for wide-area networking.

In the pure sense, DBAs available today do not allow the creation of a fully integrated voice/data network. That will occur only when the high-speed packet switches become available for handling voice, data, and video applications. At present, the use of DBAs allows only a partial integration of voice and data over the trunk transmission facilities. In that sense, a corporate network based on DBAs resembles the emerging ISDN. Each ISDN node is a composite of many individual switches, and each user location is served by either the primary rate (23B + D) or the basic rate (2B + D) interface. ISDN nodes are interconnected by a flat (nonhierarchical) backbone network consisting of T1, T2, and T3 circuits. The circuit establishments and network control are achieved through the use of the SS7 signalling scheme on a call-by-call basis. Eventually, the DBAs will evolve to provide such interfaces first on the user side, and eventually on the trunk side. The marketplace will demand such an evolution.

The network node capability is determined by the number of CS individual channels, total bandwidth (or number of T1 channels) handled, efficiency of BW utilization, and the extent of network management and control. Other selection criteria may be listed as number of channels of each type (such as voice, 9.6 Kbps, RS232, and X.21), a mixed number of channels, the available T1 bandwidth for user data, the total number of operator consoles and any restrictions on their location, the time required to reroute the path, the time to resynchronize the backbone network, the time to recover from a DBA nodal failure, the time to recover from a network console(s) failure, partitioned network management, and control and network security.

9.3 PREMISE WIRING SYSTEMS

The premise wiring systems (PWSs) constitute a major portion of the one-time costs for the corporate network systems. The 1984 divestiture of AT&T dictated that users must own their own PWSs, in addition to owning their own switching nodes. Furthermore, each user must also worry about the costs of moves, changes, and equipment reconfigurations each time employees change offices and major company reorganizations occur. The user must also consider the need for an integrated voice and data transport to prepare for the emerging ISDN and OSI environments. As a result, each user is now obligated to give much attention to PWS planning for both LAN and WAN applications. Due the nature of interconnections at each customer premise, the LAN considerations far outweigh than those for the WAN. The communications facilities that interconnect LANs are usually leased from a common carrier.

Most emerging PWSs start with the requirement for a standard information outlet (SIO) that will provide a user-changeable standard jack with six or eight pin configurations for a multivendor environment. In other words, up to two four-pairs of twisted cables can be terminated in the two-jack (popularly known as the RJ11 jack) outlet for handling all future voice and data requirements. Most popular voice and data terminals using twisted cable pairs can now be easily connected to the corporate network in a transparent manner by the use of baluns, as shown in Figure 9.2. Many other vendors sell products to replace moderately expensive cables (e.g., Apple Talk) with the existing

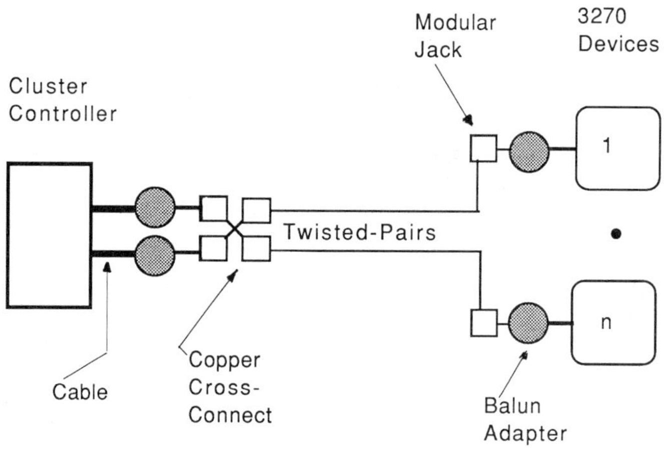

Figure 9.2. Use of Balun adapters for replacing cables with a premise wiring system based on twisted pairs.

twisted pairs. The need for expensive cables can thus be eliminated, since only twisted pairs are utilized. Data transmission speeds of up to 6 Mbps can be handled.

The other components of a PWS are the copper cross-connect (CCC) and/or fiber cross-connect (FCC) that reside in a satellite closet and are used to interface the horizontal cables to the riser cables, as shown in Figure 9.3. As shown, concentrators or multiplexers may also be needed to reduce costs for interfacing large hosts and/or interconnecting a CCC to an FCC.

The third component deals with the cables that are used in the building or on the campus for interconnecting buildings. A four-pair twisted cable terminates each SIO. Several four-pair twisted cables can be bundled together to form one or more horizontal cables on each floor. A riser cable can consist of a fiber cable handling a very large number of voice and data calls. The riser cable ultimately connects all of the premise devices to the switches, LANs, and hosts, located in an equipment room. The horizontal cable is analogous to the access line and the riser cable is analogous to the trunk, as discussed in previous chapters. The reader should thus realize the similarities between the topologies of a WAN and a PWS.

Many vendors now offer proprietary wiring systems. Northern Telecom offers the Integrated Building Distribution System (IBDS). AT&T offers the Premise Distribution System (PDS). Both of these systems are similar to the one described above. Each employs the standard number 24 AWG copper wire, with cable sizes varying from one hundred to several thousand twisted pairs. This technique is usually called the space division multiplexing (SDM). The fiber cable employs frequency division multiplexing (FDM) to concentrate several streams of data into one. In a large building, the use of fiber cables can yield significant savings when compared to the copper twisted wire cables. Specially covered plenum horizontal cables can be used for preventing smoke in case of fire.

The IBM Cabling System employs several types of cables. Type 1 data cable has two twisted pairs of number 22 AWG solid copper conductors protected by a corrugated

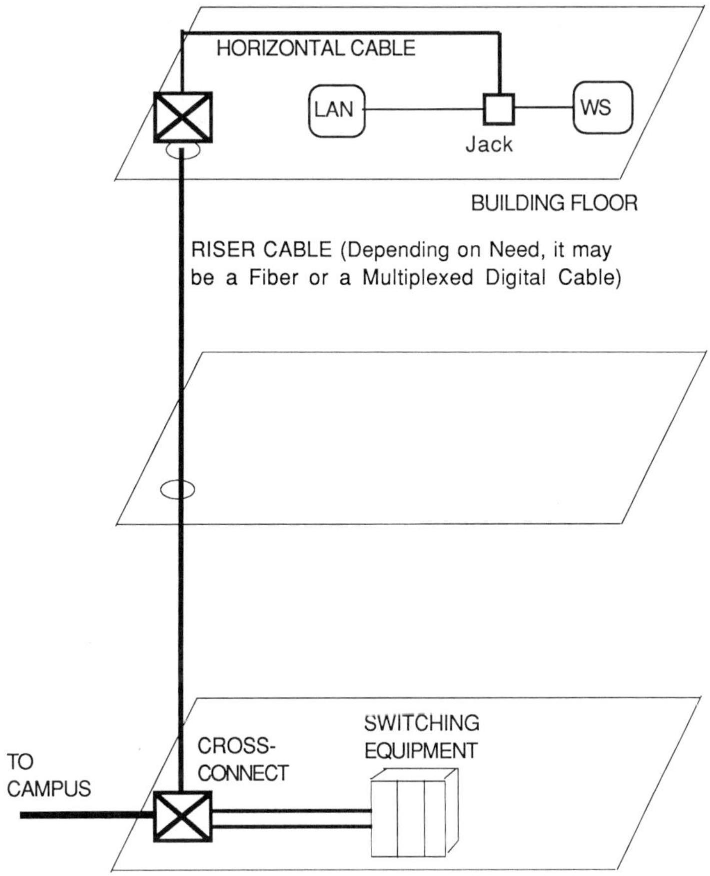

Figure 9.3. A premise wiring system based on the star topology.

metallic shield. Type 2 cable has two twisted wire pairs with a braided shield for data communications and four unshielded twisted wire pairs for voice. The Type 3 data cable has four shielded pairs. Type 5 cable, which has two optical fibers can be used for the backbone. The IBM Type 1 and 2 cables can handle data rates of up to 16 Mbps, even in noisy conditions. Type 3 cable can handle data rates of up to 4 Mbps.

Other well-known PWSs are the DEConnect System, from Digital Equipment Corporation, and the Universal Information Transport Plan (UITP) from Bell of Pennsylvania. There are still no international standards. One must be careful in selecting the vendor. The user must get at least three bids from the PWS vendors for their IVDN. The user should also insist on actual performance tests.

9.4 SYNTHESIS OF AN IVD NETWORK TOPOLOGY

The topology of an IVDN can be synthesized through the process of superimposing separate networks. In other words, each network can be separately designed using the methodology presented in the previous chapters. Later, the backbone networks of each separate network can be superimposed on top of each other, resulting in an IVDN.

The IVDNET module of the ECONETS package can be employed to synthesize an IVD network topology. In order to use this software module, a VHD file that defines the V and H coordinates, the voice traffic in millierlangs, and data traffic in bps for each location is required. In case a location has multiple sources of traffic, it should be represented as two locations, one handling only voice and the other handling only data. One will need a LINK file with zero and nonzero values to fully represent the voice and data locations, respectively. Network topologies and their monthly costs were computed for the VHD100 file, with backbone switches varying from 1 to 8, and for several combinations of design parameters. See Figure 9.4 for an illustration of a 100-node network topology obtained for four backbone switches, seven concentrators, TGF = 20, and three types of transmission facility designs. See Figure 9.5 for the plots of monthly costs for three types of IVDNs as a function of the number of backbone switches. The first type employed only the voice-grade type of circuits for ALs and trunks. The second type of IVDN employed T1 circuits for only trunks. The third type of the IVDN employed T1's for trunks and ALs. It should be stressed that all of the IVDNs were designed to handle a traffic growth factor (TGF) of 20 when considering the VHD100 file. The results show that a monthly savings of up to $400,000 and $1,000,000 can be achieved by using the second and third types of networks, respectively. One can now appreciate the reason behind the popularity of T1-based private networks.

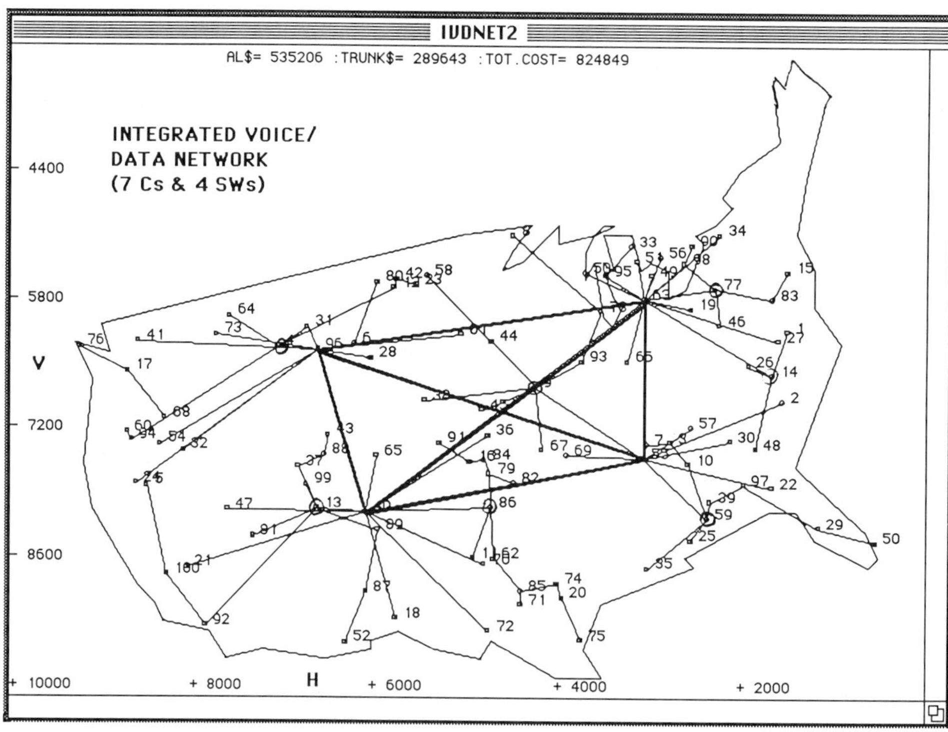

Figure 9.4. A 100-node IVDN topology with four switches and seven data concentrators.

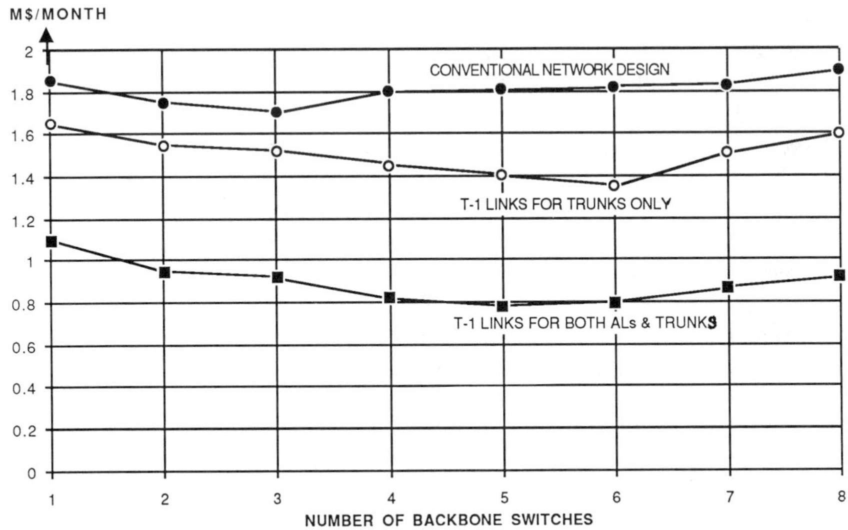

Figure 9.5. Monthly costs of transmission versus the number of optimally located switches for some useful design parameters for the 100-node IVDN.

The above example illustrates the design of an IVD network topology based on a fully connected mesh backbone network. One reason for employing a fully connected backbone network was that From-To traffic data was unavailable. The best way to optimize the topology of a backbone network is to simply employ a From-To data file. This will result in a correctly dimensioned backbone network.

Since a fully connected backbone network, as described previously, does not allow tandem connections (i.e., allowing two or more tandem switching nodes in the path of a call), it is possible to observe lightly loaded trunk bundles. For such cases, their light traffic may be routed over two T1 links in tandem, and thus reduce the IVDN costs even further. One way to achieve that goal is to make two passes at the IVDNET module. The first pass is as described above, with or without the From-To traffic data file. The second pass will involve the following steps:

1. Creating a new VHD file for only the backbone nodes;
2. Using only the internodal traffic values of the Original Trunk Traffic Matrix (see Tables 8.12 or 9.3 as examples of 2- and 5-node backbone networks, respectively); and
3. Constructing a new From-To traffic file that disallows any direct connections between certain locations.

The second pass will, therefore, create a partially connected topology for only the backbone network. These costs can then be compared with those obtained earlier for a fully connected backbone network.

Those organizations that are interested in interconnecting only their major locations via T1 facilities may employ only the second pass. In order to obtain meaningful results, one should convert the voice traffic intensities into applicable bit rates (e.g., 64 Kbps for PCM, 32 Kbps for ADPCM, and 7 Kbps for low-bit-rate voice encoders) and combined with the actual data rates from the data concentrators, packet-switching nodes, and video encoders, before running the IVDNET module. The resulting network topology graph will only show the tandem nodes and the desired tandem connections.

The second pass for hypothetical five-node backbone network can be illustrated by running any of the ECONETS software module. The designer must use all of the five nodes as locations for backbone switches. Table 9.1 defines the corresponding data base of V and H information, traffic intensities, LINK numbers, and location names for the five backbone switch nodes. It should be emphasized that the peak period traffic intensities are those derived from the From-To files described later, while assuming a fully connected backbone network. The traffic intensity values, as defined by the original VHD file are ignored. Also shown in Table 9.1 are the NLT data (i.e., LINK 0 corresponds to Tariff number 4), TGF = 1, and the parameters of the four Tariffs. The From-To file of Table 9.2 defines the traffic flows between the various nodes for the case of a

Table 9.1 EOSN/Network Design Data Base for VHD5

TOTAL NUMBER OF NODES = 5

#	-V-	-H-	LOAD	LATA	LINK	NAME
1	8492	8719	13880000	0	0	SFOCA
2	9213	7878	19800000	0	0	LAXCA
3	7029	4202	42780000	0	0	KANMO
4	8432	4033	46600000	0	0	DFWTX
5	7574	7066	22700000	0	0	SLKUT
TOTAL NODAL TRAFFIC LOAD=			145760000 BPS			

************************NLT DATA PRINTOUT *************************

LINK/NODE TYPE= 0 MAX LINK RATE=1540000 POT. NODAL RATE=56000 TARIFF#= 4

ANNUAL GROWTH FACTOR = 1

NODAL LOAD FACTOR(0 FOR FROM-TO FILE: 1 FOR ACTUAL :2 FOR POTENTIAL) = 0

***********TARIFF DATA PRINTOUT************

AVG. FIXED LOCAL LOOP CHARGES ($)= 204 50 FOR VOICE GRADE CIRCUIT (TARIFF#1)

MILEAGE BAND LIMITS	50	100	500	1000	10000
FIXED MO COSTS	58.74	122.24	164.24	324.24	324.24
COST PER MILE PER MO	2.33	1.06	0.64	0.32	0.32

AVG. FIXED LOCAL LOOP CHARGES ($)= 264 FOR 9600 BPS DDS (TARIFF#3)

MILEAGE BAND LIMITS	50	100	500	1000	10000
FIXED MO COSTS ($)	79.34	142.84	184.84	344.84	344.84
COST PER MILE PER MO	2.33	1.06	.64	.32	.32

AVG. FIXED LOCAL LOOP CHARGES ($)= 62.76 FOR 56KBPS DDS (TARIFF#3)

MILEAGE BAND LIMITS	50	100	500	1000	10000
FIXED MO COSTS($)	302	566.5	742.5	1407.5	1407.5
COSTS PER MILE PER MO	10.07	4.78	3.02	1.69	1.69

AVG. FIXED LOCAL LOOP CHARGES ($)= 2600 FOR TERRESTRIAL T1 (TARIFF#4)

MILEAGE BAND LIMITS	50	100	500	1000	10000
FIXED MO COSTS ($)	0	0	0	0	0
COSTS PER MILE PER MO	15.5	15.5	15.5	15.5	15.5

Table 9.2 Traffic Flows for a Fully Connected 5-Node Backbone Network

FROM	TO	TRAFFIC(BPS)
1	2	2500000
1	3	180000
1	4	2000000
1	5	2200000
2	1	2300000
2	3	2400000
2	4	2600000
2	5	3000000
3	1	200000
3	2	2500000
3	4	3000000
3	5	3500000
4	1	2500000
4	2	2000000
4	3	28000000
4	5	3000000
5	1	2000000
5	2	2500000
5	3	3000000
5	4	3500000

fully connected backbone network. Such a table is invoked by making the SDF parameter number 18 equal 1.

Table 9.3 shows the trunking analysis for such a network. At the beginning, it shows some additional SDF design parameters implying a ''0'' type trunk link (requiring Tariff number 4) and a full duplex trunk link. The FDX assumption implies that the direction with the higher data rate defines the number of T1 trunks to be leased. An HDX assumption would have required the addition of bit rates in both directions before computing the number of T1 trunks. This choice may be useful for tariffs that allow the leasing of both HDX and FDX T1 circuits. The next SDF parameter, number 34, implies that the DBA node (or a T1 multiplexer) allows a maximum utilization of a T1 trunk equalling 0.9. In other words, each T1 trunk is constrained to carry a maximum of 1,540,000 $*$ 0.9 bps only.

Table 9.3 also shows the Original Trunk Traffic, Modified Trunk Traffic, and the Modified T1-trunk matrices. Each of these matrices are 5 $\times$ 5 in size. The Original Trunk Traffic matrix represents the peak period traffic in bps flowing in both directions between any pair of switches. The Modified Trunk Traffic matrix represents the effective peak period traffic in bps flowing between the two switches, irrespective of direction. These values are obtained after considering the FDX or HDX nature of the trunks. The Modified Trunk matrice defines the number of T1 trunks connecting the ith node to the jth node, where both i and j vary from 1 to 5 for a matrix. It also defines each T1 trunk by the nodes it serves, its mileage, the cost of each T-1 trunk/bundle, and the total cost of the entire backbone network. It should be emphasized that the total cost of each trunk bundle is obtained by multiplying the unit cost of each trunk by the number of trunks from the Modified Trunk matrix. One should note that the AL costs are reduced to zero, since no ALs are used. Figure 9.6 illustrates the topology of the fully connected five-node backbone network, along with some pertinent cost summaries.

Table 9.4 defines the traffic flows between the various nodes for the case of a partially

Table 9.3 Trunking Analysis for a Fully Connected 5-Node Backbone Network

```
TRUNK LINK/NODE TYPE    =  0
FDX FACTOR (1 FOR FDX) =  1
MAX TRUNK UTILIZATION   = .9
TOTAL INTER/INTRA NODAL TRAFFIC= 72880000
```

ORIGINAL TRUNK TRAFFIC MATRIX AS FOLLOW

0	2500000	180000	2000000	2200000
2300000	0	2400000	2600000	3000000
200000	2500000	0	3000000	3500000
2500000	2000000	28000000	0	3000000
2000000	2500000	3000000	3500000	0

MODIFIED TRUNK TRAFFIC MATRIX IS AS FOLLOWS

0	2500000	200000	2500000	2200000
0	0	2500000	2600000	3000000
0	0	0	28000000	3500000
0	0	0	0	3500000
0	0	0	0	0

MODIFIED T1 TRUNK MATRIX IS AS FOLLOWS

0	2	1	2	2
0	0	2	2	3
0	0	0	21	3
0	0	0	0	3
0	0	0	0	0

TRUNK COST MATRIX IS AS FOLLOWS

FROM	TO	MILES	COST	TOTALS
1	2	350	8025.1	16050.2
1	3	1501	25865.6	25865.6
1	4	1482	25571.1	51142.2
1	5	598	11869.1	23738.2
2	3	1352	23556.1	47112.2
2	4	1241	21835.6	43671.2
2	5	578	11559.1	34677.3
3	4	447	9528.6	200100.6
3	5	922	16891.1	50673.3
4	5	997	18053.6	54160.8

```
*****SUMMARY OF SYSTEM COSTS*****
      ACCESS LINK COST= 0
      TOTAL TRUNK COST= 547191
      TOTAL SYSTEM COST= 547191
      CENTS PER 1MBITS= 1.04
```

connected backbone network. The original traffic flow from Node 1 to Node 3, as defined in Table 9.2, is now routed over the tandem route of 1-5-3, and the direct traffic flow between Node 1 and Node 3 is eliminated. Similarly, the original traffic flow from Node 3 to Node 1, as defined in Table 9.2, is now routed over the tandem route of 3-4-1, and the direct traffic flow between Node 3 and Node 1 is eliminated. Table 9.5 shows the corresponding trunking analysis, which implies a monthly saving of $25,866. Figure 9.7 shows the corresponding topology of the partially connected five-node network. The trunk bundle between Node 1 and Node 3 is now missing. One can repeat the above

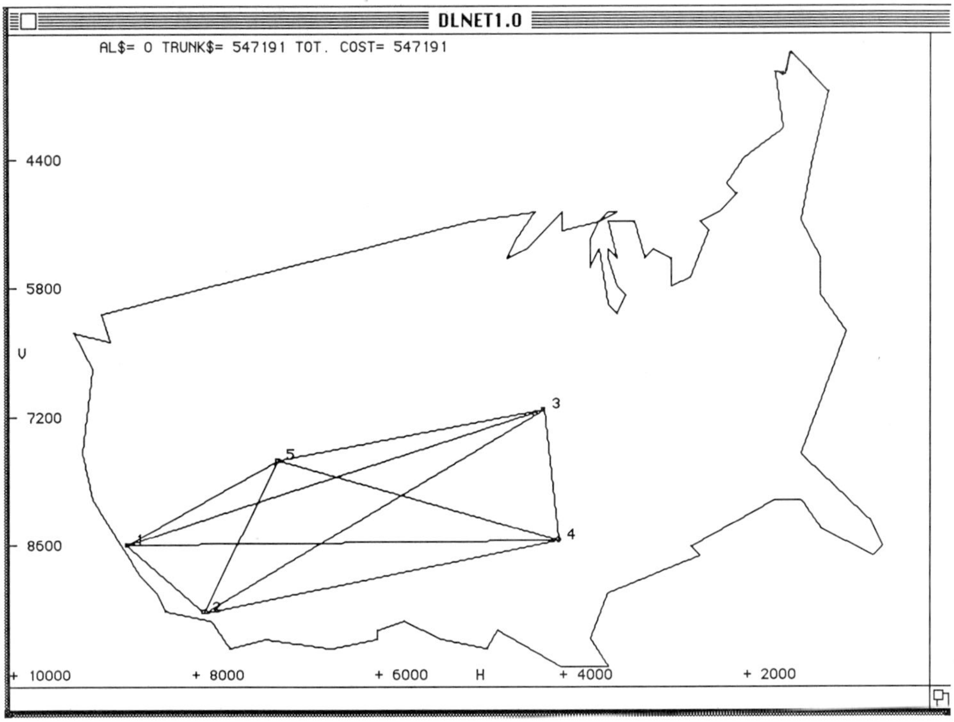

Figure 9.6. A fully connected five-node T1 backbone network.

Table 9.4 Traffic Flows for a Partially Connected 5-Node Backbone Network

FROM	TO	TRAFFIC(BPS)
1	2	2500000
1	3	0
1	4	2000000
1	5	2380000
2	1	2500000
2	3	2400000
2	4	2600000
2	5	3000000
3	1	0
3	2	2700000
3	4	3000000
3	5	3500000
4	1	2500000
4	2	2000000
4	3	28000000
4	5	3000000
5	1	2000000
5	2	2500000
5	3	3180000
5	4	3500000

Table 9.5 Trunking Analysis for a Partially Connected 5-Node Backbone Network

```
                              TRUNK LINK/NODE TYPE   =  0
                              FDX FACTOR(1 FOR FDX)  =  1
                              MAX TRUNK UTILIZATION = .9
                              TOTAL INTER/INTRA NODAL TRAFFIC= 73260000

ORIGINAL TRUNK TRAFFIC MATRIX AS FOLLOW

0          2500000    0          2000000    2380000
2500000    0          2400000    2600000    3000000
0          2700000    0          3000000    3500000
2500000    2000000    28000000   0          3000000
2000000    2500000    3180000    3500000    0

MODIFIED TRUNK TRAFFIC MATRIX IS AS FOLLOWS
0          2500000    0          2500000    2380000
0          0          2700000    2600000    3000000
0          0          0          28000000   3500000
0          0          0          0          3500000
0          0          0          0          0

MODIFIED TRUNK MATRIX IS AS FOLLOWS

0          2          0          2          2
0          0          2          2          3
0          0          0          21         3
0          0          0          0          3
0          0          0          0          0

TRUNK COST MATRIX IS AS FOLLOWS

FROM       TO         MILES      COST       TOTALS
1          2          350        8025.1     16050.2
1          3          1501       25865.6    0
1          4          1482       25571.1    51142.2
1          5          598        11869.1    23738.2
2          3          1352       23556.1    47112.2
2          4          1241       21835.6    43671.2
2          5          578        11559.1    34677.3
3          4          447        9528.6     200100.6
3          5          922        16891.1    50673.3
4          5          997        18053.6    54160.8

                    *****SUMMARY OF SYSTEM COSTS*****
                       ACCESS LINK COST= 0
                       TOTAL TRUNK COST= 521325
                       TOTAL SYSTEM COST= 521325
                       CENTS PER 1MBITS= .98
```

procedure by creating another From-To traffic table, and eliminating trunk bundles one by one until no more savings are achieved.

The above approach should be especially attractive when a T3 facility could be leased from a bulk common carrier between Nodes 3 and 4 to handle 21 T1 links at a reduced rate. Additional trunk bundles can be eliminated by routing their respective traffic over longer tandem paths.

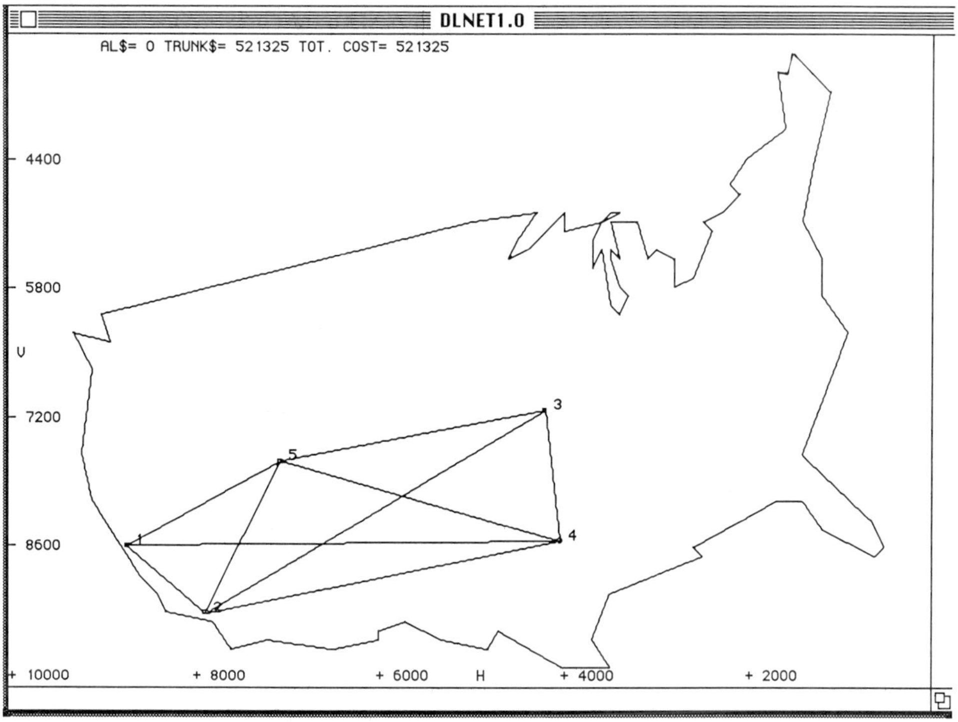

Figure 9.7. An optimized partially connected five-node T1 backbone network.

9.5 CONCLUDING REMARKS

A methodology has been presented for designing integrated voice and data networks. Several criteria for selecting (1) switching nodes for voice, data, and IVD applications, and (2) Premise Wiring Systems (PWSs) were also introduced. The use of From-To files for designing backbone networks, and the principle of topological superimposition, were also illustrated with actual cases.

10

The Overall Network Planning Process

10.1 INTRODUCTORY REMARKS

Most corporations have already invested large sums of money in their voice, data, and information processing networks. Along with their employees, these networks have become their strategic resources for maintaining a competitive position. The customer premise equipment (CPE) consists of voice terminals, dumb data terminals, PABXs, hosts, intelligent work stations (IWSs), minicomputers, personal computers (PCs), and local area networks (LANs). Whereas voice requirements are growing at about 3–4 percent a year, the PCs, desktop workstations (WSs), and LANs are increasing at the rate of 30 percent or better annually. Furthermore, most corporations would like to have the capability for introducing new products, services, and applications quickly and economically to stay abreast of the competition. This requirement is forcing them to consider corporation-wide system integration to achieve large economies-of-scale.

Since most of the existing CPEs are likely to be interconnected via proprietary networks, any migration to a new integrated corporate network is further complicated by the emergence of new standards for voice, data, and distributed DP. Standards such as ISDN, IEEE 802, ANSI's FDDI, and OSI-based network standards for distributed data processing are confusing the situation. Several new technologies are forcing these already known standards to change and will probably help create several new standards. These developments provide several alternatives to corporations for solving their complex problems. But this also causes many uncertainties in the minds of the corporate officers.

A major portion of corporate network cost is still in the form of monthly lease expenses for communications facilities connecting their many locations. Such facilities consist of voice grade, DDS, and digital T1 lines. Many corporations also own microwave and optical fiber routes. Before 1984, there was only one carrier, AT&T. Now there are many common and bulk carriers with excess fiber capacity. This has increased the number of alternatives for designing private network(s). With the availability of virtual private networking (VPN) or software-defined networking (SDN), along with new AT&T offerings such as Tariffs 12 and 15, the concept of a stable tariff is now almost dead. This is forcing corporations to follow the regulatory issues even more closely than before.

How can planners pierce this ever-thickening wall of complexity and bring some sense of order to the planning process? The following sections describe several concepts underlying the ongoing planning process and the evolving technologies and standards that can be employed to face the new realities, opportunities, and challenges. The techniques for optimizing network topologies, as described in the previous chapters, form only a small but important part of this planning process. The discussion that follows should be combined with the contents of Appendix B, dealing with evolving international and de facto standards, to understand the overall planning process.

10.2 THE PLANNING PROCESS

The concepts and the solutions offered in this section can be understood only within the framework of the planning process followed by each corporation. The planning process for corporate networking should be divided into three ongoing cycles:

1. A strategic planning cycle;
2. A tactical planning cycle; and
3. An operational planning cycle.

The strategic planning cycle deals with studying (1) the ever-changing, long-term objectives and goals of a corporation, (2) understanding the competition, and (3) ways of harnessing future developments in corporate networking to stay abreast of the competition. The strategic planning cycle generally requires the longest period (3–5 years).

The tactical planning cycle requires 1 to 3 years, and it deals with the actual design and evaluation of alternative solutions. Consequently, it requires the use of intensive model making and network design techniques.

The operational planning cycle is an integral part of the ongoing process, with a yearly check on the network management and control (NMC) techniques employed.

Based on the proceeding discussions, it can therefore be said that a typical planning process is an ongoing process that requires the services of a large number of dedicated and experienced people. The entire planning cycle is doomed to failure if a full commitment from the top management is not available.

To accomplish the above tasks, a new type of corporate thinking is needed. The traditional telecommunications and DP managers of most corporations are generally adept in the operational aspects of the planning process. New organizational structures must be in place for the guidance and rewarding the works of new thinkers and workers adept in the strategic, tactical, *and* operational aspects of the planning process. In some cases, the corporation should also reward its planners, not only for guiding its own destiny, but also influencing international standards.

10.2.1 THE STRATEGIC PLANNING PROCESS

The strategic planning group must constantly study the new technologies and evolving standards for voice, data, and DDP, for their usefulness to the corporation. In particular, any relationships between the two must be clearly understood by the strategic planners before recommending a migration path leading towards the future private network(s). One can illustrate the strategic planning process by studying recent developments for any lessons learnt.

Most of the new standards for voice and data communications derive their origins from the technologies related to VLSI, digital transmission, and packet switching. VLSI technology has also made DDP affordable to corporations. The phenomenal increase in the sale of PCs, desktop workstations (WSs), and LANs can be attributed to this fact. Most of the standards are now based on a multilayer concept that facilitates deeper understanding and structured testing of multivendor products for interoperability.

The standards for data communications (as initiated by CCITT and later incorporated by ISO) arose from the underlying need for rapid data interchange in the highly developed countries and the availability of affordable packet-switching products. This resulted in the lower four-layers of the ISO Open System Interconnect (OSI) reference model, as shown in Figure B.2.1 of Appendix B. The top three layers of the OSI reference model deal with DDP, and generally reside in host processors located in separate locations. Figure B.4.9 illustrates the reference model for Local Area Networks (LANs) as originally defined by IEEE, and now accepted by ISO. The new ANSI-initiated standard, called the Fiber Distributed Data Interface (FDDI), is based on optical fiber technology. Such a standard for Metropolitan Area Networks (MANs) will employ a data rate of 100 Mbps. Such an architecture will allow the interconnection of LANs within a metropolitan area. It is viewed as a vehicle to provide high performance networking capabilities to large corporations hungry for rapid solutions. At the present time, IEEE, ISO, and CCITT have not accepted FDDI.

IEEE is likely to accept the so-called DQDB (Distributed Queued Dual Bus) or QSPX (Queued Synchronous Packet Exchange) architecture into the IEEE 802.6 fold. This architecture can switch both voice and data packets over fiber optic media with no apparent data rate limitations. It also allows the distribution of the stations over a larger distance than those allowed by FDDI. Future developments in fast packet switching (with throughputs approaching 1 Gbps) may have a greater influence on the LAN, MAN, and WAN architectures than the fiber technology.

The standards related to the Integrated Services Digital Network (ISDN) resulted from the need for upgrading the public telephone networks (PTNs) of all nations and the availability of affordable high-capacity digital transmission facilities. Third world countries with no pressing need for increased data communications may take a long time to upgrade their PTNs. The developed countries with a need for very high-speed data interchange may find the current ISDN-related standards inadequate. Consequently, the current ISDN-related standards may see significant changes resulting from additional technologies, such as fast packet-switching and optical fibers. The best advantage going for ISDN is its foundation, built on Signaling System No. 7 (SS7) network, which will provide the small-to-medium sized corporations with an ability to provide a wider range of services quickly. Large corporations will tend to incorporate SS7 within their own private network, while employing the public ISDN facilities for off-net users. The relationships between the layers of OSI and the ISDN reference models are not clearly defined yet. The strategic planners must continue to watch for developments in this area during the coming years. Figure B.5.4 attempts to show the various layers for the voice and data applications of ISDN as things stand today.

The strategic planners have long been confused when it came to choosing either a PABX for their integrated voice and data applications or a separate LAN to handle their data needs. The sudden emergence of T1 multiplexers finally resolved the matter to some degree. The voice PABXs, data LANs, and sometime even end-user terminals are now being integrated to share the T1 backbone network. Figure 9.1 illustrates the use of the emerging dynamic bandwidth allocators (DBAs) for implementing an affordable ISDN-like private network right now. Eventually, when ISDN will be universally available, such private networks will also evolve to comply with the ISDN standards related to the Basic Rate Interface (BRI), the Primary Rate Interface (PRI) on the user side, and the trunk interfaces for accessing the public networks. This example clearly shows the manner in which the standards influence the architectures of private networks.

The DDP revolution has still a long way to go. IBM's SNA architecture was originally developed in 1974 to provide a means to allow the dumb data terminals, dispersed throughout the many locations of the corporation, to communicate with the centralized mainframe. Despite the fact that the OSI reference model was unavailable at the time, SNA also ended up employing seven layers. Figure B.6.2 illustrates the similarities and differences between the layers of the SNA and OSI reference models. In 1981, IBM unleashed another de facto standard, called the MS-DOS. The sales of their PC had just taken off and the DDP revolution got a big boost. Dumb terminals finally got a back seat. But PCs cannot live by word processing and spreadsheets alone. They must also communicate with one another and with mainframes. LANs came to the rescue. Finally IBM had to shift gears. Figure B.6.4 shows how SNA has been evolving to face the new realities of DDP.

As time progresses, IBM will also be forced to offer OSI interfaces. They already offer many such interfaces to European customers and a few for U.S. customers (e.g., NPSI for accessing the X.25 network and X.400/EM). This does not imply that SNA will disappear. Instead, the OSI-based networks will provide vehicles for internetworking, just as PTNs did for voice applications. Which network standards will be employed for strategic purposes will depend mainly on the existing investments in hardware and software products and the migration path a corporation intends to follow to meet its long-term goals. The tactical planning process can provide additional help in choosing the right technologies and standards, as discussed next.

There will be cases when a large corporation must also influence the development of a new standard to handle its operations (e.g., General Motors asking for the MAP-based network architecture of Figure B.6.7).

10.2.2 THE TACTICAL PLANNING PROCESS

Once the long-term migration path has been defined by the strategic planning process, the corporation may decide to actually define a set of viable solutions to achieve its long-term goals and objectives. It is the duty of the tactical planning group to provide concrete networking solutions. This is the group that actually quantifies the relationships between the products or services provided by the corporation and the performance of its communications and information processing networks. Based on these analyses, it assesses the applicability of several viable technologies and various levels of system integration for increasing the productivity of the worker and/or providing better products/services at a lower cost than before. The tactical planning process consists of the following well-defined tasks or steps:

1. Modeling the current environments related to CPEs, applications/services, associated levels of performance provided to the users, and their relationships to productivity levels;
2. Modeling the new environments in terms of additional CPEs, applications, and desired levels of performance;
3. Developing a life cycle traffic growth model for each of the applications/ services;
4. Assessing the new technologies and associated standards available for applicability;
5. Developing viable designs for one or several integrated networks with life cycle cost and performance models. Chapters 2–9 of this book describe this process;
6. Developing viable cutover models, ensuring uninterrupted service to existing users and modifying the life cycle cost models accordingly. This task must include facility planning; and
7. Selecting the most desirable or least risky solution, and selling it to management.

The first step of the tactical planning process requires the longest period to complete, especially if it has never been attempted before. The second and third steps require the full cooperation from the top corporate managers. The fourth, fifth, and sixth steps are greatly facilitated by the availability of time-proven modeling, analysis, and design tools. The success of the seventh step is obviously dependent upon (1) a clear understanding of the problem, (2) the depth of the task accomplished, and (3) faith in one's approach or solutions. The use of meaningful return-on-investment (ROI) models also helps. Quick, dogmatic, and revolutionary solutions will never accomplish the goals of the corporation.

The effectiveness of this group is directly dependent upon the availability of time-proven tools that are based on algorithms, and not on artificial intelligence. Many of the traditional tools required large mainframes for providing "very accurate" answers. Such an accuracy was found to be needless since the tariff and traffic environments usually change while going from the network synthesis stage to the implementation stage. Furthermore, the network topologies are generally independent of the actual tariff employed. This was clearly shown in Chapter 8. The choice of mainframes was natural, since only large computers could handle the huge amount of data associated with tariffs and traffic. However, the use of bypass facilities, the availability of discounts to large users, and the leasing of private lines from bulk carriers has destroyed the constancy of tariffs. Simplified models of tariffs can now be constructed for computing the approximate costs of any networking alternative during the early stages of network planning. Such models allow the use of powerful desktop computers employing Intel's 80386 or Motorola's 68020 processing capabilities. Graphs detailing network topologies and text detailing link/trunk costs can be stored as files for later retrieval, editing, and printout. Due to the interactive nature of the activity, one can now evaluate a large number of alternatives within days instead of months. Several technologies can be quickly evaluated for application. Several network topologies can be superimposed to synthesize integrated voice, data, and video networks. Chapters 5 through 9 illustrate this process in some detail.

These days, there is a movement that is tilting towards the use of expert systems for tactical network planning. Such an approach seems completely inappropriate, in view of the availability of meaningful network design algorithms. Expert systems are only useful for highly complex situations such as patient diagnosis or network fault prediction/location. Using expert systems for network planning is like using a lawn mower for shaving one's beard. The use of expert systems will also tend to encourage the use of mainframes again, and thus slow down the tactical planning process.

Some planners employ computer simulations for designing networks. There are several problems with this approach. Simulation results are only valid when the network system is in a state of statistical equilibrium. For a typical large network system, it takes a long time to reach that state. The simulation of large network systems also requires the use of a mainframe. Finally, most simulation studies cannot handle a real system operation without spending a huge amount of money and effort on the development of the simulation program and running that program on a mainframe. For example, the operation of an automatic call distribution system requires the scheduling of agents every half hour. This creates large system transients, whose effects last for hours, thus preventing the achievement of any statistical equilibrium. It is easier to learn about ACD subtleties from the on-going operations of a real ACD system. That is the responsibility of the operational planning process, discussed in the next subsection.

10.2.3 THE OPERATIONAL PLANNING PROCESS

Ongoing network planning is one of the many services provided by a well-designed network management and control (NMC) system. Major vendors (AT&T in particular) and ISO are converging towards defining nine critical NMC functions:

1. Configuration management, dealing with network inventory, topology, and directory on a real-time basis, preferably using a relational data base;
2. Fault management, dealing with network status supervision, end-to-end testing, diagnostics, fault-reporting/trouble ticket creation/repairing, back-up, and reconfiguring;
3. Performance management, dealing with performance definitions, monitoring, trending, and thresholding, on a real-time basis;
4. Accounts management, dealing with planning of network-related budgets, allocating costs to end users, and verifications of network system goals;
5. Security management, dealing with establishing and maintaining criteria for network access management, and reconfiguration control for any number of well-defined partitions;
6. Network planning, dealing with periodic network optimizations, the modeling of network contingencies for disaster recovery, and ongoing development of strategies to help the user to achieve short-term and long-term goals;
7. Operations management, dealing with efficient running of all operations centers, including the tasks of staffing, training, and controlling information flows;
8. Programmability, dealing with the management and enhancement of existing software packages and the development of new custom software packages; and
9. Integrated system control, dealing with a capability to manage and control a multivendor network transparently, from any user-friendly console.

The objective of the operational planning group is to maintain a cost-effective private network despite the ongoing perturbations occurring within the corporate network(s), dealing with traffic, outages, and catastrophes. Since mid-1980s, the field of NMC has been gaining the undivided attention of not only sophisticated large users and vendors, but also the ISO.

Large users have already discovered that a sophisticated NMC system is essential for operating a cost-effective, multivendor, multiapplication private network. Unfortunately, each vendor started out with a proprietary NMC system good only for its own subnetwork. Since its NMC system could not communicate with the NMC systems of other vendors, the multivendor network could not be operated effectively. A great deal of confusion resulted. Several vendors have now started to introduce products based on the concepts of network elements (NEs), elemental management systems (EMSs), for handling multivendor environments, and some services based on the evolving OSI message protocol. AT&T's Unified Network Management Architecture, DEC's Enterprise Network Management, and IBM's Netview are some of the examples.

Since the completion of the basic OSI framework, the ISO has started to define the NMC functions performed at each OSI layer. ISO Draft International Standards (DIS) 7498-4, as drafted in October 1988, defines the OSI Management Framework. It contains the Common Management Information Protocol (CMIP) and Common Management Information Services (CMIS). CMIP is the basic protocol that defines the format of messages that will be passed within and between multivendor systems for implementing services. CMIS defines services dealing with how the standard messages are to be passed within and between multivendor systems or network elements. Since the ISO has not yet dealt with the extent and type of services (i.e., with the content of NMC messages), a lot of work has yet to be accomplished. In order to influence this work, several vendors have joined together to form the OSI / Network Management (OSI/NM) Forum. Its objective is also to achieve multivendor network interoperability in the shortest possible time. Consequently, their NMC protocol will start with the X.25 / IEEE 802.3 transport layers and the upper layers' functionality, as expressed by ISO's CMIP / CMIS DIS.

Another group of vendors that market TCP / IP (Transmission Control Protocol/Internet Protocol) products is also interested in influencing OSI NMC-related standards. Vendors of ISDN and FDDI products have also started to worry about the OSI/NMC standards. During the next decade, users will start seeing NMC products conforming to the OSI standards. That will enable large sophisticated users to fully manage and control their networks. The real benefit of such a control will be a full understanding of the network behavior under varying operational environments. This, in turn, will encourage the development of expert systems (ESs) that will enable the NMC team to not only achieve self-healing at all levels of the network, but also to help provide the desired level of performance at the lowest possible cost. Since such an approach is fully automated, the cost of providing a full spectrum of NMC services will be reduced drastically when compared to the existing costs.

10.3 FINAL CONCLUSIONS

This book devoted its first six chapters to describing concepts, tools, and algorithms for the analysis and design of topologies that characterize most modern private networks.

Chapters 7, 8, and 9 concentrated on providing actual examples of voice, data, and integrated voice/data network designs. Chapter 10 dealt with the overall planning process for creating a meaningful network architecture for the enterprise. In this final chapter, it was recommended that each corporation must hire a new set of employees to concentrate on the strategic, tactical, and operational network planning processes, as ongoing activities. Corporations must learn to reward these planners for innovative thinking and bold initiatives. In some cases, it may be necessary to influence a new network standard or architecture. Such was the case when General Motors chose to back the seven layer Manufacturing Automation Protocol (MAP). They realized the strategic importance of their new private network to their future profitability. One can, however, say with emphasis that no single architecture dogma, or utopia will solve all problems. Each corporation will find that it requires a unique solution for its unparalleled set of problems. Each corporation will also find a growing dependence on internetworking to achieve a cost-effective, multivendor solution. This will require the synthesis of networks with increasingly mixed topologies. The network planners should find the tools and algorithms in this book useful for evaluating any hypothesis regarding a new technology or networking solution. But they must never forget that the analysis and design tools provided in this book are only a small part of the overall network planning process.

Appendix A

Derivation of M/D/N Equations

Let

$$C = A/N = \text{Average server utilization} \tag{A.1}$$

and

$$\tau = (t/t_s - \nu) \qquad 0 \le \tau < 1 \tag{A.2}$$

where

A = traffic intensity in erlangs
N = the number of servers
t = any real number for delay time
t_s = fixed service time.
ν = integer (t/t_s)

One can now define the Prob(Delay > t) as follows:

$$\text{Prob(delay} > t) = \sum_{i=0}^{N-1} \sum_{w=1}^{\infty} V_i \text{ Exp } (-A(w - \tau)[R] \tag{A.3}$$

where

$$V_i = \sum_{j=0}^{i} \mu_j \tag{A.4}$$

$$R = \left\{[A(w - \tau)]^{[N(\nu+w)+N-i-1]}\right\} \Big/ [N(\nu + w) + N - i - 1]! \tag{A.5}$$

$$\mu_j = \text{Exp } (-A) \left\{ \sum_{i=0}^{N} \mu_i (A^j/j!) + \sum_{i=N+1}^{N+j} \mu_i [A^{j-i+N}/(j - i + N)!] \right\} \tag{A.6}$$

$= \text{final value of } \mu_j$

The final values of μ_j are computed by substituting the values of μ_j previously obtained, again and again, until the values of μ_j become stable from one iteration to another. This technique is only successful if "j" is selected large enough to satisfy the relationship $\mu_{j+1} = C * \mu_j$. Furthermore, one must select the first set of μ_j values as follows:

$$\mu_j = \left[(N * C)^j / j!\right] \Bigg/ \left\{ \left[(N * C)^N / (1 - C)N!\right] + \left[\sum_{i=0}^{N-1} (N * C)^i / i!\right] \right\} \ldots \text{for } j \leq N \tag{A.7}$$

$$\mu_j = \left[(N^N * C^j) / N!\right] \Bigg/ \left\{ \left[(N * C)^N / (1 - C)N!\right] + \left[\sum_{i=0}^{N-1} (N * C)^i / i!\right] \right\} \ldots \text{for } j > N \tag{A.8}$$

The published literature dealing with traffic theory does not yet provide simple expressions for the first two moments of the time spent in the M/D/N type queue in the form that these are available for the M/M/N type queue, as shown in Equations 4.9c and d. The author has simulated the M/D/N type queue for a large number of values pertaining to server load and N. He has discovered that the following normalized (for $T_s = 1$) expressions agree very well with the simulation results.

$$\text{AVG}(T_q) = 1 + \left[P_0 / 2(N - A)\right] \tag{A.9}$$

$$\text{VAR}(T_q) = \left[1/(N - A)^2\right] * \left[P_0/3 - P_0^2/12\right] \tag{A.10}$$

It is interesting to note that Equations A.9 and A.10 reduce to corresponding Equations 4.5 and 4.6 for the single server (or $N = 1$). The value of P_0 for $N = 1$ reduces to the average server load, represented by the variable ρ. The value of P_0 in Equations A.9 and A.10 must be solved using Equation A.3 and the method described in this Appendix. One can also use the curves of Figure A.1 to obtain the values of P_0 for some useful values of C and N. It is also interesting to note that the values of P_0 for the M/M/N queue are only about 6 percent higher than those shown in Figure A.1 for identical values of N and C. The reader can now use the analytical tools presented in this book to study the behavior of the worst case (i.e., an M/M/N queue) and the best case (i.e., an M/D/N queue), and thus interpolate the performance of all practical systems that generally fall in between the two extremes.

NOTE: The above method is attributed to Everett (1953). His method, which is now quite suitable for computations on the new desktop computers, was first described in a simple manner in an unpublished *Manual of Techniques for Traffic Analysis*, prepared at the request of Mr. A. A. Collins, President of Collins Radio Company, in 1968. One of author's colleagues, Dr. J. C. Shah, is credited with researching the original papers on M/D/N type queues.

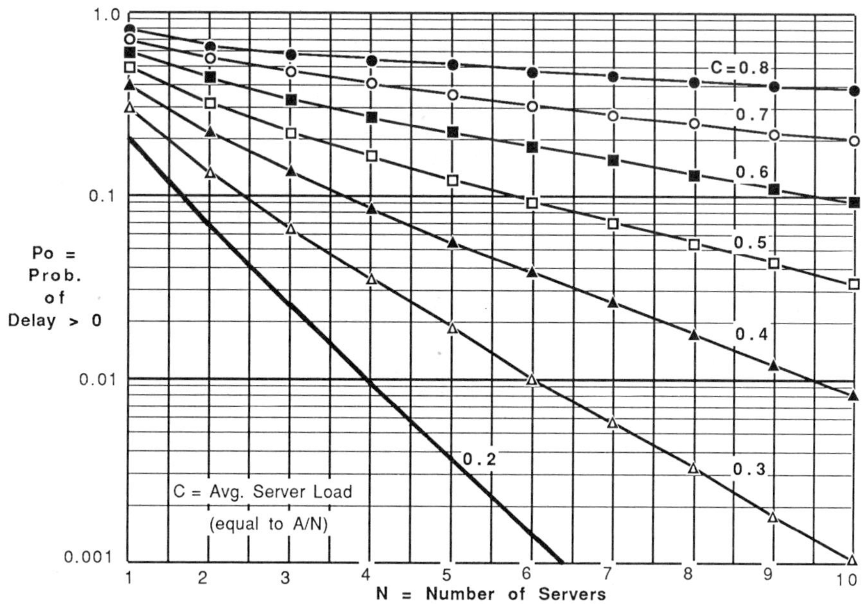

Figure A.1. Probability (delay > 0) or P_0 versus N and some useful values of C.

Appendix B

Evolving International and De Facto Standards for Networking

B.1. THE MAKING OF INTERNATIONAL STANDARDS

B.1.1 INTRODUCTION

One of the many functions of the United Nations is to promote the development of international standards for telecommunications. To achieve this goal, the International Telecommunications Union (ITU) was created. Two committees, known as CCITT and CCIR, handle the development of standards for telephone/telegraph (or wire-related) and radio systems. This chapter will describe (1) the activities of CCITT, and (2) the features of several popular standards.

Each member nation sends their representatives to the CCITT-sponsored meetings held in Geneva and other cities at regular intervals. Once every three years, a meeting is held for the final approval of new standards under consideration. The national CCITT sponsored subcommittees also meet in their own countries for finalization of national recommendations. There are many other standards-making bodies in technologically advanced countries (e.g., the U.S.A.) and regions (e.g., the Common Market countries of Europe) that coordinate with the CCITT activities. Many Bell Operating Companies (BOCs) and vendors also participate in the CCITT activities as observers. Users of the developed countries are not yet allowed to participate in the CCITT activities, but they do participate in the activities of other national standards-making bodies.

B.1.2 THE MAJOR STANDARDS-MAKING BODIES

The Consultative Committee for International Telephone and Telegraph (CCITT) and Consultative Committee for International Radio (CCIR) have already been mentioned. The International Standards Organization (ISO) represents the international computer and terminal manufacturers. The interests of the Electronic Industry Association (EIA) are self-evident. The American National Standards Institute (ANSI) oversees the interests of the U.S. manufacturers. The National Bureau of Standards and the Institute of Electrical and Electronic Engineers (IEEE) also promote the development of standards for American industries. On behalf of the BOCs and OCCs, the newly formed Exchange Carriers Standards Association (ECSA) is also interested in the development of standards that are suitable for their administrations. All of the above-mentioned bodies do not operate in a vacuum. They cooperate with one another all of the time. Some of the

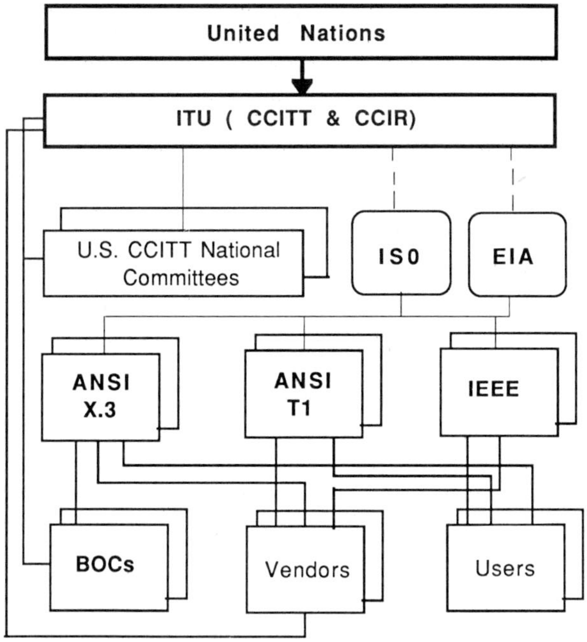

Figure B.1.1. An illustration of the standards-making process.

standards may be named differently, but equivalences exist between different nomenclatures employed by different industries of different countries. The relationships between the above bodies during the standards-making process is illustrated in Figure B.1.1.

CCITT is composed of 18 study groups (SGs). Each study group has a responsibility for a particular area, such as voice, data, or integrated services. These SGs are described as follows:

1. SGI influences standards dealing with telegraph and telematic services.
2. SGII handles telephone system operations.
3. SGIII handles tariff principles.
4. SGIV handles transmission maintenance of international circuits.
5. SGV handles protection against dangers and EM disturbances.
6. SGVI handle protection of and specifications for cable sheathes and poles.
7. SGVII handles public data networks.
8. SGVII handles telegraphy and related DTE (Data Terminal Equipment).
9. SGIX handles telegraph transmission quality.
10. SGX handles telegraph switches.
11. SGXI handles telephone switching and signalling.
12. SGXII deals with telephone transmission performance.
13. SGXIII deals with automatic and semiautomatic public telephone networks.
14. SGXIV deals with facsimile telegraph transmission and equipment.
15. SGXV deals with transmission systems.

16. SGXVI deals with telephone networks.
17. SGXVII deals with data transmission over telephone and telex networks.
18. SGXVIII deals with digital networks, such as ISDN.

The relationships between the CCITT, ISO, and U.S. standards organizations is shown in Figure B.1.2 and Figure B.1.3. Two of the major standards organizations in the U.S., ANSI and ECSA, jointly sponsor the T1 Technical Committees such as T1C1, T1D1, T1M1, T1Q1, T1X1, and T1Y1 to cover the standards related to Network/CPE, ISDN, NMC, Network Performance, Carrier-to-Carrier Interfaces, and other subjects such as SONET (as related to a synchronous optical fiber network transmission standard for data rates in excess of 50 Mbps). See Figure B.1.4 for an organization of the T1 Steering Council.

An interesting aspect of the standards-making process is the need for a complete consensus of all of the members of the ITU before creating a new standard. As a consequence, the standards-making process is quite slow when compared to the way that the technology is changing in the developed nations. The disparity between the developing nations and the relatively few developed nations (e.g., the U.S.A., Sweden, Japan, and Germany) makes the process look very slow. Futhermore, the presence of PT/T (Public Telephone/Telegraph) monopolies most everywhere, versus the free economies in developed nations also creates a disparity between the two interests. This gives rise to the creation of many de facto standards in developed nations. The making of the original T1 carrier structure as implemented by AT&T and the IBM's Systems Network Architecture (SNA) within the U.S.A. could not wait for the CCITT-sponsored standards. Whereas the North American T1 standard has now been accepted as a CCITT standard, IBM's SNA will probably remain out of the CCITT fold forever.

Some of the most well-known international and de facto standards will be described

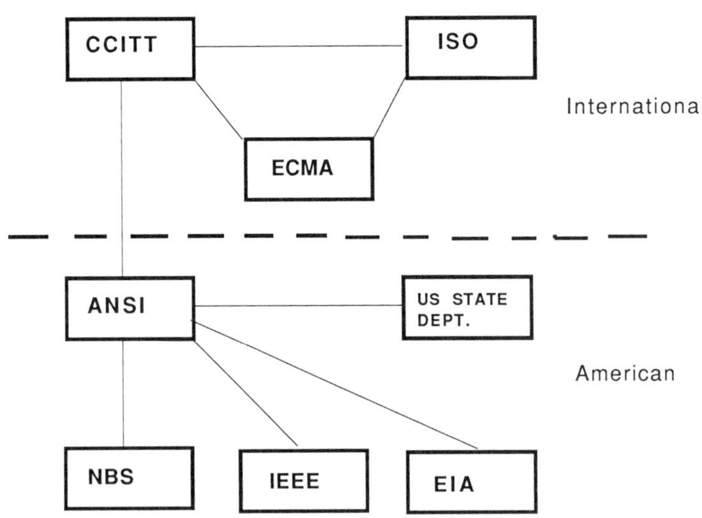

Figure B.1.2. Relationship between the international and U.S. standards-making bodies.

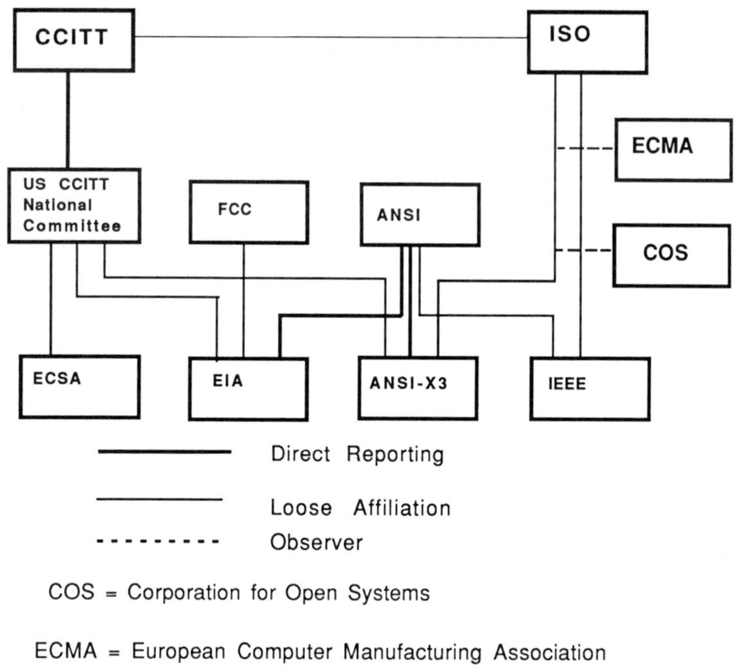

Figure B.1.3. Another view of the interplay between the international and U.S. standards-making bodies.

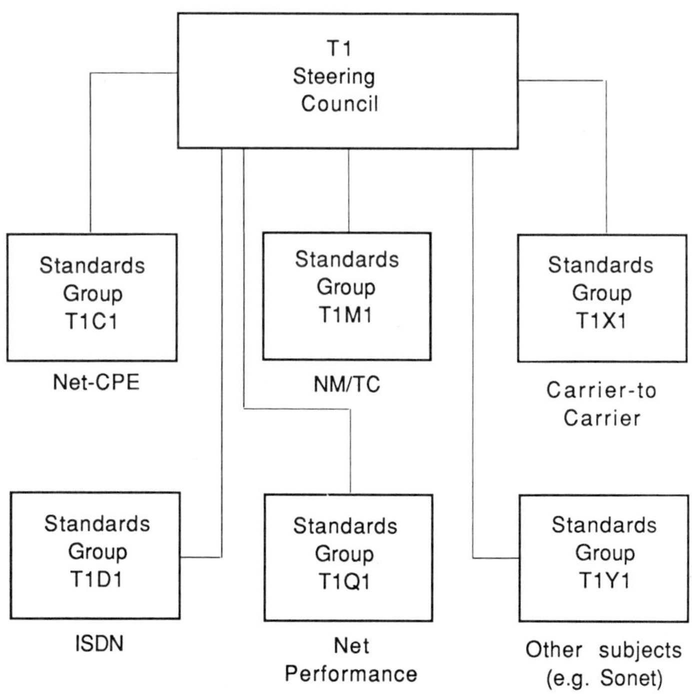

Figure B.1.4. The structure of the ANSI T1 committee.

in the following sections of this chapter. Section B.2 will describe the ISO's Open System Interconnect (OSI) reference model, which will be used to see how a given standard compares to an international, multilayer model. The standards described in Sections B.3, B.4, B.5, and B.6 will deal with voice, data, integrated voice and data, and distributed data processing (DDP), respectively.

B.2 REFERENCE MODEL FOR OPEN SYSTEM INTERCONNECTION

B.2.1 INTRODUCTION

The increasing use of Very Large Scale Integration (VLSI) techniques in the implementation of communications and data processing equipment has decreased the costs of hardware to such an extent that it is now possible to distribute the intelligence in an integrated fashion throughout the network in a very cost-effective manner. Each network node can be characterized by a collection of unique functions that define the NMC capability, or the intelligence of a network system, and therefore the associated network architecture, as defined previously. Using the well-known technique of successive decomposition or iterative modeling, it should be easy to model any collection of communications and data processing functions handled by a network node. The international standards bodies such as ISO and CCITT have provided just such a modeling technique based on the concepts of standards and protocols. A collection of communications and data processing functions is generally known as the standard. A communications protocol defines the communication procedures and data encodings used to allow information exchange between two compatible network nodes or systems. A data processing protocol deals with the presentation and application control. Most of the standards-related protocols (international and de facto types) can be represented in a consistent fashion. An obstacle to understanding the standards or protocols is their diversity and mundane complexity.

Reading about a protocol is never an exciting event. One way to deal with their diversity is to model a standard in terms of hardware/software layers. When two compatible network nodes communicate with one another to provide service, communication occurs between layers. Two adjacent layers within the same network node are separated by appropriate interfaces. Two different protocols can then be compared by studying the corresponding software layers.

B.2.2 OPEN SYSTEM INTERCONNECTION (OSI) REFERENCE MODEL

The International Standards Organization (ISO) created a standard (ISO 7498) for a reference model for distributing communications and data processing functions within a node in 1983. CCITT has also accepted this model as a basis for its evolving standards. A reference model is an attempt to represent the layers of communications and data processing functions within a network node and the required interprocess communication. See Figure B.2.1 for the seven-layered reference model for OSI designed to foster unfettered interoperability between hardware/software products manufactured by different vendors. The lower four layers—physical, data link, network, and transport represent the basic network service functions. These functions are distinct from the higher three: the session, presentation, and application layers. These three higher layers mainly deal with the data processing functions. Since this section is discussing only the OSI reference

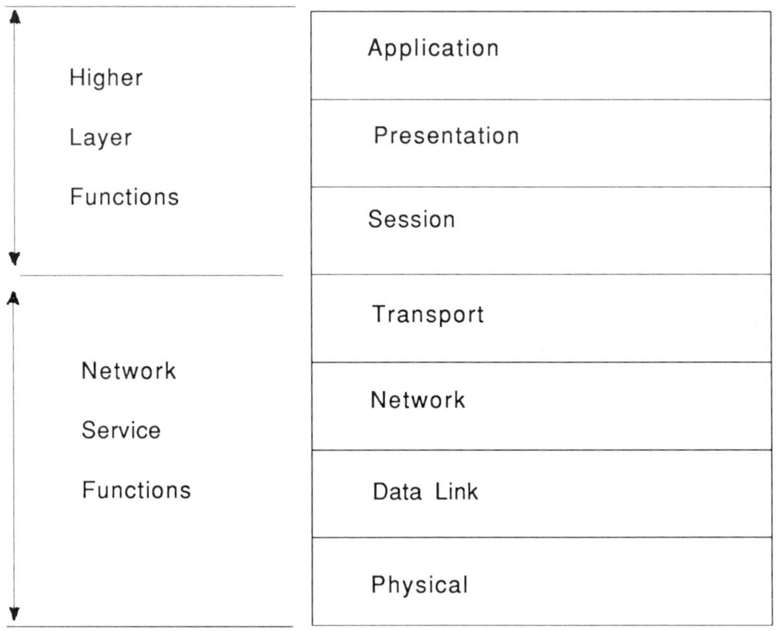

Figure B.2.1. ISO's Open System Interconnect (OSI) reference model.

model, no attempt will be made to define the actual standards or protocols covered by each layer. The succeeding sections will deal with only the major standards for voice, data, integrated voice/data, and distributed DP environments, respectively.

The physical layer deals with the following attributes:

1. Electical (e.g., data rates and voltage levels);
2. Functional/procedural (e.g., definitions of data, timing, and control signals); and
3. Mechanical environments (e.g., the type of connector and the number of pins).

The data link control (DLC) layer performs the following functions:

1. Data link connection maintenance;
2. Link activation and deactivation processes;
3. Map data units to data link units;
4. Multiplex data units to multiple physical connections;
5. Delimit data link units;
6. Error detection, recovery, and notification to end users; and
7. Network identifications (ID) and parameters exchange with peer DLC parties within the compatible network nodes.

The network layer provides the following capabilities:

1. Providing network address and end-point ID;
2. Multiplexing of network connections to DLC;

3. Segmenting/blocking of data streams;
4. Selecting the appropriate service;
5. Selecting the desired service-quality;
6. Error detection and recovery and end user notification;
7. Expediting data transfers; and
8. Connecting resets with loss of data and end user notification.

The transport layer has three phases:

1. A call establishment phase, dealing with transit set-up and network service, throughput, delay, and error characteristics and data unit sizing;
2. A data transfer mode, dealing with blocking/segmenting and multiplexing of connections, end-to-end data flows control, data unit ID, connection ID, error detection and error recovery, and expedited data; and
3. A termination phase, dealing with the cause of termination and the ID of termination connection.

The session layer handles the following tasks:

1. Session connection;
2. Session establishment and release/synchronization;
3. Normal and expected data exchange;
4. Quarantine service; and
5. Exception reporting.

The presentation layer provides the following capabilities:

1. Data transformations required for information processing (IP);
2. Data formatting into either ASCII, binary, EBCDIC, numeric, or Graphic form; and
3. Syntax selections.

The application layer provides the following services necessary for any application:

1. Identifying participating users;
2. Establishing the authority to communicate;
3. Testing of privacy mechanisms, if required;
4. Allocating costs or billing and testing quality-of-service;
5. Recovering from error and allocating responsibilities;
6. Selecting a dialog discipline; and
7. Transferring actual information between partners.

A study of the various layers shows that the higher the layer, the more complex is the software. It should also be obvious that, the lower the layer, the more dependent it is on the physical hardware, Furthermore, it should be obvious that the application layer does not specify a given application—it represents only the interfaces provided for running the various applications/services handled by the network system. A network of compatible nodes represents an intelligent system quite similar to a single computer. Together, the interconnected network nodes provide a set of services to a distributed population of users while sharing distributed data bases and computing resources. See Figure B.2.2 for an illustration of interlayer communications between interconnected nodes. In some quarters, the OSI reference model is referred to as the OSI architecture. This concept should not conflict with the one introduced in Section 1.7, if one considers the various layers as DP/DC entities and how these are related to one another. In the sections that follow, several popular standards (of the international or de facto type), and their similarities to the OSI reference model, will be discussed.

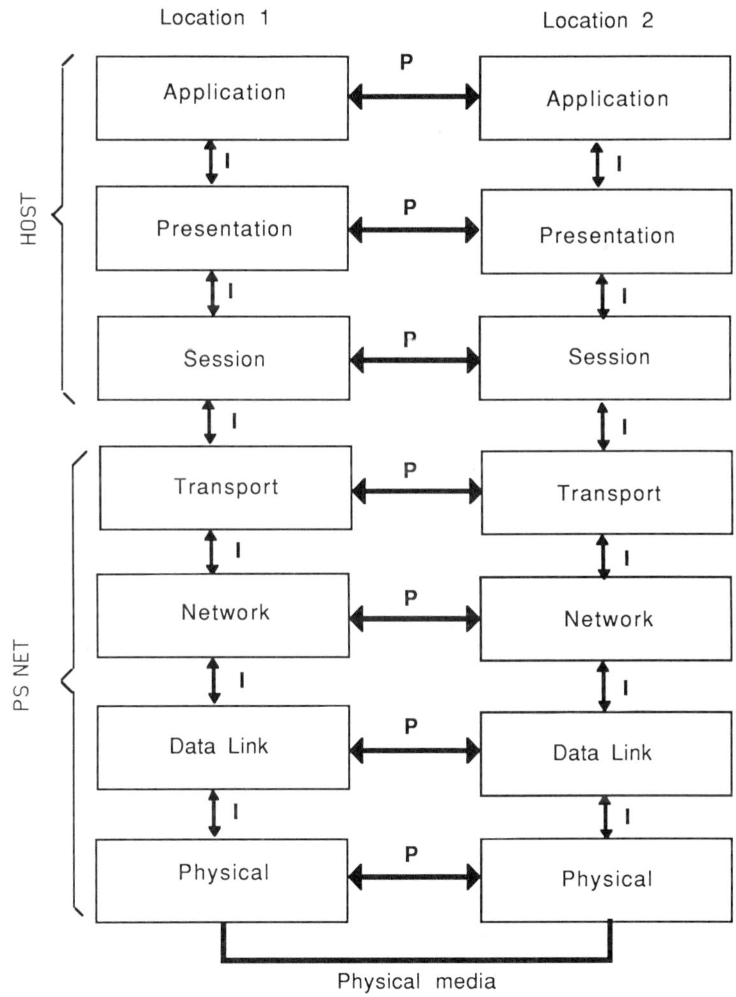

Figure B.2.2. An illustration of the interaction between two nodes based on the OSI reference model.

B.3 STANDARDS RELATED TO VOICE COMMUNICATIONS

B.3.1 INTRODUCTORY REMARKS

Most of the telephone networks in the developed countries such as the U.S.A., Canada, and Japan were developed before the international standards were finalized. The first attempt to interwork two voice networks was made after World War II when the gateways in New York and London were connected together to perform the necessary handshaking and translating of the signaling functions over the transatlantic cable. The networks in North America and European countries have evolved rather independently.

The CCITT Green Book describes most of the international standards dealing with voice communications, which primarily employ the star topology at all levels of hierarchy. Therefore, most of well-known standards will be divided by transmission and switching considerations only. For transmission, radio and wireline media will be considered. Both the analog and digital switches will be considered.

Whereas a PS data communication network standard can generally provide the OSI network services using the lowest four layers, namely physical, link, network, and transport, most existing circuit switched (CS) voice networks employ only the lowest physical layer, at this time. These networks must evolve to use the higher layers of OSI services in order to handle other services on an integrated basis. A new set of standards that are the basis of a new network architecture, Integrated Services Digital Network (ISDN) attempts to achieve that goal. ISDN will be discussed in Section B.5.

B.3.2. RADIO TRANSMISSION SYSTEMS

B.3.2.1 Analog Techniques

No important standards worth mentioning here for this category.

B.3.2.2 Digital Techniques: Some Standards

1. CCIR Rec. 382-3 for 2 GHz Band, 90 Mbps rate, and 1344 PCM channels
2. CCIR Rec. 383-2 for 4 GHz Band, 90 Mbps rate, and 1344 PCM channels
3. CCIR Rec. 384-3 for 6 GHz band, 140 Mbps rate, and 1344 PCM channels
4. CCIR Rec. 497-2 for 8 GHz band, 70 Mbps rate, and 960 PCM channels

B.3.3 WIRELINE TRANSMISSION SYSTEMS: SOME STANDARDS

1. CCITT Rec. G.120, G.121, G.123, G.133 define transmission performance
2. CCITT Rec. G.101, G.111, G.113, G114 define the transmission loss plan

B.3.3.1 Analog Techniques: Some Standards

1. CCITT Rec. G.232 for FDM Group with 12 Voice Channels @ 60-108 KHz
2. CCITT Rec. G.233 for FDM SGroup with 60 Voice Channels @ 312-552 KHz
3. CCITT Rec. G.232 for FDM MGroup with 300 Voice Channels @ 812-2044 KHz

4. CCITT Rec. G.232 for FDM SMGroup with 900 Voice Channels @ 8516-12388 KHz

B.3.3.2 Digital Techniques: Some Standards

1. CCITT Rec. G.733 for T1 @ 1.544 Mbps with 24 PCM channels (N. America)
2. CCITT Rec. G.732 for CEPT @ 2.048 Mbps with 32 (30 data, 2 supv) PCM channels
3. CCITT Rec. G.743 for T2 @ 6.312 Mbps with 96 PCM channels (N. America)
4. CCITT Rec. G.742 for CEPT @ 8.448 Mbps with 120 PCM channels
5. CCITT Rec. G.752 for T3 @ 44.736 Mbps with 672 PCM channels (N. America)
6. CCITT Rec. G.751 for CEPT @ 34.368 Mbps with 480 PCM channels
7. CCITT Rec. (NA) for T4 @ 274.176 Mbps with 4032 PCM channels (N. America)
8. CCITT Rec. G.751 for CEPT @ 139.264 Mbps with 1920 PCM channels

B.3.4 SWITCHING SYSTEMS

B.3.4.1 World Numbering Plan

This numbering plan allocates the country codes to each fifteen digit international telephone number. Such a plan was necessary to allow international dialing.

B.3.4.2 CCITT Signalling System #5 for Intercontinental Telephone Cable

B.3.4.3 CCITT Signalling System #6 for Public Telephone Networks

Such a scheme was implemented within the Bell Systems and several other public telephone networks prior to its replacement by the CCITT Signalling System #7 for ISDN.

B.3.4.4 CCITT Signalling System #7 for Public Telephone Networks (for ISDN-based public telephone networks)

B.3.4.5 Q.22/Q.23 Recommendations for Push-Button Telephone Sets (Vol. GVI-1)

B.3.4.6 Q.35 Recommendations for Characteristics of Various Tones (e.g., Ringing, Busy, Congestions, and Warning): (Vol. GVI-1)

B.3.4.7 Q.40-45 Recommendations for the Characteristics of Transmission Plans: (Vol. GVI-1)

B.3.5 NETWORK PLANNING AND TRAFFIC ENGINEERING

1. CCITT Rec. E.520 for the use of Erlang B formula for sizing trunk bundles.
2. CCITT Rec. E.541 for the use of P.01 for the final choice route trunk.

B.4 STANDARDS RELATED TO DATA COMMUNICATIONS

B.4.1 INTRODUCTION

Whereas the need for data communications for a distributed data processing (DDP) system is apparent, the need for data communications for a voice network is not that obvious. Practically all of the existing voice communications systems use some form of data communications for signaling and NMC functions. In many cases, signaling data are sent in-band on the same circuit that is employed for voice. In some parts of the system, special order-wire circuits are used to transport alarms and diagnostic data from one node to another. Most existing voice communications systems employ a proprietary protocol for data communications. An international standard for accessing a public data network (PDN) is described in the following subsections.

B.4.2 V.SERIES STANDARDS FOR ANALOG MODEMS

Nowadays, everybody is familiar with the modulator/demodulator (MODEM) devices employed to connect personal computers (PCs) with mainframes. Modems are also employed to synthesize multidrop links to allow several communications controllers (each controlling several CRT devices) to communicate with a mainframe. In most systems, modems constitute a large investment. For that reason, many intelligent PABXs and hosts employ modem pooling (or sharing) to reduce the costs of communications hardware. Modems are evolving into very intelligent devices, thanks to VLSI technology. This intelligence is provided to achieve a desired amount of NMC functionality for a PDN.

See Table B.4.1 for a list of V.Series standards for analog modems or their maintenance and/or performance monitoring. Of course, some vendors sell modems that are based on proprietary protocols and used within a private data network.

B.4.3 X.SERIES STANDARDS FOR WIDE AREA PUBLIC DATA NETWORKS

Many assume that X.25 is the only international standard that is required to define a Public Data Network (PDN). Actually, a large collection of international standards is used to plan and implement a PDN. See Table B.4.2 for a list of the related X.Series standards required to define a PDN. The X.1 standard specifies all of the user classes of services provided by a PDN. See Table B.4.3 for a list of such services.

Figure B.4.1 illustrates a topology of a typical PDN based on the X.Series standards. Any packet-mode data terminal equipment (DTE) based on the X.25 standard can directly access a PDN node by interfacing with its associated data communication circuit-termination equipment (DCE). A nonpacket mode terminal (any one of existing terminals such as ASCII, BSC, and SDLC) must be serviced by another PDN node called the packet assembler and disassembler (PAD). To define the various interfaces and protocols between the PAD and PDN node, X.3, X.29 and X.28 were developed. A PAD is generally assumed to handle the X.25 protocol.

The three software layers for the X.25 protocol for accessing a PDN are illustrated in Figure B.4.2. These three layers are identical to the ones described for the ISO's OSI

Table B.4.1 Some Standards for Analog Transmission and Modems

Standard	Description
V.3	International Alphabet#5 for Information Exchange
V.15	Acoustic Couplers for Data Communications
V.16	Standards Description for Analog Modems
V.19	DTMF Modem for PTNs
V.20	Extended DTMF Modem for PTNs
V.21	300baud Modem for PTNs
V.22	1200bps Modem (FDX)
V22bis	300baud (Async), 1200/2400bps (Async or Synchronous)
V.23	600/1200bps Modem for PTNs
V.24	Definitions for DTE-DCE Information Exchange Circuits
V.26	1200/2400bps Modem for 4-W Circuits
V.26bis	1200/2400bps Modem for PTNs
V.27	4800bps Modems for 4-W Circuits
V.27bis	4800bps Modem with Equalizer Standard for PTNs
V.27ter	2400/4800bps Modem for PTNs
V.28	Definition for the Unbalanced Interface
V.29	9600bps Modem for 4-W Leased Lines
V.32	9600bps Modem (FDX-Synchronous) for 2-W Leased/Dial-Up
V.35	48000bps Modem for 60-108KHz Group Circuits
V.36	48000-72000bps Modem for 60-108KHz Group Circuits
V.37	96000-168000bps Modem
V.40	Parity Error Indicators
V.41	Error Controllers
V.50	Transmission Quality
V.52/53	Bit Error Patterns/Limits

standard, as described in Section B.2. The exact correspondence between the corresponding layers of the two standards resulted from the close association between the CCITT and ISO. Some of the well-known international and national (e.g., the CCITT and EIA) recommendations are specified in the following paragraphs.

1. The first layer deals with the electrical, functional, and mechanical aspects of transmission:

Table B.4.2 Some Standards for Public Data Networks

Standards	Description
X.1	International User Classes of PDN Services
X.2	International User Facilities in PDNs
X.3	Packet Assembly/Disassembly (PAD) Device
X.4	International Alphabet#5 for PDNs
X.20	Start/Stop DTE-DCE Interface in PDNs
X20bis	V.21 Modem for PDNs
X.21	Synchronous DTE-DCE Interface in PDNs
X.21bis	DTE-DCE Interface for Synchronous V.Series Modems
X.24	DTE-DCE Information Exchange Circuits (IECs)
X.25	DTE-DCE Interface for DTEs Operating in Packet Mode
X.26	Electrical Characteristics of Unbalanced Double Current IECs
X.27	Same as above for Balanced IECs
X.28	DTE-DCE Interface for Start/Stop DTEs Accessing PADs
X.29	Exchange of Control and User Data between DTE and PAD
X.32	Dial-Up PDN Service (< 9600bps)
X.60	Common Channel Signaling for Synchronous Data Service
X.70	Terminal and Transit Control Signaling for Start/Stop Data Service over International Circuits between Asyncronous PDNs
X.71	Same as above for Synchronous PDNs
X.75	Data Exchange on International Trunks
X.95	Network Parameters in PDNs
X.96	Call Progress Signals
X.121	International Numbering Plan for PDNs

1.1. CCITT Rec. X.27 (or V.11) and EIA RS-422 specify the balanced electrical interface with a data rate between 0 and 10 Mbps. It is interoperable with RS-423 and X.26 (or V.10). Distance limitations are 4000 feet @ 100 Kbps and 40 feet @ 10 Mbps. CCITT V.28 and EIA RS-232C represent the first generation and CCITT X.26 (or V.10) and EIA-423 represent the second generation of standards for 0 to 300 Kbps rate and shorter distances dealing with the unbalanced interfaces.

Table B.4.3 PDN Services Related to the X.1 Recommendation

Class-of-Service (COS)	Description of Services Offered
1. Asynchronous, Circuit Switched	1.1 @300baud, 11-Unit Code, Start/Stop
	1.2 @50-200baud, 7.5 to 11 Unit, Start/Stop
2. Synchronous, Circuit Switched	2.1 @600bps Using Data Terminals
	2.2 @2400bps Using Data Terminals
	2.3 @4800bps Using Data Terminals
	2.4 @9600bps Using Data Terminals
	2.5 @48000bps Using Data Terminals
3. Synchronous, Packet Switched	3.1 @2400bps with Packet Mode Terminals
	3.2 @4800bps with Packet Mode Terminals
	3.3 @9600bps with Packet Mode Terminals
	3.4 @48000bps with Packet Mode Terminals

1.2. CCITT Rec. V.24 and EIA RS-232C represent the first generation and CCITT Rec. V.24 and EIA RS-449 represent the second generation of standards dealing with the functional definitions of data, timings, and control signals.
1.3. ISO 2110 and EIA RS-232C represent the first generation standard for a 25-pin connector. ISO 4902 and EIA RS-449 represent the second generation standard for the 37-pin connector that is interoperable with RS-232C.

2. The second layer deals with the logical link control used for establishing, maintaining, and terminating a session between the DTE and DCE. It is identical to the HDLC protocol as specified in the X.25 standard for accessing a PDN. The IEEE 802.2 standard for LANs is also similar to the X.25 HDLC with some differences as shown in the following paragraphs. IBM's SNA employs a synchronous data link control (SDLC) protocol that is similar to HDLC. In all cases, each LLC protocol involves the use of a frame, as shown in Figure B.4.8.

3. The third layer deals with the creating of packets with the calling and called node addresses, multiplexing packets from DLCs onto a network link, selection of service and its quality, error checking and recovery, and other related functions. Again, this is identical to the network layer of the CCITT X.25 protocol.

Figure B.4.3 illustrates the basic interfaces between the DTE, DCE, and the PDN at a very high level. Figure B.4.4 illustrates the physical and other standard layers of the interfaces between the DTE, DCE, and the PDN node. Figure B.4.5 shows two classes

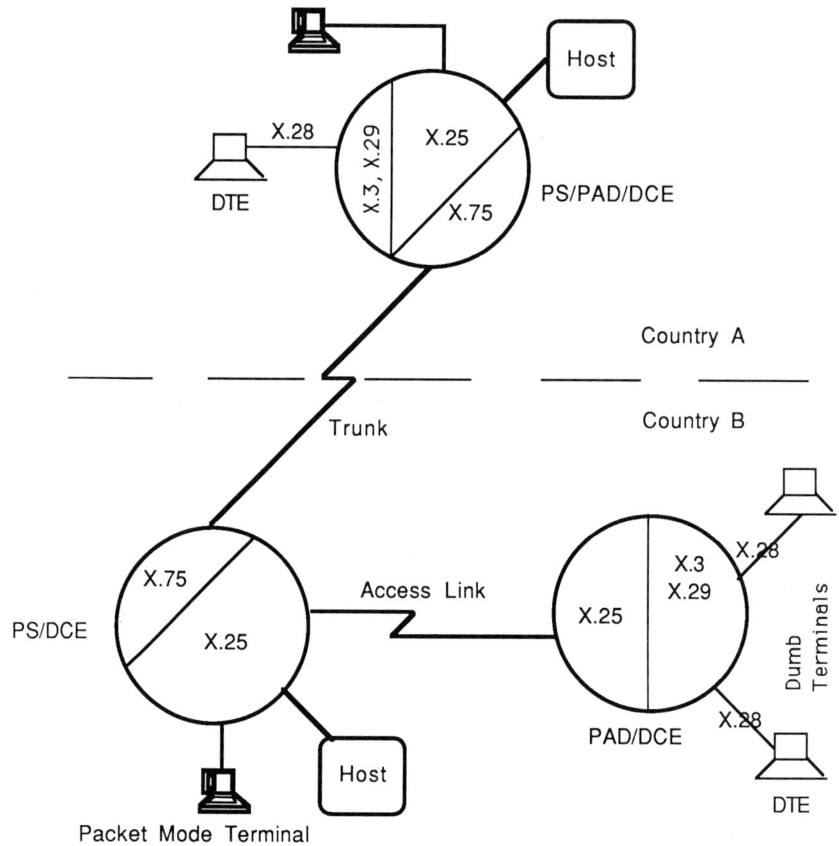

Figure B.4.1. A public data network topology based on some well-known standards.

of DTE to DCE link access protocols (LAPs). The LAP A deals with the async-response-mode (ARM). The LAP B deals with the balanced mode of operation, where either the DTE or DCE can initiate the process of data exchange.

The X.25 protocol employs both the virtual circuit (VC) and the datagram (or the connectionless) concepts for data interchange. For the VC approach, a logical VC (involving PDN nodes and links) is established before allowing any data packet to move. This allows the sharing of the physical resources among many logical VCs. The connectionless approach will also allow the sharing of PDN's physical resources by using the dynamic routing technique described in Chapter 1. The VC and the connectionless concepts also allows the sharing of the physical link connecting a X.25-based host and the PDN among up to 4096 VCs. The resulting resource sharing is perhaps the most attractive feature of X.25 protocol. See Figure B.4.6 for an illustration of the VC operations related to the X.25 protocol.

Several types of data packets and data frames are required to make a PDN provide the data communications service. The structures of a call request packet, call accept

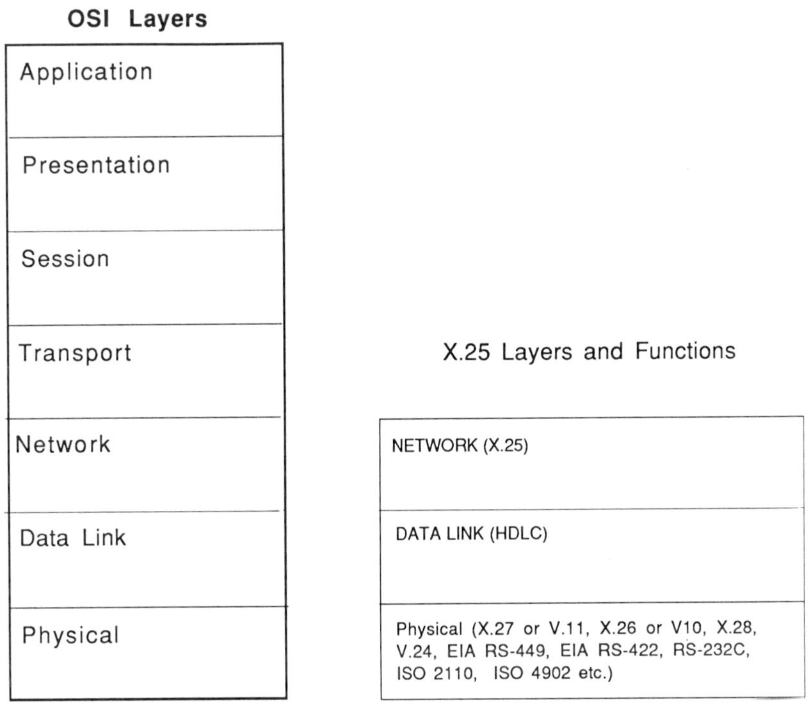

Figure B.4.2. The X.25 layers and their relationships to the OSI reference model layers.

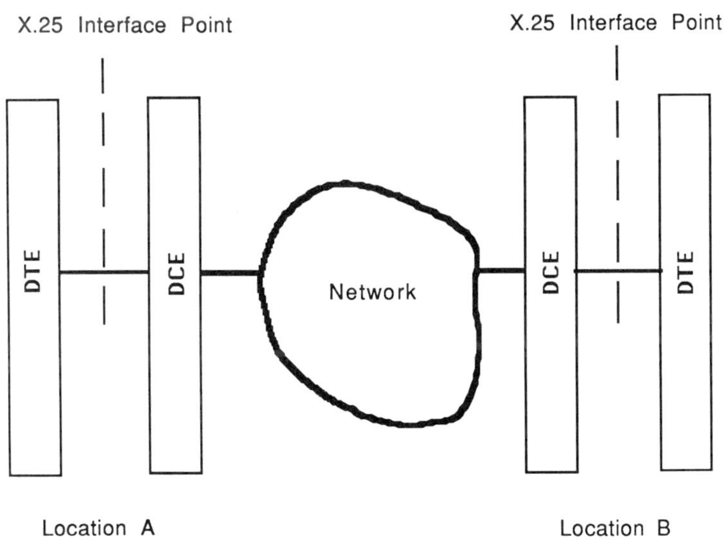

Figure B.4.3. An illustration of basic X.25 DTE-DCE interfaces.

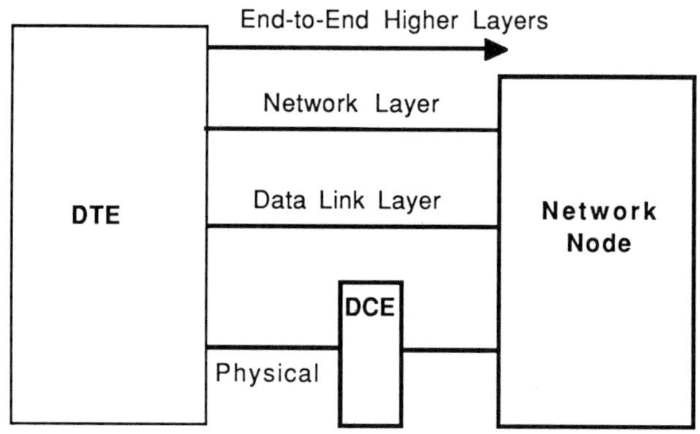

Figure B.4.4. An illustration of the X.25 layers involved in basic network interfaces.

packet, information packet, supervisory frame, unnumbered frame, and the information frame are shown in Figures B.4.7 and B.4.8.

B.4.4 IEEE STANDARDS FOR LOCAL AREA NETWORKS (LANs) FOR DATA COMMUNICATION

In order to provide low-cost LANs to the computer community, the IEEE Computer Society established a committee in 1980 to create LAN standards. Thus began the IEEE 802 project. Although the original goal was to establish a single standard, the effort ended up with several meaningful standards to face the reality of several evolving LAN architectures.

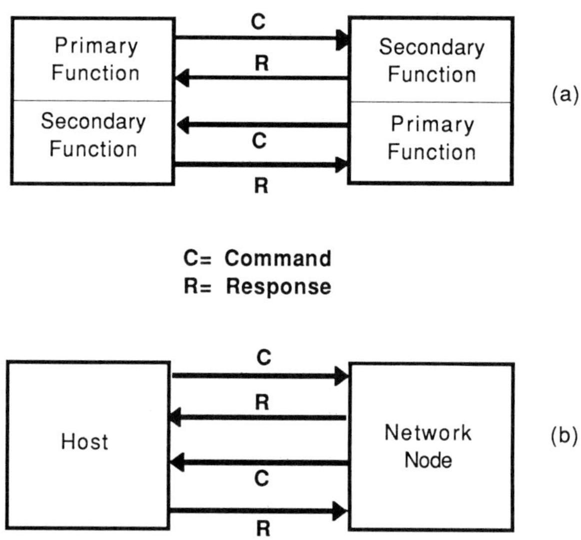

Figure B.4.5. Two classes of DTE-DCE link access procedures involved.

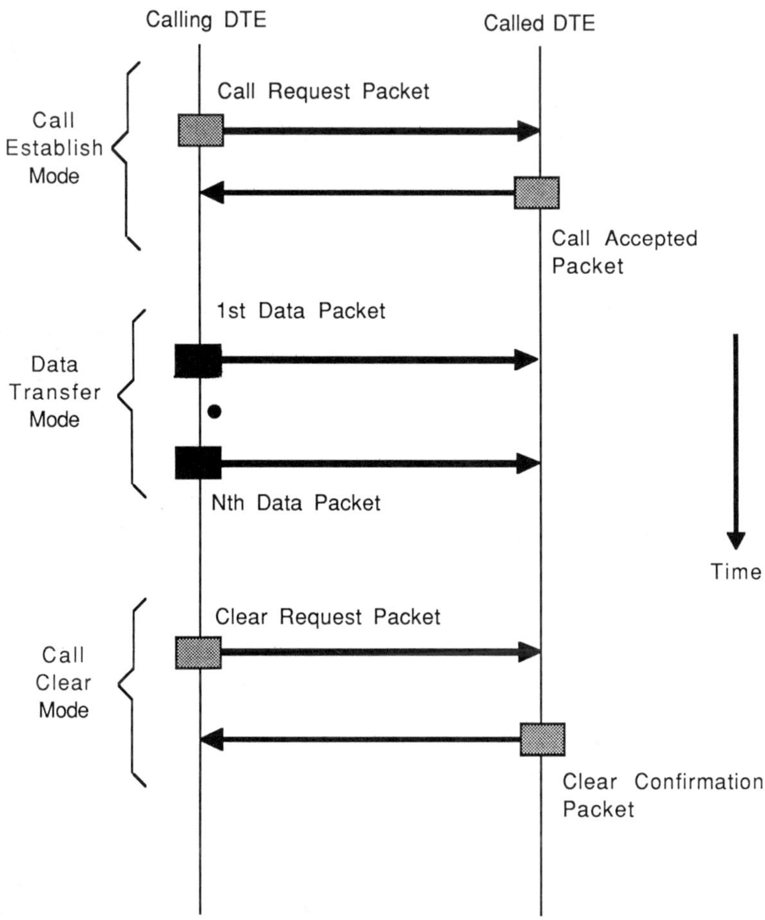

Figure B.4.6. An illustration of the X.25 data exchange operations in a virtual circuit mode.

The work of the IEEE 802 committee is currently divided into the following subcommittees, each dealing with a specific LAN standard:

1. IEEE 802.1 dealing with Higher Layer Interface (HILI) standard.
2. IEEE 802.2 dealing with the Logical Link Control (LLC) standard.
3. IEEE 802.3 dealing with the Carrier Sense Multi-Access with Collision Detection (CSMA/CD) standard.
4. IEEE 802.4 dealing with the token bus standard.
5. IEEE 802.5 dealing with the token ring standard.
6. IEEE 802.6 dealing with the Metropolitan Area Network (MAN) standard.

At the time of writing, work has been completed on the 802.2, 802.3, 802.4, and 802.5 standards, and IEEE has approved initial standards. Work on 802.1 is still pro-

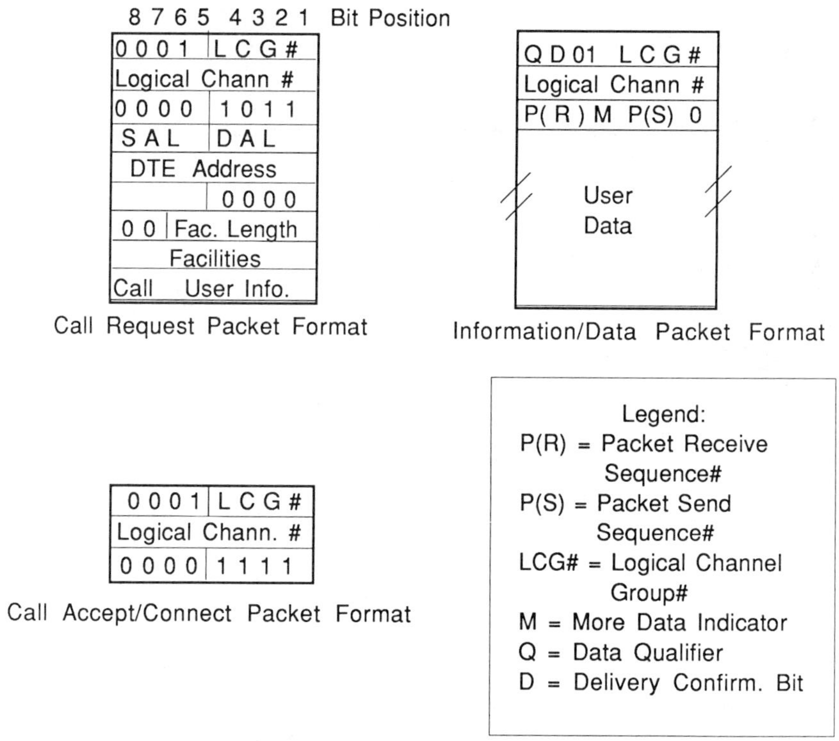

Figure B.4.7. An illustration of the various X.25 packet formats.

gressing. It deals with higher layer interfaces for each of the other standards, internetworking, global addressing, and network management and control (NMC). Other study groups are always being added to consider other standards for LANs such as a IVDLAN (SG 802.9). The acceptance of the IEEE 802 standards has been remarkably quick and widespread. The National Bureau of Standards (NBS) which issues Federal Information Processing Standards (FIPS) for U.S. government procurements, has issued a FIPS for the CSMA/CD and LLC standards. The other FIPS based on IEEE 802 standards will soon follow. The ISO has also adopted the IEEE 802 documents in their entirety as a draft proposal. This clearly shows how the various standards organizations cooperate. Another influential standards organization, ECMA, is also looking into the IEEE 802 standards for a basis for its own LAN-related standards.

Each of the 802.3, 802.4, and 802.5 standards represents a unique LAN architecture and will employ the 802.1 and 802.2 standards to interface with the higher layer protocols (e.g., transport, session, presentation, and application) and to handle the LLC layer, respectively. Whereas the higher layers are the same as those described for the OSI Reference Model (see Section B.2), the LLC for a LAN is somewhat different from the LLC for the X.25 protocol for WANs. The 802.2 LLC layer must support the multiaccess nature of the bus/ring media while sharing the link access with the medium access control (MAC) and providing a minimum amount of network layer functions. As

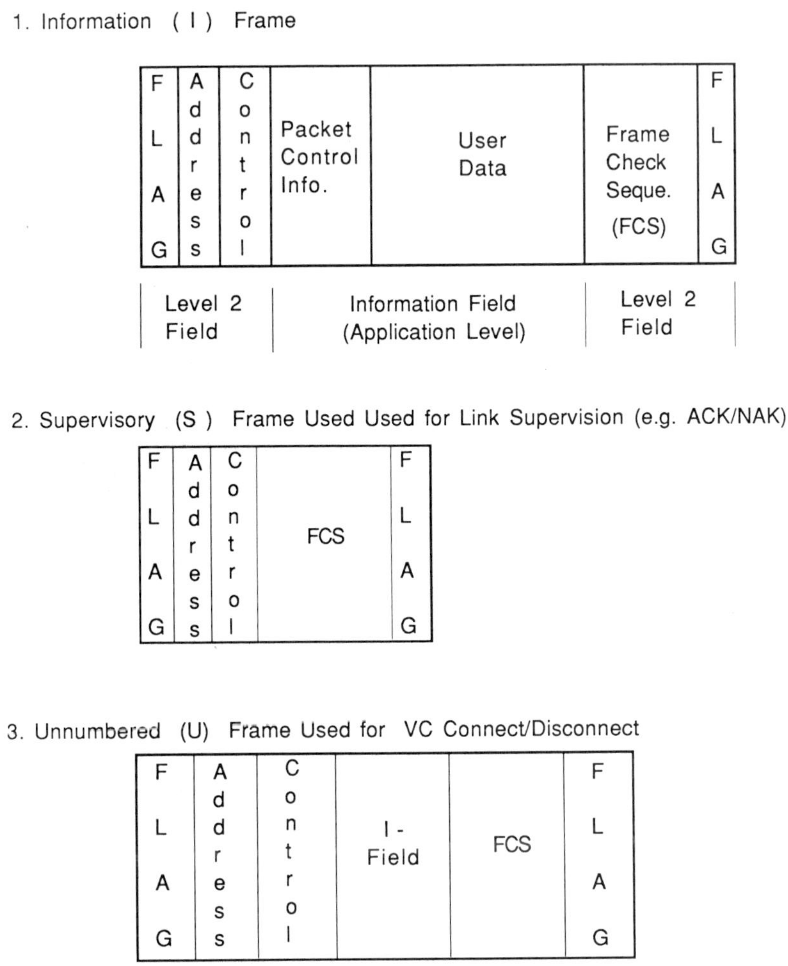

Figure B.4.8. An illustration of the various X.25 frame formats.

a minimum, LLC must perform error control, flow control, and frame sequencing, just as HDLC does. Since there are no intermediate switching nodes, a LAN does not need a formal network layer. The remaining parts of the network layer, if any, can be incorporated into the LLC layer. All of the IEEE 802 standards are capable of handling both the datagram (or connectionless) and the virtual circuit (VC) services. Furthermore, all of the IEEE standards employ the asynchronous TDMultiplexing technique for the sharing of common media.

See Figure B.4.9 for an illustration of a reference model for LAN standards and its comparison with the OSI reference model.

At the present time, the FDDI standard is being developed under the ANSI jurisdiction. IEEE and ISO are mere observers. The FDDI standard will provide a capability to interconnect (1) very-high-speed mainframes located in adjacent areas, and (2) low-

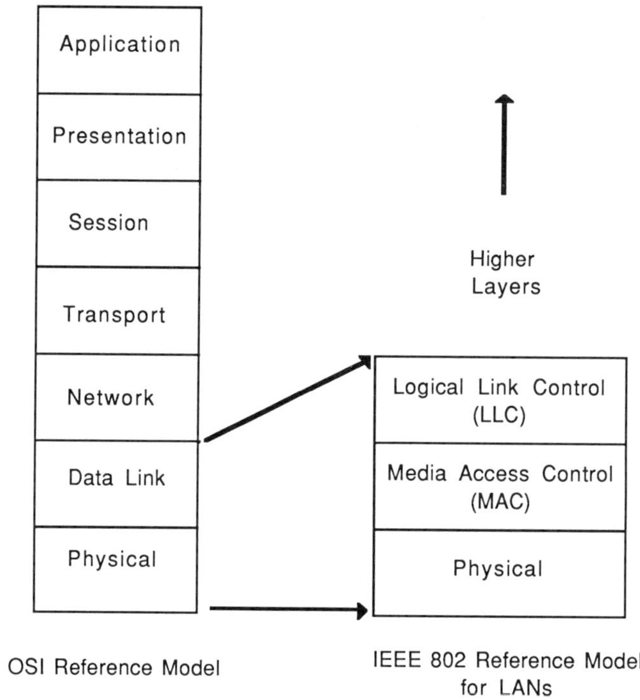

Figure B.4.9. Reference models for the IEEE 802 and OSI standards.

speed LANs located within a metropolitan area. Many vendors supply proprietary bridges, routers, and gateways to interconnect LANs. Some vendors have started to interconnect LANs via X.25-based backbone networks. At present, IEEE is leaning towards the DQDB (Distributed Queued Dual Bus) architecture for the IEEE 802.6 Standard. See Chapter 10 for a discussion on the subject.

Since the media access control layer provides the unique character to the IEEE 802 standard, one should discuss the standards protocols associated with it.

B.4.4.1 CSMA/CD (802.3) Standard for the Bus Topology

For the CMSA/CD scheme, a station wishing to transmit first senses the presence of a carrier to see if another station is transmitting. If the medium is busy, station will defer transmission of its packet/message until the carrier drops and a guard band time interval (also called the interframe gap) have elasped. If the medium is idle, the station may transmit. During its transmission the station also tests for a collision with the transmission from another station. Such a collision is caused by the propagation-time separations between the stations. If there is a collision, the transmissions will be garbled and full of errors. If the station detects a collision, it will cease transmission and jam or reinforce the detected collision with a 48-bit pattern, and then defer transmission attempts until a randomized retry time interval has passed. This standard is an evolved version of the original Ethernet product as developed by the Xerox Corporation.

B.4.4.2 Token Bus (802.4) Standard for the Bus Topology

According to this scheme, all stations on the bus receive the superposition of all other station transmissions, as well as their own. Control is passed via a token (equivalent to a bit pattern). A station must receive a full message preamble before it can pass the token onward or transmit its own packet/message. This interval equals the time for energy to propagate from one end of the bus to the other end plus the time it takes for each station to perform the required signal processing. Such an interval is named as the interface delay.

B.4.4.3 Token Ring (802.5) Standard for the Ring Topology

In this scheme, each station receives from a single station and transmits only to a single station. Control is passed via a token which is similar to a bit pattern. The performance of a token ring is determined mainly by the propagation time, which defines the length of time to propagate energy from one station through all of the other stations and back to itself. Each station interface unit is characterized by the time it takes to detect the token and decide whether to pass the token to the next station or transmit a packet/message. The interface time is therefore equal to one bit time (for a one-bit token) only, since the full message preamble can be handled by a subsequent serial-to-parallel conversion process.

B.4.5 SOME DE FACTO STANDARDS FOR DATA COMMUNICATIONS

A large number of private data networks employ proprietary protocols for LAN and WAN applications. Despite their widespread use, no attempt has been made to represent these in terms of a multilayer protocol in a manner similar to that for the OSI reference model. Only a brief description of these protocols will be given.

B.4.5.1 Datapoint's ARCNET

About 8000 LANs currently utilize the Attached Resource Computer network (ARCnet) architecture. The topology of an ARCnet consists of hubs and resource interface modules (RIMs) as illustrated in Fig. B.4.10 and it utilizes the token passing (of both the ring and bus variety) access scheme over a 2.5 Mbps bus for a large number (<255) of intelligent devices and computers. Hubs and RIMs are interconnected via buses and all connected devices listen to transmissions. Each hub provides connections (via an internal ring) to RIMs serving PCs, file servers, and printers according to a physical star topology. Bridges and gateway are also available to interface with other ARCnets, and SNA/SDLC, HDLC or X.25 networks. A 20 Mbps version of ARCNET is also planned.

B.4.5.2 AT&T's ISN

Since its divestiture in 1984, AT&T has been developing several data communications networking products for BOC and business applications. Two products, Datakit and Information Systems Network (ISN), are now available. Their unique data networking architecture is based on the concepts of shared virtual packet, shared high-capacity contention bus, and fast packet switching. Several LANs based on ISN can be intercon-

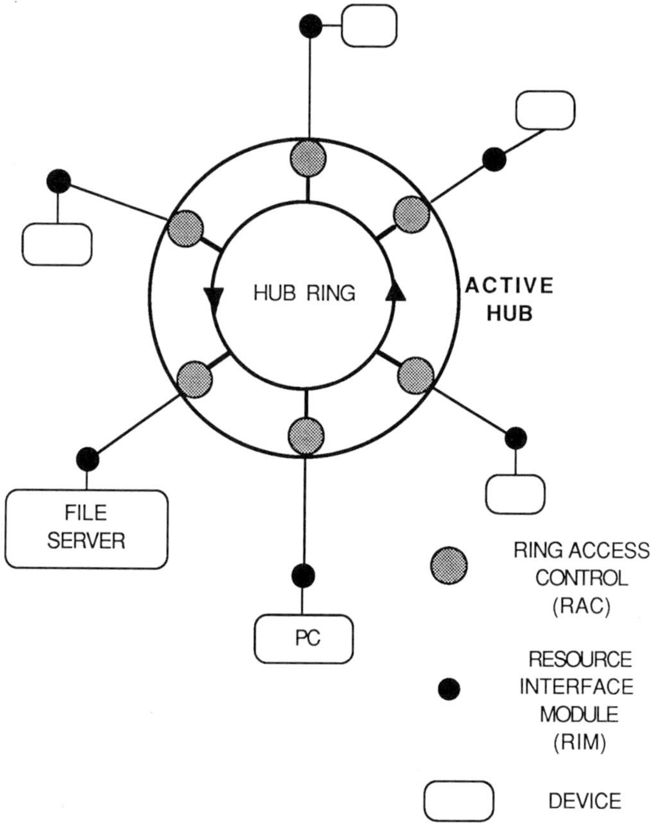

Figure B.4.10. Mixed topology for Datapoint's ARCNET LAN architecture.

nected to achieve a WAN. It employs the concept of VCs to establish end-to-end paths within the network. See Figure 4.9 for an illustration of an ISN/Datakit-based LAN. It consists of several input/output modules, a fast packet switch, a controller, and high-capacity contention and broadcast buses. Most modules transmit on the contention bus and listen to the broadcast bus. The switch receives on the contention bus and transmits on the broadcast bus.

A concentrator is offered for interfacing with most of the existing terminal types. Gateways to X.25 PDN are also available. The high-speed packet switch can also service a multiple of 56-Kbps, T1, or fiber optical transmission lines, while introducing very little end-to-end delays. Several of these PSs can be interconnected via high-speed trunk bundles according to a mesh topology to achieve a high-capacity nationwide WAN. The resulting WAN should obey the multistar, multicenter topology, as illustrated in Figure B.4.11. The constraint to employ only the star topology usually results in high transmission costs with widely scattered CPEs. However, the transmission costs may be minimized when only pure LANs based on the ISN/Datakit architecture are interconnected to create a nationwide WAN. This implies that CPEs are located within a building or a campus, allowing the use of privately owned cables.

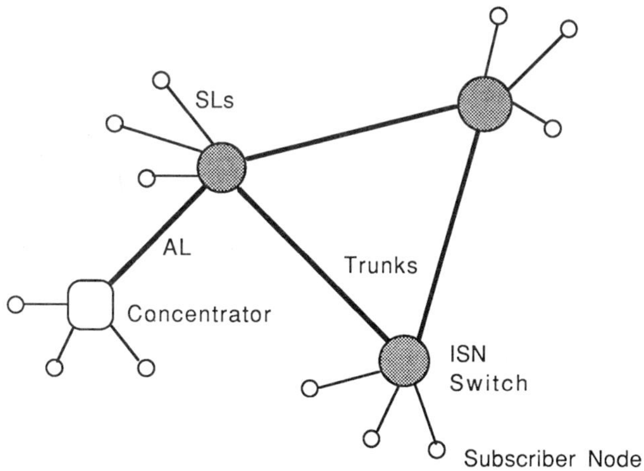

Figure B.4.11. Mixed topology for AT&T's ISN and Datakit architectures.

B.4.5.3 Doelz's ESR/Virtual Packet Network

The Extended Slotted Ring architecture employs two types of nodes. A desktop box (called the Elite) provides packet handling and pure statistical packet-multiplexing and demultiplexing (i.e., a PAD function) as illustrated in Figures 4.10 and 4.11. The large node (called the Esprit) provides high-speed packet switching functions for interchanging packets between a number of netlinks. Each Elite can service up to 16 different types of terminals. A fully shared transmission link (or netlink) can serve several Elites according to the DL topology. Several Esprits can be connected with one another, according to a mesh topology, to provide a WAN. See Figures B.4.12 and B.4.13 for an

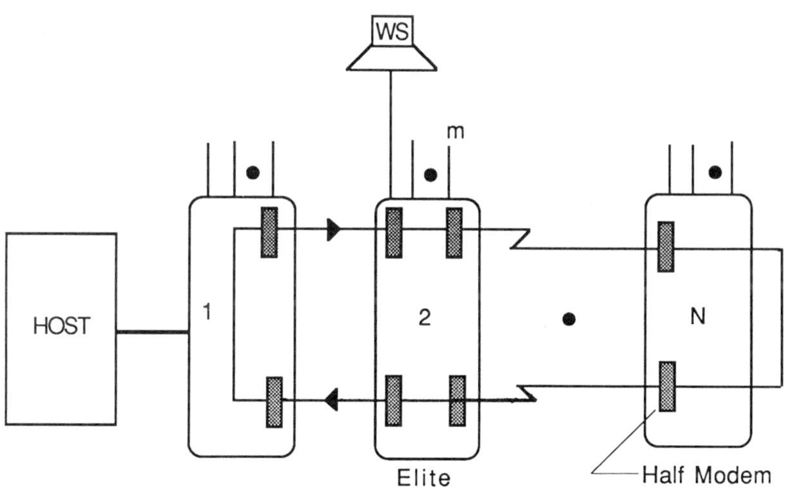

Figure B.4.12. An illustration of the extended slotted ring employed by a netlink in Doelz architecture.

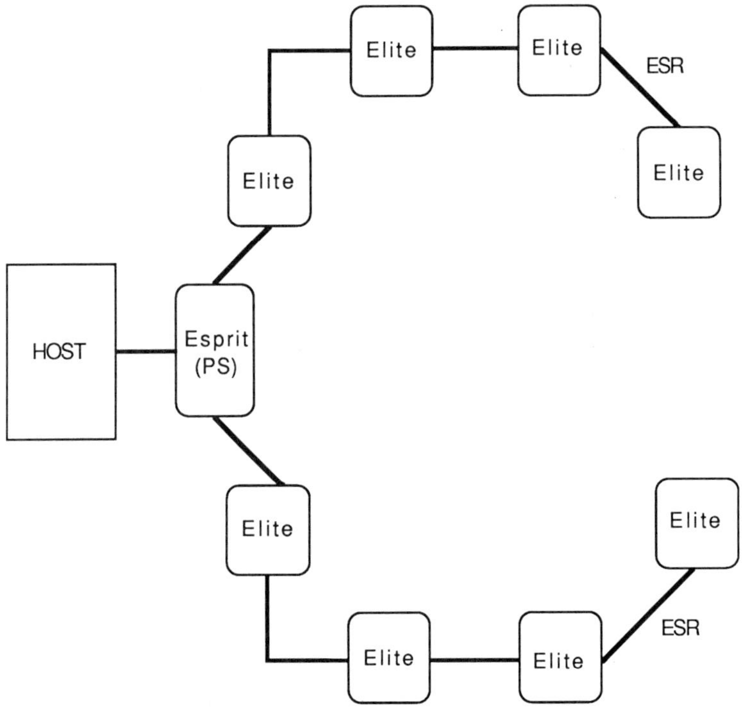

Figure B.4.13. A directed netlink topology employed in Doelz architecture.

illustration of the extended slotted ring used in synthesizing a netlink and a typical directed netlink topology respectively. The originating Elite node establishes a virtual circuit (VC) within the network before data packets can flow between the two ends. The ESR architecture packetizes every stream of data/message into very small proprietary packets and provides nodal bypass to achieve (1) protocol transparency, and (2) very small end-to-end queuing delays. Gateways are provided to interface with X.25 PDNs.

B.5 STANDARDS RELATED TO INTEGRATED VOICE AND DATA NETWORKS

B.5.1 INTRODUCTION

The evolving standards dealing with the concept of Integrated Services Digital Network (ISDN) are the subject of this section. ISDN framework was created to provide a direction for the evolution of existing public telephone networks in such a manner as to provide many additional services, such as data, video, and leased variable bandwidth circuits. Most of the confusion regarding ISDN stems from the fact that ISDN is only a concept. Depending upon the set of ISDN-related standards, one can come up with a new variation in the ISDN concept. Basically, there are four set of standards that deal with ISDN access interfaces, the basic-rate, primary-rate, the broadband ISDN, and Signaling System #7 (SS7).

B.5.2 THE BASIC-RATE ISDN ACCESS INTERFACE

This interface involves the use of a digital subscriber loop (DSL) for accessing the ISDN central office. A DSL provides 2 B channels, each at 64 Kbps and 1 D channel rated at 16 Kbps. Whereas a B channel can be utilized for the full-duplex (FDX) CS voice or data service, the D channel is employed for signaling and/or FDX PS and telemetry. See Figure B.5.1 for a detailed illustration of the ISDN interfaces on both sides of the network termination interface (NTI) unit. The U interface is defined by the ISDN service provider and the T interface is defined by the CPE. The U interface utilizes the existing two-wire (or one-pair) subscriber line with the restriction that its length must be the usual range of five–seven kilometers. The T interface utilizes a four-wire connection that should not exceed one kilometer. The NTI unit provides line termination, layer 1 multiplexing/ management, timing, power transfer, and interface termination functions. As shown in Figure B.5.2, additional ISDN terminals can be served through the use of a single NT2 unit using the star topology. In order to interface existing non-ISDN terminals (e.g., based on the RS-232C physical layer), a terminal adapter (TA) unit will be required. This results in an additional interface called the R interface.

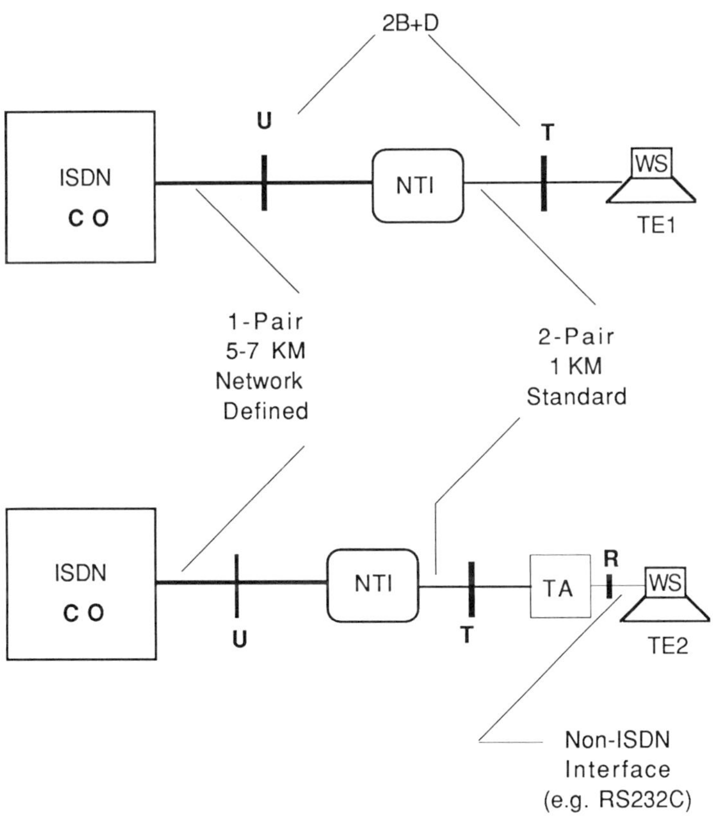

Figure B.5.1. Basic rate interface for a single user of ISDN.

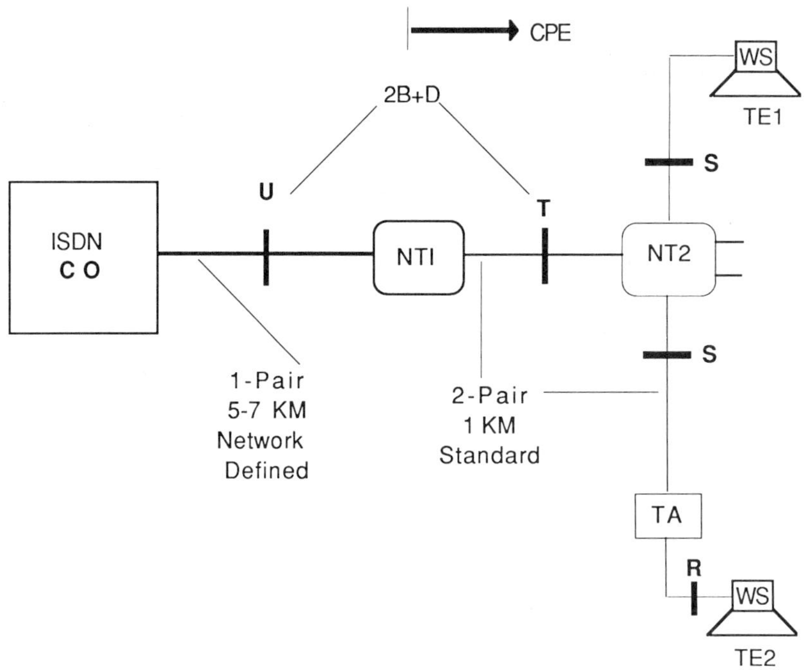

Figure B.5.2. Basic rate interface for several users of ISDN.

B.5.3 THE PRIMARY-RATE ISDN ACCESS INTERFACE

This interface involves the use of a T1-based extended digital subscriber line (EDSL) for accessing the ISDN central office. An EDSL provides 23 B channels, each at 64 Kbps, and 1 D channel also rated at 64 Kbps. Whereas a B channel can be used for the FDX CS voice or data service, the D channel is utilized for FDX PS. The signaling functions are handled through the I.451 and SS#7 standards and some NMC related functions are accomplished through the use of an extended superframe (ESF), discussed earlier in Chapter 1. See Figure B.5.3 for an illustration of the associated T and S interfaces.

B.5.4 THE BROADBAND ISDN ACCESS INTERFACE

This interface is a relatively new concept. When fully finalized in the ECSA's T1Y1 standards group, it will provide dynamically configurable channels or packets at rates up to 150 Mbps transmitted via a synchronous optical network (SONET) interface. Additional interface rates of up to 600 Mbps are also being discussed. This interface differs from the Basic and Primary interfaces in the fact that, instead of using synchronous fixed-channel techniques, a universal, extremely flexible optical transport mechanism for voice data, and high-density digital video based on the Asynchronous Transfer Mode (ATM) is provided.

Additional ISDN channel types have been defined: the 64 Kbps PS E-channel is recommended for use with the SS7, the 384 Kbps CS/PS H0-channel is recommended for

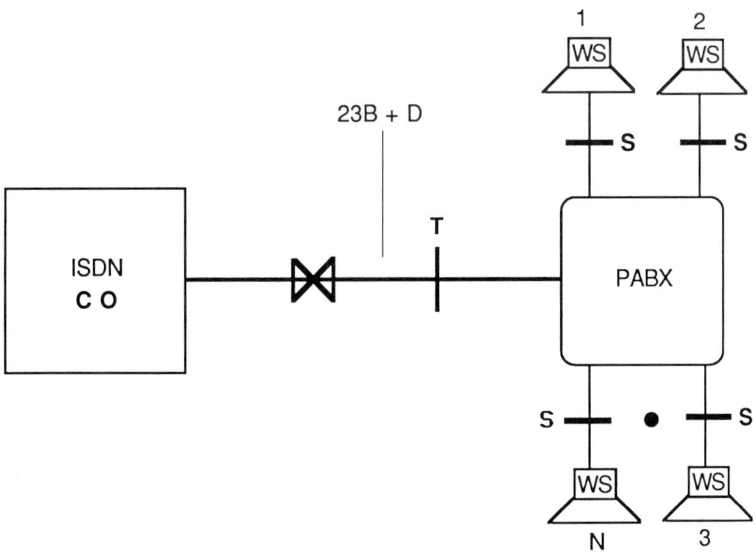

Figure B.5.3. Primary rate interface for an ISDN compatible PBX.

HS facsimile data, or slow-motion video, the 1.536 Mbps CS H11- and 1.920 Kbps CS H12-channels are recommeded for high-speed data, facsimile, and full-motion video.

B.5.5 OSI-TYPE LAYERS FOR ISDN AND RELATED STANDARDS

Study Group XI is concerned with the Q. recommendations and SG XVIII is concerned with the I. recommendations for ISDN. Both the Q. and I. recommendations are concerned with the signalling and interfaces of the first and third ISDN layers. The link control layer is derived from the X.25 data link protocol. The applicable CCITT standards for the lower three layers are shown in Figure B.5.4. The first layer handles the physical and electrical interfaces. The applicable standards can be included in the set I.43X, where X is any integer. This layer is concerned with only the raw bit transport. Separate standards for the basic, primary, and the broadband interfaces will exist. The CCITT recommendation Q.921 is for the ISDN link control. It is almost identical to the second layer of X.25, except that only the extended addressing mode is used and the address field defines the required service (e.g., voice, data, or NMC) provided by the ISDN access switch to the CPE. The Q.921 recommendation for the link control is also called the LAPD standard for ISDN. The third layer (network related) is defined by the Q.431 recommendation.

It specifies the OSI layer 3 functions for ISDN and the means for establishing, maintaining, and terminating network connections across an ISDN and between CPEs. Therefore, it controls the set up of calls across an ISDN, using structured messages. The messages flow on out-of-band channels and consist of digital symbols. The message set can be easily extended to handle future growth of service.

The manner in which an ISDN provides voice- and data-related service on a wide-area basis is illustrated in Figure B.5.5. Several ISDN COs are connected to a SS7-based

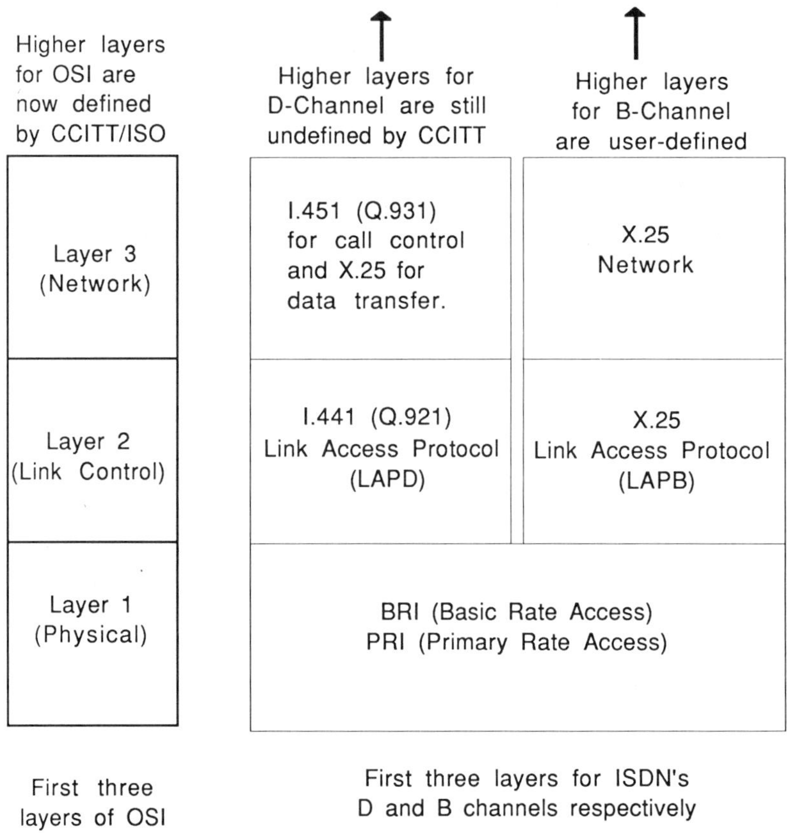

Figure B.5.4. Reference models for ISDN and OSI standards.

mesh network to provide switched and nonswitched services using the separate public telephone and data networks. Figure B.5.6 illustrates the composition of an ISDN CO. Line termination (LT), control point (CP), PS, and CS modules are provided to interface with the SS7, PTN, PDN, U, T and primary rate signalling points. Figure B.5.7 illustrates the hierarchy and the topology of a typical SS7 subnet. Such a network is a replacement of AT&T's original SS6 signaling network, which employed 2.4 Kbps trunks and 180-bit messages. By the end of 1988, AT&T was scheduled to convert their old signaling network to SS7. The use of 56 (or eventually 64 Kbps) trunks and 256-bit message provides a tremendous improvement and growth potential. When finally completed, AT&T's SS7 network will consist of over 172 4ESSs and 14 signal transfer points (STPs) to handle up to 12 million call attempts per hour (versus 2 million call attempts in SS6). Properly designed large private networks will also employ a SS7 subnetwork for signaling and control and interworking with the public ISDN-based networks. Some medium-sized corporations may find the cost of a SS7 subnetwork too high. In those cases, it is possible to employ some hardware interfaces that allow the creation of SS7-type subnetwork using 9.6 Kbps trunks. This approach will not only reduce the cost of

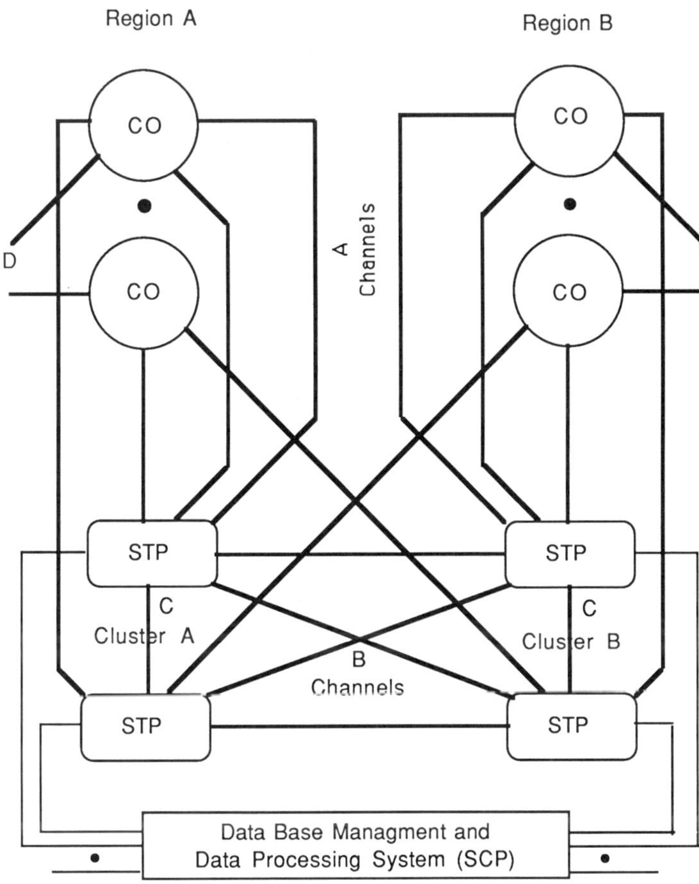

Figure B.5.5. Mixed topology of a SS7 network employed by ISDN.

the total network but also allow unfettered interworking with other private and public networks.

B.6 STANDARDS RELATED TO DISTRIBUTED DATA PROCESSING

B.6.1 OSI-RELATED STANDARDS

Since the acceptance of the OSI Reference Model standard ISO 7498 in 1983, the ISO subcommittee SC16 has been engaged in the arduous task of developing standards for the OSI protocols. CCITT, which cooperates with ISO, is also engaged in the development of relevant standards for PDNs. Both ISO and CCITT create standards that obey the OSI reference model and foster interconnection between multivendor computer systems. Figure B.6.1 illustrates two example sets of standards that have been accepted for some unique environments (e.g., LANs and WANs). Figure B.6.1 does not imply that the effort for developing standards is finished. Additional standards and protocols are

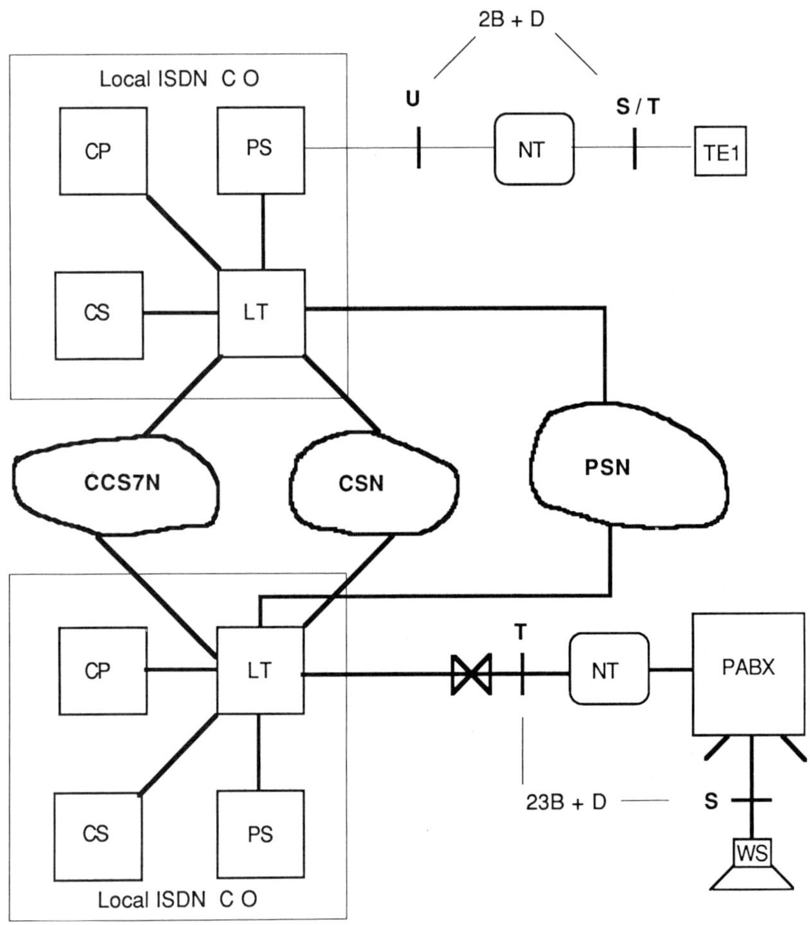

Figure B.5.6. Components of an ISDN Central Office.

still needed to satisfy the requirements of other meaningful environments. It is interesting to observe that X.25 forms the basis of the first three WAN layers. ISO has added a few additional standards for LANs and connectionless (or datagram) protocols. Although the CCITT standards appear different from those developed under the auspices of ISO, there are many similarities. For example, the CCITT standard X.400 for electronic messaging requires only the basic activity subset of the ISO session protocol ISO 8327 and as represented by the CCITT standard X.225. In general, the scope of ISO and CCITT are to handle all DDP environments and the PDN-based services, respectively. Eventually, these two subsets of standards will merge together to provide users with a single set of fully developed standards.

B.6.2 IBM's SYSTEM NETWORK ARCHITECTURE (SNA)

IBM first announced its proprietary SNA in 1974. It is a collection of hardware and software modules for distributing the computing resources and handling the needs for

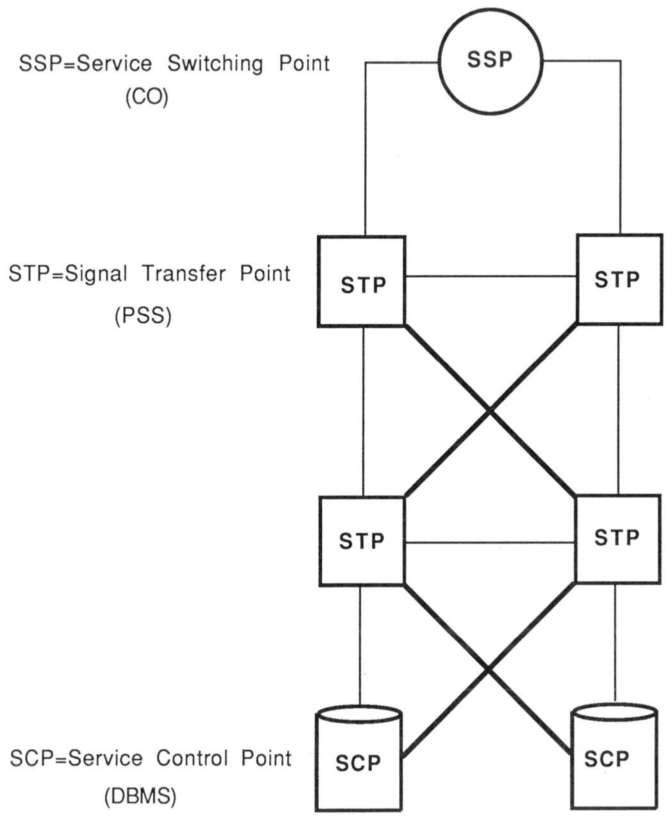

Figure B.5.7. Hierarchy and topology of a typical SS7 Network.

their large user base. It can be termed as a de facto standard due to its omnipresence in the U.S.A. Since its inception, SNA has been continuously evolving. Its architecture can also be represented by seven layers. Since SNA was developed before the availability of the OSI reference model, there are several differences between the SNA and OSI models. See Figure B.6.2 for a comparison between the two models. Each SNA node consists of at least a physical unit (PU). Some PUs are assigned the responsiblity of a system service control point (SSCP). Some PUs also consist of a Logical Unit (LU) through which an operator or an application program accesses the network. Each LU is associated with a network address and contains a service manager and one or more half-sessions. One half-session of an LU can be connected to another half-session of another LU in another SNA node to form a full session. On a logical level, the SNA network consists of interconnected LUs, resulting in a distributed network. Therefore, an SNA network can consist of intelligent workstations and a large mainframe that controls a large SNA domain. Originally, an SNA-based network was always centrally controlled by the mainframe. However, since the recent introduction of the LU type 6.2, peer-to-peer communication can occur if the communicating programs are written on the LU6.2 platform. Interestingly, LU6.2 can be spread across several layers—a fact that violates

	ISO	CCITT
Application (Layer 7)	CASE 8650. FTAM 8571	DIRECTORY SVC. X.400 EM
Presentation (Layer 6)	PROTOCOL 8823 SYSTEM ASN.1	SYNTAX X.409
Session (Layer 5)	8327	X.225
Transport (Layer 4)	8073	X.224
Network (Layer 3)	X.25 8208/8878. CLNP 8473	X.25
Data Link (Layer 2)	X.25 LAPB 7776. LAN LLC 8802	X.25 (LAPB)
Physical (Layer 1)	DTE-DCE. LAN 8802	V.24(RS.232C) V.35, X.21

ASN.1 = Abstract Syntax Notation 1
CASE = Common Application Service
CLNP = Connectionless Network Protocol
EM = Electronic Mail
FTAM = File Transfer and Access Management
LAPB = Link Access Protocol (Balanced)
LLC = Logical Link control

Figure B.6.1. Notable equivalences between OSI and CCITT standards.

the clean-separation-between-layers-requirements of ISO. It was for this fact that ISO will not accept LU6.2 as an OSI application layer protocol. See Figure B.6.3 for some typical SNA network topologies. Since SNA supports BSC and SDLC, conventional multidrop topologies based on minimal spanning trees are very popular with the SNA networks. LU6.2 will encourage the formation of clusters based on the mesh topology in the future.

IBM is committed to preserving its market share and serving the needs of SNA users for interworking. Figure B.6.4 illustrates the manner in which SNA has evolved since 1974. IBM's approach to interworking is to offer specialized gateways. See Figure B.6.5 for the NCP-PS interface (NPSI) employed to interface with the public PS networks. Due to the disparity between the SNA and OSI application layers, the DISOSS to X.400

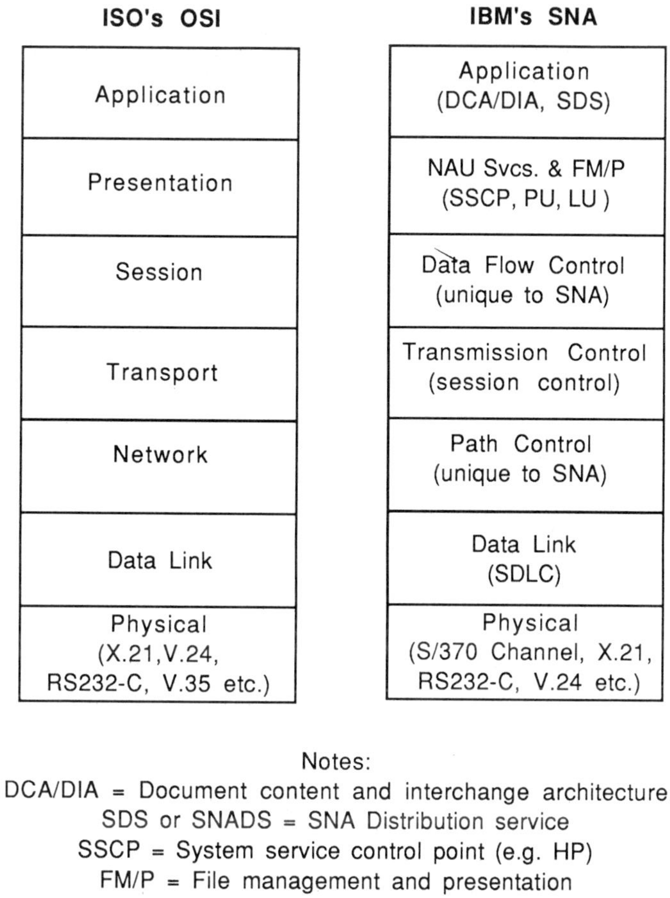

Figure B.6.2. Reference models for IBM's SNA and ISO's OSI standards.

gateway will be quite complex. All of the lower layers will have to communicate with their counterparts. It may need another mainframe or a supermini to do the job of a gateway. In any case, it is the user that will bear the extra cost of the interworking.

B.6.3 DEC's DIGITAL NETWORK ARCHITECTURE (DNA)

DNA is also a collection of hardware and software products, and it was first announced by Digital Equipment Corporation in 1975. Whereas SNA is generally characterized by hierarchical control, DNA was developed to provide a truly distributed control. See Figure B.6.6 for the evolution of the various DNA layers and the applicable standards. One should note that the DNA Phase IV application layer handles the interfaces to the network management and control functions in an integrated fashion. As time progresses, DEC plans to fold all of the OSI layers and the related standards into DNA Phase V while providing a proprietary alternative at each layer of the DNA protocol. It is inter-

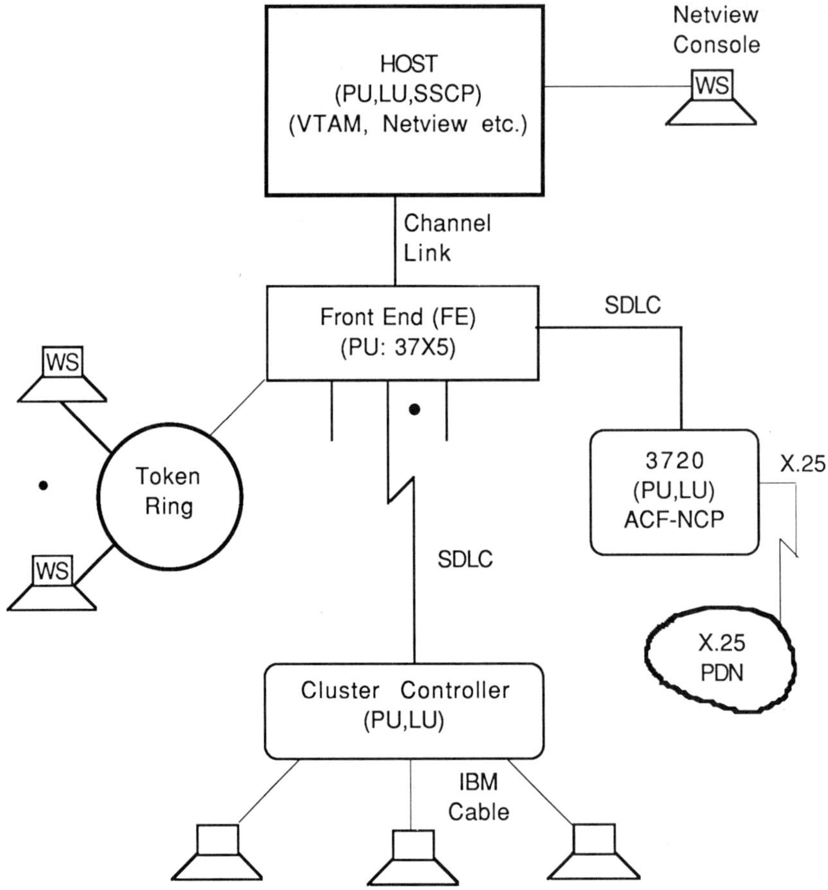

Figure B.6.3. Topology of a typical network based on SNA.

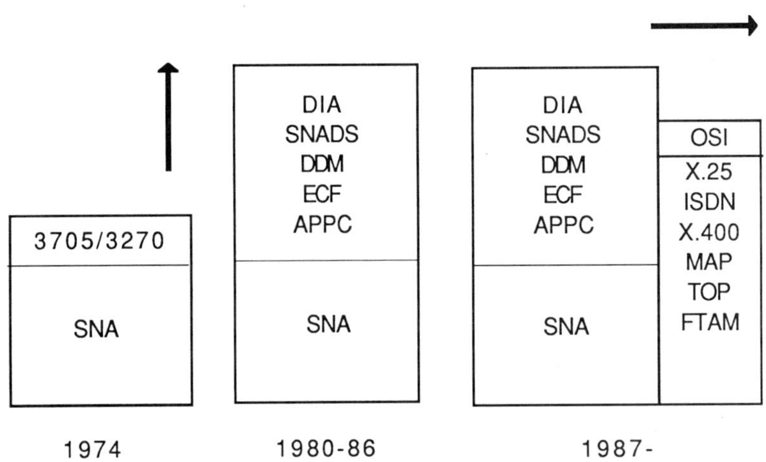

Figure B.6.4. An illustration of SNA evolution.

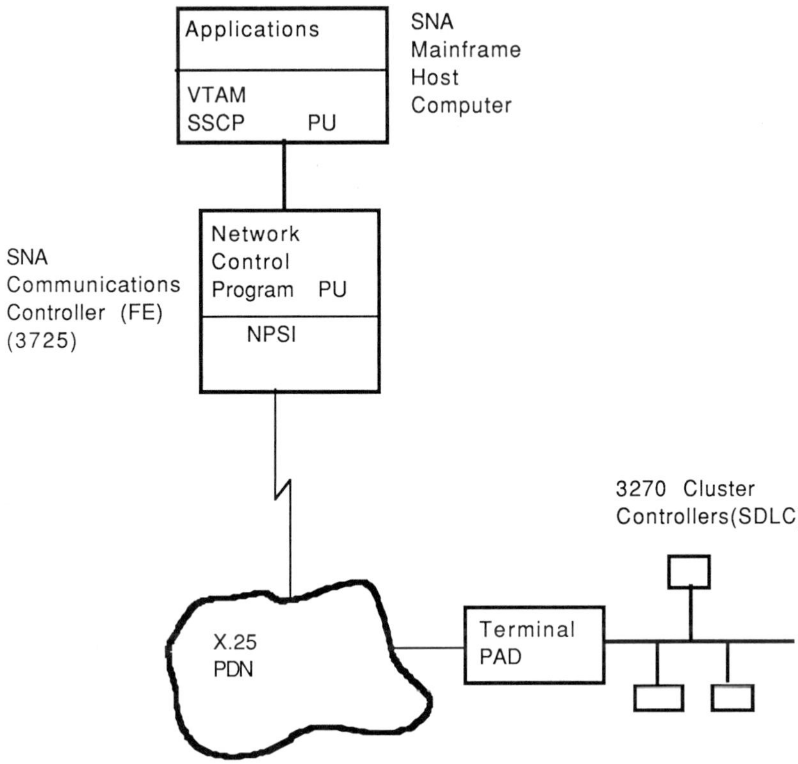

Figure B.6.5. Interworking between SNA and PDN using the NPSI interface.

esting to note that while DEC is committed to PC connectivity via the LANs, it favors only the IEEE 802.3 (CSMA/CD) standard and not the token ring. In this matter, DEC agrees with the Technical Office Protocol (TOP) architecture that will be described later. Many DECnet products are available to achieve multivendor connectivity. A typical DNA topology will tend to be identical to the one applicable for a PPS network.

B.6.4 MANUFACTURING AUTOMATION PROTOCOL (MAP) ARCHITECTURE

A seven-layered protocol to help achieve factory automation has been under development by several firms, notably General Motors Corporation. The MAP architecture is primarily based on the OSI reference model applicable for the LAN environment. Due to the response time constraints, the Token Bus (IEEE 802.4) standard was chosen for the first physical layer. The applicable standards for each of the OSI layers are illustrated in Figure B.6.7. The bus topology applies for a MAP-based network.

B.6.5 TECHNICAL OFFICE PROTOCOL (TOP) ARCHITECTURE

A seven-layered protocol to help achieve office automation and productivity increases among technical workers has been under development by several firms, spearheaded by

OSI Layers	DNA Layers	DNA Layered Functions	DNA/OSI Layers	DNA/OSI Layered Functions
Application	User ——— Network Managmnt.	File Transfer, Remote Access, Down-Line System Load, Remote Command File Submission, Virtual Terminals	Msg. Router (X.400 Gtwy) - - - - - OSI Apps. DNA User DNA Mngt	OSI FTAM, CCITT X.400 Messaging
Presentation	Network Application		OSI Prestns DNA Net Apps.	
Session	Session Control	Task-to-Task	OSI Session (OSAK) DNA Session Control	OSI Session
Transport	End-End Communications		OSI Transport (VOTS)	Task-to-Task
Network	Routing	Adaptive Routing	OSI Network	OSI Internet, Data Link Mgmnt. Adaptive Routing X.25
Data Link	Data Link	DDCMP (pt. to pt. & Multi Pt.) | X.25 | CSMA/CD	OSI DATA LINK	HDLC MAP 802.3 (see Appendix B)
Physical	Physical Link		OSI Physical	RS232, IEEE LAN 802.3 (see Appendix B)

Figure B.6.6. DEC's DNA layers/functions, as related to OSI layers.

the Boeing Corporation. The TOP architecture is primarily based on the OSI reference model applicable for the LAN environment. The office environments dictated the choice of the CSMA/CD (IEEE 802.3) standard for the lowest physical layer. The applicable standards for each of the OSI layer are illustrated in Figure B.6.8. The bus topology also applies for a TOP-based network.

ISO's OSI	MAP
Application	NM/TC, DS MMFS,FTAM
Presentation	NULL
Session	ISO Session Kernal
Transport	ISO Transport C4
Network	ISO CLNS
Data Link	IEEE 802.2
Physical	IEEE 802.4

Notes:
NM/TC = Network Management and Technical Control
DS = Directory Service : C4= Class4
MMFS = Manufacturing Message Format Standard of EIA
FTAM = File Transfer and Access Management
CLNS = Connectionless Network Service (also = Datagram)

Figure B.6.7. LAN layers for MAP.

B.6.6 GOVERNMENT OPEN SYSTEM INTERCONNECTION PROCUREMENT (GOSIP) ARCHITECTURE

National Bureau of Standards (NBS) is planning to introduce a new OSI-based architecture for distributed data communication/processing for all U.S. government agencies during 1988. Thus far, the U.S. government agencies have been procuring data communication products that satisfy the Transmission Control Protocol/Internet Protocol (TCP/IP). Most vendors still manufacture products based on TCP/IP for the government. The transition is bound to be painful, considering the installed base of TCP/IP. See Figure B.6.9 for an illustration of the four-layered TCP and the two-layered IP protocols. See Figure B.6.10 for the standards related to GOSIP layers for both the LAN and WAN environments. All of the network topologies for LAN and WAN environments discussed in earlier chapters will apply to a GOSIP-based network.

ISO's OSI	TOP
Application	FTAM
Presentation	NULL
Session	ISO Session Kernal
Transport	ISO Transport C4
Network	ISO CLNS
Data Link	IEEE 802.2
Physical	IEEE 802.3

Notes:
FTAM = File Transfer and Access Management
CLNS = Connectionless Network Service (also = Datagram)
C4= Class4

Figure B.6.8. LAN layers for TOP.

B.6.7 OPEN NETWORK ARCHITECTURE (ONA)

ONA is an FCC requirement and part of the ongoing Computer Inquiry III process. The FCC requires each public network to be redesigned to make the entry of Enhanced Service Providers (ESPs) as equitable as possible. ESPs are information processing/providing (e.g., telephone answering, voice mail, directory service) vendors. The seven Regional Bell Holding Companies (RBHCs) have recently filed their plans for ONA, which defines (1) the Basic Service Arrangement (BSA) for accessing the public network through various types of virtual/dedicated lines, and (2) the Basic Service Elements (BSEs), as provided by each RBHC. Each ESP can then combine such BSEs to provide unique sets of their services. With time, ONA will employ all of the ISDN interfaces and ESPs will be forced to employ the OSI Reference Model for their subnets to maintain high profitability.

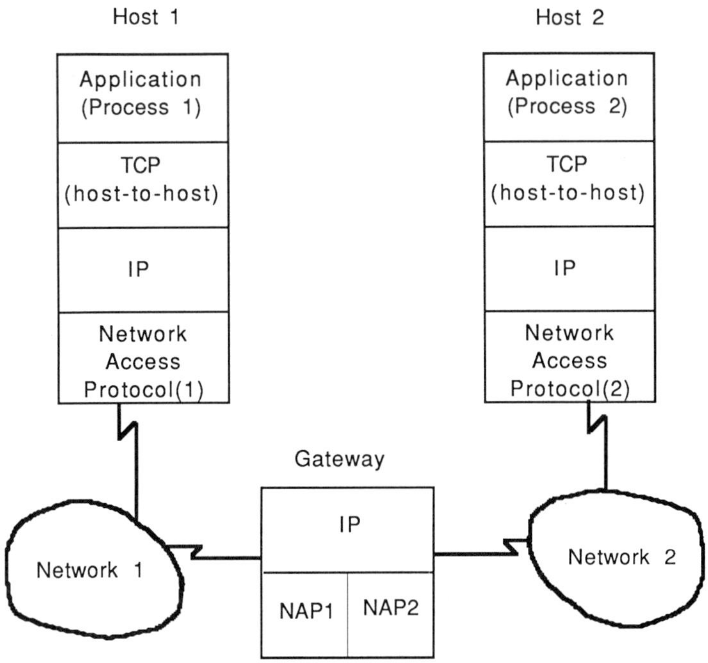

Notes:
TCP = Transmission Control Protocol
NAP = Network Access Protocol
IP = Internet Protocol

Figure B.6.9. A reference model for TCP/IP.

OSI Layers	GOSIP's Present and Future Protocols
Application	X.400 EM, FTAM, Virtual Terminal, Directory Service, Network Management, Association Control Svc. Element
Presentation	OSI Presentation Protocol
Session	OSI Session Protocol
Transport	Transport Protocol Class 4 & Connectionless Transport Protocol
Network	Connectionless Network Service + Dynamic Routing, X.25, ISDN, Connection Oriented Network Service + Internetwork Protocol
Data Link	X.25 HDLC LAP B, IEEE 802.3, 802.4, 802.5, FDDI MAC, ISDN,
Physical	BASEBAND, BROADBAND, Shielded Twisted Pair, FDDI Physical, Military Standard 114-A, RS-232

Figure B.6.10. Layers and protocols related to GOSIP.

Glossary

ACD	automatic call distribution
ACDT	ACD type
ACF	access coordination function or advanced communication function
ACK	acknowledgement
ACPM	average cost per minute (of a voice call)
ACS	advanced communication service
ADPCM	adaptive differential pulse code modulation
AI	artificial intelligence
AL	access line
ALT	access line type (it defines the link capacity, allowable data rate on the link, potential data rate per terminal, and the applicable tariff)
ALMF	access link multiplexing factor (equals the number of VG circuits/AL)
ALU	arithmetic logic unit
AM	amplitude modulation
AMA	automatic message accounting
AMEF	average message expansion factor
AMLC	asynchronous multiline controller
ANSI	American National Standards Institute
AQT	average queuing time
ARM	async response mode
ARPA	Advanced Project Research Agency
ARS	automatic route selection
ASK	amplitude shift keying
ASYNC	asynchronous mode of communication
AT&T	American Telephone & Telegraph Co.
ATA	American Transportation Association
ATAT	average turn-around time
ATDM	asynchronous time division multiplexing
ATDMA	asynchronous time division multiple access
ATP	analysis type (for computing response times)
AWG	American wire gauge
Balun	BALanced to UNbalanced impedance matching device used as a cable to twisted-wire pair adapter in a premise wiring system
BBN	Bolt, Beranek, and Newman
BBTF	back bone trunking factor
BC	bridged channel (for a multidrop link)
BCC	blocked cell cleared
BCD	blocked call delayed

BCH blocked call held
BER bit error rate
BHR busy hour
BINS Barclay's integrated network system
BLC bridged local channel
BLKL block length in bytes
BLU basic link unit
BNR Bell Northern Research
BOC Bell Operating Company
BOP bit-oriented protocol
bps bits per second
BR block redundancy associated with each encoded block
BRA basic rate area
BRI basic rate interface (for ISDN)
BSA basic service arrangement
BSC binary synchronous communication
BSE basic service element
BST bid-to-start time
BTL Bell Telephone Laboratories
BW bandwidth
B8ZS bit 8 zero suppression
CAC customer administration center
CAS centralized attendant service
CATV cable television
CC common control
CCC copper cross connect or clear channel capability
CCIR International Radio Consultative Committee
CCIS common channel interoffice signaling (AT&T terminology)
CCITT International Telegraph and Telephone Consultative Committee
CCM customer coordination/control management
CCP communication and control processor
CCS common channel signaling (also equivalent to CCITT's SS7) or hundred call seconds
CCSA common control switching arrangement (AT&T nomenclature)
CCU communication control unit
CDC Control Data Corp.
CDMA code division multiple access
CEC Commission for European Communities
CLLI Common Language Location ID
CLNS connectionless network service
CLT central limit theorem (for statistical analysis)
CMIP common management information protocol
CMIS common management information services
CMST capacitated minimal spanning tree
CNCC customer network control center
CNCP Canadian National Canadian Pacific
CNS communication network systems
CO central office
COG center of gravity
COS class of service or communications operating system

CP	communications processing
CPE	customer premise equipment (e.g., telephone, PABX)
CPU	central processing unit
CRC	cyclic redundancy code
CRT	cathode ray tube (also used for a dumb computer terminal)
CS	circuit switching (also circuit switched)
CSA	customer service area
CSABDS	customer service area bands
CSATBL	customer service area table
CSMA/CD	carrier-sense-multiple-access with collision detection
CSMT	capacitated (or constrained) minimal spanning tree
CSS7	common-channel signaling system #7 (also equivalent to SS7)
CSU	channel service unit
CSV	customer service vehicle
CU	control unit
CV	customer vehicle
CVSDM	continuously variable slope delta modulation
DAMA	demand-assignment multiple access
DAP	device access protocol
DARPA	Defense Advanced Research Project Agency
DB	decibel
DBA	data base access or dynamic bandwidth allocator
DBM	data-base-management
DCA	distributed communications architecture
DCE	data circuit-termination equipment or data communication equipment
DCP	distributed communications processor
DCPSK	differentially coherent phase shift keying
DCS	distributed computing system or data circuit switches
DDBM	distributed-data-base-management
DDCMP	digital data communication message protocol
DDD	direct distance dialing
DDP	distributed data processing
DDS	dataphone digital system
DEC	Digital Equipment Corp.
DENOM	denominator
DF	disk file or dividing factor as used in ROI analysis
DFO	digital fiber optics
DIS	draft international standard (per ISO)
DISTF	distance factor
DL	directed link associated with Doelz networks
DLA	delayed acknowledgment
DLC	data link control
DM	delta modulation
DMA	direct memory access
DMW	digital microwave
DNA	distributed network architecture of NCR and DEC or Digital Equipment Corp. Network Architecture
DNHR	dynamic non-hierarchical routing
DNIC	data network identification code
DP	data processing

DQDB	distributed queued dual bus
DREQ	delay required
DS	directory service
DSA	digital service area
DSDDS	Dataphone switched digital data system
DSI	digital speech interpolation (e.g., TASI)
DSU	data service unit
DTE	data terminal equipment
DTMF	dual-tone multifrequency
DTN	data transporting network
DTP	daily traffic profile (for ACD)
DTS	digital tandem switching
DUC	data unit control
DUV	data under voice
ECC	economical call cost (for voice application)
ECD	expected call duration
ECSA	Exchange Carriers Standards Association
E(D)	expected delay
EDS	electronic data systems
EEC	European Economic Community
EFT	electronic fund transfer
EIA	electronic industry association
EIN	European Informatic Network
EIS	electronic information system (AT&T)
EM	electromagnetic or electronic mail
EMA	enterprise management architecture of DEC
EMD	expected message duration
EMS	elemental management systems
EO	end office
EOF	end-of-file
EOSN	end-of-service node (e.g., PBXs, PADs, data concentrators, key telephones, sets, etc.)
EPL	expected polling latency
EPSCS	enhanced private switched communication system (AT&T offering)
EPSS	experimental packet switching service
ERL	erlang
ERLB	erlang-B
ERLC	erlang-C
ES	expert system
ESF	extended super frame
ESP	enhanced service provider
ESR	Extended Slotted Ring of Doelz Inc.
ESRA	en-route station routing approach (a traditional approach for public telephone networks)
ESS	electronic switching system
ETN	electronic tandem network (AT&T offering for analog systems)
EW	Esau-Williams network design algorithm
FADS	force administration data system (ACD related)
FAX-PAK	facsimile-packet switching system
FCC	Federal Communications Commission or fiber cross connect

FCS	frame check sequence based on CRC
FDDI	fiber data distribution interface (IEEE standard)
FDM	frequency division multiplexing
FDMA	frequency division multiple access
FDX	full duplex
FDXFACT	full duplex factor regarding trunks
FE	front end
FIDN	fully integrated digital network
FIFO	first-in-first-out
FIPS	federal information processing standard
FM	frequency modulation
FSK	frequency shift keying
FTAM	file transfer and access management
FTD	from-to-data
FTF	from-to-file (equivalent to FTD file)
FTP	file transfer protocol
FX	foreign exchange
GOS	grade of service
GOSIP	government open system interconnect protocol
GTE	General Telephone and Electronics Corp.
HD	high density
HDLC	high-level data link control
HDX	half duplex
HEHO	head-end-hop-off
HF	high-frequency
HILI	higher layer interface
HLC	high-level center
HMD	half modem delay (ms)
HP	host processor
HPR	host port rate (bps)
HTT	host think time in seconds
HU	high usage circuit in a hierarchical network
IATA	International Airline Transportation Association
IBC	inter bridge channel (for multidrop links)
IBDS	integrated building distribution system
IBM	International Business Machines Corp.
ICPB	information characters per block
ICS	Info Call Service
IEC	inter exchange carrier for interLATA services
IEC	information exchange circuit in a PDN
IED	information exchange distribution
IEEE	Institute of Electrical and Electronics Engineers
IES	Infoexchange Service
IGS	InfoGram Service
IMA	input message acknowledgement
IML	input (inquiry) message length in bytes
IMP	interface message processor
IMS	information management system (IBM offering)
IMT	intermachine trunks
INAP	Infogram network access protocol

IOC	inter office (or POP) channel for inter-LATA use
IP	information processing (also inter-process)
IPL	initial program load
IPSS	international packet switching service
IPT	idle polling time
IR	inquiry response or interest rate
IRC	international record carriers or inter rate center
ISDN	Integrated Services Digital Network
ISN	information system network of AT&T (similar to Datakit)
ISO	International Standards Organization
ITU	International Telecommunications Union
IVD	integrated voice and data
IVDN	integrated voice/data network
IWS	intelligent work station
LAMBDA	avg. message arrival rate per VMPT circuit per second (also λ)
LAN	local area network
LAP	link access procedure
LAPB	link access procedure, balanced (for X.25)
LAPD	link access procedure–D channel (for ISDN)
LATA	Local Access and Transport Area
LC	local channel as a part of the total end-to-end circuit
LCC	local communication controller
LCM	line concentrator module
LD	long distance or low density
LEC	local exchange carrier for intraLATA services
LIFO	last-in-first-out (queuing system)
LL	local loop
LLC	logical link control
LNA	local network architecture
LR	link rate
LSI	large-scale integration
LSO	local service office
LT	line termination
LTU	line termination unit
LU	logical unit
MA	multiple access
MAC	media access control
MAN	Metropolitan Area Network
MAP	Manufacturing Automation Protocol
MATDM	message-asynchronous time division multiplexing
MCI	Microwave Communications, Inc.
MCMD	multicenter, multidrop
MCMS	multicenter, multistar
MD	multidrop (also see CMST or MST)
MF	*multifrequency* signaling scheme or a PCM *multiframe*
MFD	main frame distribution
MIPS	million instructions per second
MIS	management information system
MJU	multipoint junction union
MMFS	manufacturing message format standard

MMOS	message multiplex operating system
MODEM	modulation-demodulation
MPL	multipoint private line
MPT	multipoint
MS	message switching
MSK	minimum shift keying
MST	minimal spanning tree
MSU	multiplex service unit
MT	magnetic tape
MTBCF	mean time between catastrophic failures
MTBEB	mean-time-between-error bursts
MTBF	mean time between failures
MTBMF	mean time between major failures
MTBNF	mean time between nodal failures
MTBSF	mean time between subsystem failures
MTKU	maximum trunk utilization factor or maximum trunk utilization factor for data networks
MTS	message telecommunications service
MTTR	mean-time-to-repair
MUTBL	MegaCom Usage Table
MW MULDEM	microwave multiplexer-demultiplexer
N-SMLC	network synchronous multiline controller
NAK	no acknowledgement
NAU	network addressable unit
NBS	National Bureau of Standards
NCC	network control center
NCPM	number of calls per month
NCU	number of control units per virtual multipoint circuit
NE	network element
NIM	network interface machine
NLT	node-link tariff data for each location or link (it defines the link capacity, allowable link rate, and the applicable tariff)
NM/TC	network management and technical control
NMC	network management control (also network management center)
NMC/TC	network management control/technical control
NMP	network management protocol (CCITT, ISO)
NNX	network numbering exchange
NPA	numbering plan area
NPL	National Physical Laboratories of U.K.
NPSI	NCP packet service interface (IBM)
NSC	Network Systems Corporation
NSP	network service protocol
NTI	network trunk interface (ISDN)
NUM	numerator
NWF	nodal weight factor
OCC	other common carriers
ONA	open network architecture
OS	operating system
OSI	open systems interconnection
OSRA	originating station routing approach

PABX	private automatic branch exchange
PAD	packet assembly/disassembly
PAM	phase amplitude modulation
PATDM	packet-asynchronous time division multiplexing
PBX	private branch exchange
PC	primary center or personal computer
PCM	pulse code modulation
PCTG	Programmable Channel Termination Group
PDM	pulse duration modulation
PDN	public data network
PDS	premise distribution system or premise wiring system
PEXD	probability exceeding DREQ
PGK	propagation constant (also Kpg)
PLP	private line pricing
PLU	primary logic unit
PM	phase modulation
PMST	poll-to-message-start-time
PMT	packet-mode terminal
PNW	potential nodal weight
POI	point-of-interface (applies to IEC/OCC exchange)
POP	point-of-presence (applies to IEC/OCC exchange)
POS	point-of-scale
PPL	Pluribus private line
PPM	pulse position modulation
PPS	private packet service
PPX	private packet exchange
PRI	primary rate interface (for ISDN)
PS	packet switching (also packet switched)
PSE	packet switching exchange
PSI	Pluribus satellite interface message processor
PSK	phase shift keying
PSN	packet switched network
PSS	packet switching system
PSV	public service vehicle
PTN	public telephone network
PTS	public telephone system
PTT	Postal, Telephone, and Telegraph
PT/T	Postal, Telephone/Telegraph
PU	physical unit (for SNA)
PV	principal value
PWM	pulse-width-modulation
PWS	premise wiring system or premise distribution system
QOS	quality-of-service
QSPX	queued synchronous packet exchange (also see DQDB)
RAC	ring access control for Datapoint's ARCnet
RAM	random access memory
RBHC	Regional Bell Holding Company (such as U.S. West, SouthWestern Bell)
RC	regional center
RCA	Radio Corp. of America

RCP	Royal Canadian Police
RFP	request for proposal
RHO	regional hub office
RIM	resource interface module for Datapoint's ARCnet
RJE	remote job entry
RJEP	remote-job-entry protocol
RML	response message length in bytes
R_{mph}	message per hour rate
ROE	remote order entry
ROI	return-on-investment
ROM	read-only memory
RSTBL	rate step table (for designing ACDs)
RTIU	remote terminal interface unit
RU	request unit
S/F	store-and-forward
SBS	Satellite Business Systems
SC	sectional center
SCC	specialized common carrier
SSCP	system service control point
SCMD	single-center, multidrop
SCP	service control point (or a DB management node)
SCSS	single-center, single-star
SDF	system design file
SDLC	synchronous data link control
SDM	space division multiplexing
SDN	software defined network (see also VPN)
SG	study group (associated with CCITT)
SIO	standard information outlet as defined by CCITT
SITA	Societe Internationale de Telecommunications Aeronautiques
SL	subscriber line
SMDR	station message detail report
SMLC	synchronous multiline controller
SNA	system network architecture
SNAP	standard network access protocol
SNRC	Successive Node Route Control
SOM	start of message
SONET	synchronous optical network
SOW	statement of work
SPC	stored program control
SPLS	shared private line service
SPNS	switched private network service
SQR	square root
SSB	single-side band
SSCP	system service control point
SS#7	signaling system number 7 (also SS7)
STP	signal transfer point (or an HS packet switch)
SW	switch
SWIFT	Society for Worldwide Interbank Financial Telecommunications
T-carrier	digital transmission system (e.g., T1)

TA	terminal adapter (for ISDN)
TAC	Telenet access controller
TASI	time asynchronous speech interpolator
TC	toll center
TCA	time-consistent average
TCAM	telecommunications access method (IBM)
TCO	Telenet central office
TCP/IP	transmission control protocol/internet protocol
TCPM	total cost per minute of a voice call
TCTS	Trans-Canada Telephone System
TDC	terrestrial digital circuit
TDM	time division multiplexing
TDMA	time division multiple access
TDX	time-division-exchange
TEHO	tail-end-hop-off
TELCO	telephone company
TGF	traffic growth factor
THI	terminal handler interface
TIH	terminal interface handler
TIP	terminal interface processor
TK	trunk
TKMF	trunk multiplexing factor (equals the number of VG circuits)
TKLT	trunk link type
TM	transaction module
TMC	total monthly cost of a network
TN	transport network
TOD	time-of-day
TOP	technical office protocol
TP	Telenet processor
TS	time-sharing
TST	time-space-time
TTTN	tandem tie-trunk network
TTY	teletypewriter
TWX	teletypewriter exchange service
UBER	undetected bit error rate
UITP	universal information transport system
UNMA	unified network management architecture (AT&T)
UPR	user port rate (bps)
URTBL	use-rate-table (for designing ACD)
VAN	value-added network
VC	virtual circuit
VCL	virtual-circuit-link or virtual circuit and link data for each location
VCU	video compression unit
VG	voice-grade
VGL	voice grade line/link
VGC	voice grade circuit
VHD	vertical (V) and horizontal (H) coordinate data of each location
VLSI	very-large-scale-integration
VMPT	virtual multipoint circuit associated with Doelz network

VPN	virtual private network (see also SDN)
VRU	voice-response-unit
VSAT	very small aperture terminal (for Satellite systems)
VTAM	virtual telecommunications access method (IBM)
VTP	virtual terminal protocol
WAN	wide-area-network
WATS	wide area telecommunications service
WB	wide-band
WC	wire center that serves the customer premise equipment or co
WPM	words per minute
WS	workstation
WUI	Western Union International
WUTBL	WATS Usage Table

Bibliography

Andrews, R. 1967. No. 2 ESS service features and call processing plan. *BSTJ*, vol. 48, October 1967.

AT&T. 1966. *Rural Electrification Administration*: Telephone Engineering and Coordination Manual, Section 325.

Bell Telephone Laboratories. 1977. *Engineering and Operations in the Bell System*. New York.

CCITT. 1968. *Economic Studies at the National Level in the Field of Telecommunications*: (1964–1968), Chapter VI, CCITT.

Chandy, K. and Lo, T. 1973. The capacitated minimum spanning tree, *Networks*, vol. 3, no. 2, pp. 173–182.

Collins, A. A. and Pederson, R. D. 1973. *Telecommunications: A Time for Innovation*. Dallas: Merle Collins Foundation.

Doelz, M. L. and Sharma, R. L. 1984. Extended Slotted Ring Architecture for a Fully Shared and Integrated Communication Network, *Proc. MIDCON 1984 Conference* at Dallas, Texas.

Doelz, M. L. 1985. Using the basic Equation 4.11 he has derived a large collection of expressions valid for actual applications of ESR. The listed Equations, 4.12, 4.13, and 4.23–4.30 represent only a few generalized applications.

Eldin, A. 1975. Basic Traffic Theory, *Seminar on Traffic Engineering and Network Planning, Organized by ITU*, New Delhi, India, November 24 to December 5, 1975.

Esau, L., and Williams, K. 1966. On teleprocessing system design. *IBM Syst. Jrnl.*, vol. 5, no. 3, pp. 142–147.

Everett, J. L. 1953. State probabilities in congestion problems characterized by constant holding times, *J. Operations Research*, 1, pp. 279–285.

Feller, W. 1957. *An Introduction to Probability Theory and its Applications*. New York: John Wiley.

Frank, H., Frisch, I., Chu, W., and Van Slykes, R. 1971. Optimal design of centralized computer networks, *Networks*, vol. 1, no. 1, pp. 43–57.

Frank, H. and Chu, W. 1972. Topological optimization of computer network. *Proc. IEEE*, vol. 60, November 1972, pp. 1385–1397.

Fraser, A. 1983. Towards a universal data transport system. *IEEE J. on Selected Areas in Communications*, Vol. SAC-1, No. 5, November 1983.

Hoffman, H. 1983. Extended Erlang C: Traffic engineering for queuing with overflow, *Business Communications Review*, July–August 1983.

Karnaugh, M. 1972. Multipoint network layout program, IBM Internal Report No. RC3723.

Kershenbaum, A. and Chu, W. 1974. A unified algorithm for designing multidrop teleprocessing networks, *IEEE Trans. Communications*, vol. COM-22, no. 11, pp. 1762–1772.

Kruskal, J. 1956. On the shortest spanning subtree of a graph and the travelling salesman problem, *Proc. Amer. Math. Society*, vol. 7.

Kurose, J., Schwartz, M., and Yemini, Y. 1984. Multiple-access protocols and time-constrained communications. *ACM Computing Surveys* Vol. 16, No. 1, pp. 43–70.

Martin, J. 1972. *System Analysis for Data Transmission*. New York: Prentice-Hall, Inc.

Parzen, E. 1965. *Stochastic Processes*. San Francisco: Holden-Day, Inc.

Prim, R. 1957. Shortest connection networks and some generalizations, *B.S.T.J.*, vol. 36, pp. 1389–1401.

Schwartz, M. 1987. *Telecommunications Networks*. New York: Addison-Wesley.

Sharma, R. L. 1968. Analysis of a scheme for information organization and retrieval from a Disc File, *Proc. IFIP 1968 Conference* at Edinburgh, Scotland.

Sharma, R. 1972. U.S. Patent No. 3,703,006; Algorithm for Synthesizing Economical Data Communications Network. Issued on Nov. 14, 1972.

Sharma, R. and M. El-Bardai. 1970. Sub-Optimal Communications Network Synthesis. *Proc. 1970 Intl. Conference on Communications*, at San Francisco.

Sharma, R. L., De Sousa, P. T., and Ingle, A. D. 1982. *Network Systems*. New York: Van Nostrand Reinhold.

Sharma, R. L. 1983. Design of an economical multidrop network topology with capacity constraints, *IEEE Trans. Communications*, vol. COM-31, no. 4, pp. 590–591.

Siemens Aktiengesellschaft. 1970. *Telephone Traffic Theory*. Munich, FRG.

Smith, R. and Smith, J. 1962. Loss and delay in telephone call queuing systems, *ATE Journal*, vol. 18, no. 1, pp. 18–30.

Stallings W. 1984. Local networks. *ACM Computing Surveys* Vol. 16, No. 1, pp. 3–41.

Stuck, B. 1983. Calculating the maximum data rate in local area networks. *Computer*, May 1983, pp. 72–76.

Syski, R. 1960. *Introduction to Congestion Theory in Telephone Systems*. Edinburgh and London: Oliver and Boyd.

Whittaker, B. 1975. Analysis and optimal design of a multiserver, multiqueue system with finite waiting space in each queue, *B.S.T.J.*, vol. 54, no. 3, pp. 595–623.

Wolman, E. 1965. A fixed optimum cell-size for records of various lengths. *JACM*, vol. 12, no. 1, pp. 53–70.

Index

Index